Sitabhra Sinha, Arnab Chatterjee, Anirban Chakraborti, and
Bikas K. Chakrabarti

Econophysics

An Introduction

WILEY-VCH

WILEY-VCH Verlag GmbH & Co. KGaA

Sitabhra Sinha, Arnab Chatterjee,
Anirban Chakraborti, and
Bikas K. Chakrabarti

Econophysics

The Authors

Sitabhra Sinha *is professor of theoretical physics at the Institute of Mathematical Sciences (IMSc), Chennai, India. He received his doctorate from the Indian Statistical Institute, Kolkata, for research on nonlinear dynamics of neural network models in 1998. Following postdoctoral positions at the Indian Institute of Science, Bangalore, and Cornell University, New York City, he joined IMSc in 2002. His research interests include complex networks, nonlinear dynamics of biological pattern formation, theoretical/computational biophysics, and the application of statistical physics for analyzing socio-economic phenomena. He was an International Fellow of the Santa Fe Institute (2000–2002).*

Arnab Chatterjee *is a postdoctoral researcher at the Condensed Matter and Statistical Physics Section, The Abdus Salam International Centre for Theoretical Physics at Trieste, Italy. He was awarded his Ph.D. from the Saha Institute of Nuclear Physics, Kolkata, India, where he worked on dynamic transitions in Ising models, and also on the application of statistical physics to varied interdisciplinary fields such as complex networks and econophysics. Dr. Chatterjee has studied structural properties of the transport networks and has developed the kinetic models of markets. In recent years he has worked on percolation models and even on problems related to stock market crashes, resource utilization, and queuing.*

Anirban Chakraborti *has been an assistant professor at the Quantitative Finance Group, Ecole Centrale Paris, France, since 2009. He received his doctorate in physics from Jadavpur University in 2003. Following postdoctoral positions at the Helsinki University of Technology, Brookhaven National Laboratory, and Saha Institute of Nuclear Physics, he joined the Banaras Hindu University as a lecturer in theoretical physics in 2005. Statistical physics of the traveling salesman problem, models of trading markets, stock market correlations, adaptive minority games and quantum entanglement are his major research interests. He is a recipient of the Young Scientist Medal of the Indian National Science Academy (2009).*

Bikas K. Chakrabarti *is a senior professor of theoretical condensed matter physics at the Saha Institute of Nuclear Physics (SINP), Kolkata, and visiting professor of economics at the Indian Statistical Institute, Kolkata, India. He received his doctorate in physics from Calcutta University in 1979. Following postdoctoral positions at Oxford University and Cologne University, he joined SINP in 1983. His main research interests include physics of fracture, quantum glasses, etc., and the interdisciplinary sciences of optimisation, brain modelling, and econophysics. He has written several books and reviews on these topics. Professor Chakrabarti is a recipient of the S.S. Bhatnagar Award (1997), a Fellow of the Indian Academy of Sciences (Bangalore) and of the Indian National Science Academy (New Delhi). He has also received the Outstanding Referee Award of the American Physical Society (2010).*

The Authors

Dr. Sitabhra Sinha
Institute of Mathematical Sciences
CIT Campus
Chennai, India

Dr. Arnab Chatterjee
The Abdus Salam International Centre for
Theoretical Physics ICTP
Condensed Matter and Statistical Physics Section
Trieste, Italia

Dr. Anirban Chakraborti
Ecole Centrale Paris
Laboratoire de Mathématiques Appliquées aux
Systèmes
Châtenay-Malabry, France

Prof. Bikas K. Chakrabarti
Saha Institute of Nuclear Physics
Centre for Applied Mathematics
and Computational Science
Kolkata, India

Cover Design
Cover-Image © Kit Wai Chan (Fotolia.com)
Adam Design, Weinheim, Germany

Library of Congress Card No.: applied for

British Library Cataloguing-in-Publication Data:
A catalogue record for this book is available
from the British Library.

**Bibliographic information published by the
Deutsche Nationalbibliothek**
The Deutsche Nationalbibliothek lists this
publication in the Deutsche Nationalbibliografie;
detailed bibliographic data are available on the
Internet at http://dnb.d-nb.de.

Typesetting   le-tex publishing services GmbH,
Leipzig

Printed on acid-free paper

ISBN 978-3-527-40815-3

Contents

Econophysics. Sitabhra Sinha, Arnab Chatterjee, Anirban Chakraborti, and Bikas K. Chakrabarti
Copyright © 2011 WILEY-VCH Verlag GmbH & Co. KGaA, Weinheim
ISBN: 978-3-527-40815-3

Preface

Physicists have long had a reputation of meddling in areas outside the restricted domain of physics. In some cases, the non-traditional approaches that physicists bring can result in new insights, and in due course, this new inter-disciplinary area of research can become a recognized part of science. Biophysics has been the most spectacular success in this mold; astrophysics and geophysics have also emerged in the last century from the forays of physicists into disciplines outside their own.

However, the use of physics techniques to understand economic phenomena, leading to the development of a new field dubbed "econophysics" (for better or worse), marks a new departure for physicists beyond their traditional domain. Economics, and for that matter, the social sciences in general, are often thought to be unamenable to the analytical tools used in the natural sciences, in particular, physics. As critics of the quantitative approach to socioeconomic issues are always ready to remind us: social sciences deal with issues related to human beings, whose behavior do not follow any reproducible patterns nor can they be reduced to equations. Thus, the early attempts of physicists to address economic questions have been invariably met with, at best, bemused curiosity, or worse, total indifference from the social science community.

However, the ongoing worldwide economic and financial crisis has rudely awakened everyone to the weak theoretical foundations on which the science of economics rests. It has also highlighted alternative approaches to understand social issues, especially econophysics. The past decade has seen an ever-growing number of physicists tackling problems arising in an economic or financial context, under the premise that while individual human behavior is not predictable, those of a large collection of economic agents are likely to follow discernible patterns. The increasing volume and quality of the contributions by econophysicists have now reached a critical level where they cannot be ignored either by mainstream economics or physics communities. We have reached the point where a beginner will find it difficult to figure out where to start learning about the field and even a specialist working in this area can no longer keep track of all the exciting developments that are happening in it.

Econophysics. Sitabhra Sinha, Arnab Chatterjee, Anirban Chakraborti, and Bikas K. Chakrabarti
Copyright © 2011 WILEY-VCH Verlag GmbH & Co. KGaA, Weinheim
ISBN: 978-3-527-40815-3

Thus, we felt that this is an appropriate time to write a book introducing the main currents of research in econophysics to a broad and non-specialist scientific audience. While the text will possibly be most useful for students and researchers in physics and mathematics interested in socioeconomic phenomena, we hope that economists and social scientists would also find it helpful as an introductory glimpse of econophysics. With this aim in view, we have kept assumptions of requisite background knowledge for reading the book at a minimum. The unusually extensive appendices introduce almost all the basic physics concepts that are necessary for a person not trained in physics to read this book. While it only serves as the first step towards the long journey of acquiring expertise in this rapidly developing inter-disciplinary area, we would be happy if our book serves to give the reader an overall idea of the present state of research in econophysics and hopefully inspire them to join in this effort to use the techniques of physics to understand how society works.

The choice of topics discussed in the book has necessarily been colored by our own areas of interest in econophysics. We do realize that, as econophysics is a rapidly growing field, we have not included many references and have not discussed the contributions of several scientists. We apologize that we could not give equal representation to all the areas of research in econophysics. We have tried to include important references to the literature whenever we have thought that the original material would be indispensable, and have sometimes omitted continued attribution and references which would have been very distracting otherwise.

The idea of writing a book introducing econophysics was suggested to us by our dynamic and enthusiastic editor, Anja Tschoertner of Wiley-VCH. Despite numerous delays and false starts on the way, her patience and support have helped to bring the project to a successful completion. This book is a collective effort among authors across two different continents and the principal responsibility for writing the various chapters were distributed as follows: S. Sinha for Chapters 1, 3, 4, 5, 6, 10 and 11, A. Chatterjee for Chapters 7 and 8, A. Chakraborti for Chapters 2 and 9 and B.K. Chakrabarti for Chapter 1 and the Appendices.

Several friends and collaborators have provided invaluable assistance. In particular, we would like to thank Soumyajyoti Biswas and Anulekha Dutta for transcribing and checking the lecture notes corresponding to the material included in the appendices, and Rajeev Singh for proofreading the entire manuscript and making numerous suggestions on how to improve it. We would also like to thank all our collaborators, including Frédéric Abergel, Pratip Bhattacharyya, Anindya Sundar Chakrabarti, Guido Germano, Asim Ghosh, Els Heinsalu, Antti Kanto, János Kertész, Kimmo Kaski, Subhrangshu Sekhar Manna, Sugata Marjit, Manipushpak Mitra, Jukka-Pekka Onnela, Raj Kumar Pan, Marco Patriarca, Srutarshi Pradhan, Srinivas Raghavendra, M.S. Santhanam, Nisheeth Srivastava, Robin Stinchcombe, Marko Sysi-Aho, Ioane Muni Toke and S.V. Vikram. We would also like to acknowledge helpful discussions with Ginestra Bianconi, Indrani Bose, Arnab Das, Deepak

Dhar, J. Doyne Farmer, Mauro Gallegati, Brian Hayes, Abhijit Kar-Gupta, Thomas Lux, Krishna Maddaly, Matteo Marsili, Pinaki Mitra, Pradeep K. Mohanty, Peter Richmond, Abhirup Sarkar, Parongama Sen, H. Eugene Stanley, Dietrich Stauffer, S. Subramanian, Yarlagadda Sudhakar and Victor M. Yakovenko among others.

Chennai   *Sitabhra Sinha*
Trieste   *Arnab Chatterjee*
Châtenay-Malabry   *Anirban Chakraborti*
Kolkata, January 2010   *Bikas K. Chakrabarti*

1
Introduction

"Economic theorists, like French chefs in regard to food, have developed stylized models whose ingredients are limited by some unwritten rules. Just as traditional French cooking does not use seaweed or raw fish, so neoclassical models do not make assumptions derived from psychology, anthropology, or sociology. I disagree with any rules that limit the nature of the ingredients in economic models".
– George A. Akerlof, *An Economic Theorist's Book of Tales* (1984)

Over the past couple of decades, a large number of physicists have started exploring problems which fall in the domain of economic science. The common themes that are addressed by the research of most of these groups have resulted in coining a new term "econophysics" as a collective name for this venture. Bringing together the techniques of statistical physics and nonlinear dynamics to study complex systems along with the ability to analyze large volumes of data with sophisticated statistical techniques, the discoveries made in this field have already attracted the attention of mainstream physicists and economists. While still somewhat controversial, it provides a promising alternative to, and a more empirically based foundation for the study of economic phenomena than, the mainstream axiom-based mathematical economic theory.

Physicists have long had a tradition of moving to other fields of scientific inquiry and have helped bring about paradigm shifts in the way research is carried out in those areas. Possibly the most well-known example in recent times is that of the birth of molecular biology in the 1950s and 1960s, when pioneers such as Schrödinger (through his book *What is Life?*) inspired physicists such as Max Delbruck and Francis Crick to move into biology with spectacularly successful results. However, one can argue that physicists are often successful in areas outside physics because of the broad-based general nature of a physicist's training, rather than the applicability of physical principles as such in those areas. The large influx of physicists since the late 1990s into topics which had traditionally been the domain of economists and sociologists have raised the question: does physics really contribute towards gaining significant insights into these areas? Or, is it a mere fad, driven by the availability of large quantities of economic data which are amenable to the kind of analytical techniques that physicists are familiar with?

Econophysics. Sitabhra Sinha, Arnab Chatterjee, Anirban Chakraborti, and Bikas K. Chakrabarti
Copyright © 2011 WILEY-VCH Verlag GmbH & Co. KGaA, Weinheim
ISBN: 978-3-527-40815-3

The coining of new terms such as econophysics and sociophysics (along the lines of biophysics and geophysics) have hinted that many physicists do believe that physics has a novel perspective to contribute to the traditional way of doing economics. Others, including the majority of mainstream economists, have been dismissive until very recently of the claim that physics can have something significant to contribute to the field. Physics is seen by them to be primarily a study of interactions between simple elements, while economics deals exclusively with rational agents, able to formulate complex strategies to maximize their individual utilities (or welfare).

However, even before the current worldwide crisis revealed the inadequacies of mainstream economic theory, economists had realized that this new approach of looking at economic problems cannot be simply ignored, as indicated, for example by the entry of the terms "econophysics" and "economy as a complex system" in the *New Palgrave Dictionary of Economics* (Macmillan, 2008). The failure of economists by and large to anticipate the collapse of markets worldwide in 2008 over a short space of time has now led to some voices from within the field of economics itself declaring that new foundations for the discipline are required. The economists Lux and Westerhoff in an article published in *Nature Physics* in 2009 [1] have suggested that econophysics may provide such an alternative theoretical framework for rebuilding economics. As Lux and other economists have pointed out elsewhere [2], the systemic failure of the standard model of economics arises from its implicit view that markets and economies are inherently stable. Similar sentiments have been expressed by the econophysicist Bouchaud in an essay in *Nature* published in the same year [3].

However, worldwide financial crises (and the accompanying economic turmoil) are neither new nor as infrequent as economists would like to believe. It is therefore surprising that mainstream economics have ignored, and sometimes actively suppressed, the study of crisis situations. The famous economist Kenneth Arrow even tried to establish the stability of economic equilibria as a mathematical theorem; however, what is often forgotten is that such conclusions are crucially dependent on the underlying simplifying assumptions, such as, perfectly competitive markets and the absence of any delays in response. It is obvious that the real world hardly conforms to such ideal conditions. Moreover, the study of a wide variety of complex systems, e.g., from cellular networks to the internet and ecosystems, over the past few decades using the tools of statistical physics and nonlinear dynamics has led to the understanding that inherent instabilities in dynamics often accompanies increasing complexity.

The obsession of mainstream economics with the ideal world of hyper-rational agents and almost perfect competitive markets has gone hand in hand with a formal divorce between theory and empirical observations. Indeed, the analysis of empirical data has ceased to be a part of economics, and has become a separate subject called econometrics. Since the 1950s, economics has modeled itself more on mathematics than any of the natural sciences. It has been reduced to the study of self-consistent theorems arising out of a set of axioms to such an extent that it is probably more appropriate to term mainstream economics as *economathematics*, that is mathematics inspired by economics and that too, having little connection to reality. This is strange for a subject that claims to have insights and remedies for one of the most important spheres of human activity. It is a sobering thought that decisions made by the IMF and World Bank which affect millions of lives are made on the basis of theoretical models that have never been subjected to empirical verification. In view of this, some scientists (including a few economists) have begun to think that maybe economics is too important to be left to economists alone. While a few have suggested that econophysics may provide an alternative theoretical framework for a new economic science, we think that the field as it stands is certainly an exciting development in this direction, and intend to give an introduction to it here.

Before describing in this book how physicists have brought fresh perspectives on understanding economic phenomena in recent times, let us point out here that despite the present divorce of economics from empirical observation, there has been a long and fruitful association between physics and economics. Philip Mirowski, in his book, *More Heat Than Light* [4] has pointed out that the pioneers of neoclassical economics had indeed borrowed almost term by term the physics of 1870s to set up their theoretical framework. This legacy can still be seen in the attention paid by economists to maximization principles (e.g., of utility) that mirrors the framing of classical physics in terms of minimization principles (e.g., the principle of least action). Later, Paul Samuelson, the second Nobel laureate in economics and the author of possibly the most influential textbook of economics, tried to reformulate economics as an empirically grounded science modeled on physics in his book *Foundations of Economic Analysis* (1947). While the use of classical dynamical concepts such as stability and equilibrium has also been used in the context of economics earlier (e.g., by Vilfredo Pareto), Samuelson's approach was marked by the assertion that economics should be concerned with "the derivation of operationally meaningful theorems", that is those which can be empirically tested. Such a theorem is "simply a hypothesis about empirical data which could conceivably be refuted, if only under ideal conditions". Given the spirit of those times, it is probably unsurprising that this is also when the engineer-turned-economist Bill Philips (who later became famous for the Philips curve, a relation between inflation and employment) constructed the Moniac, a hydraulic simulator for the national economy (Figure 1.1) that modeled the flow of money in society through the flow of colored water. The mapping of macroeconomic concepts to the movement of fluids was a direct demonstration that the economy was as much a subject of physical inquiry as other more traditional subjects in physics.

This was however the last time that physics would significantly affect economics until very recently, as the 1950s saw a complete shift in the focus of economists towards proving existence and uniqueness of equilibrium solutions in the spirit of mathematics. A parallel development was the rise of mathematical game theory, pioneered by John von Neumann. To mathematically inclined economists, the language of game theory seemed ideal for studying how selfish individuals constantly devise strategies to get the better of other individuals in their continuing endeavor to maximize individual utilities. The fact that this ideal world of paranoid, calculating hyper-rational agents could never be reproduced in actual experiments carried

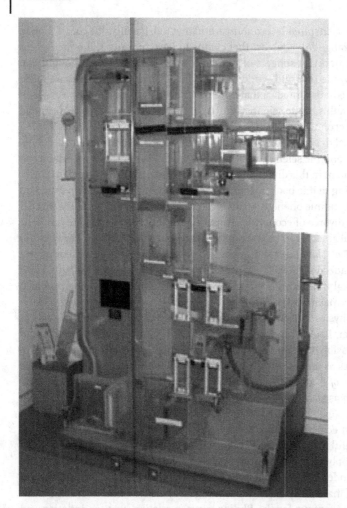

Figure 1.1 The economy machine. A reconstruction of the Moniac (at the University of Melbourne), a hydraulic simulator of a national economy built in 1949 by A.W.H. Phillips of the London School of Economics, that used the flow of colored water to represent the flow of money. It is currently again being used at Cambridge University for demonstrating the dynamic behavior of an economic system in economics first-year lectures. [Source: [5], Photo: Brett Holman]

out with human subjects where "irrational" cooperative action was seen to be the norm, could not counter the enthusiasm with which economists embraced the idea that society converges to an equilibrium where it is impossible to make someone better off without making someone else worse off. Further developments of rational models for interactions between economic agents became so mathematically abstract, that an economist recently commented that it seems (from an economic theorist's point of view) even the most trivial economic transaction is like a complicated chess game between Kenneth Arrow and Paul Samuelson (the two most famous American economists of the post-war period). The absurdity of such a situation is clear when we realize that people rarely solve complicated maximization equations in their head in order to buy groceries from the corner store. The concept of bounded rationality has recently been developed to take into account practical constraints (such as the computational effort required) that may prevent the system from reaching the optimal equilibrium even when it exists.

It is in the background of such increasing divergence between economic theory and reality that the present resumption of the interrupted dialogue between physics and economics took place in the late 1980s. The condensed matter physicist Philip Anderson jointly organized with Kenneth Arrow a meeting between physicists and economists at the Santa Fe Institute that resulted in several early attempts by physicists to apply the recently developed tools in nonequilibrium statistical mechanics and nonlinear dynamics to the economic arena (some examples can be seen in the proceedings of this meeting, *The Economy as an Evolving Complex System*, 1988) [6]. It also stimulated the entry of other physicists into this inter-disciplinary research area, which, along with slightly later developments in the statistical physics group of H. Eugene Stanley at Boston University, finally gave rise to econophysics as a distinct field, the term coined by Stanley in 1995, in Kolkata. Currently there are groups in physics departments around the world who are working on problems related to economics, ranging from Japan to Brazil, and from Ireland to Israel. While the problems they work on are diverse, ranging from questions about the nature of the distribution of price fluctuations in the stock market to models for explaining the observed economic inequality in society to issues connected with dynamical fluctuations of prices as a consequence of delays in the propagation of information, a common theme has been the observation and explanation for scaling relations (or power laws). Historically, scaling relations have fascinated physicists because of their connection to critical phenomena; but more generally, they indicate the presence of universal behavior. Indeed, the quest for invariant patterns that occur in many different contexts may be said to be the novel perspective that this recent incursion of physicists has brought to the field of economics, and that may well prove to be the most enduring legacy of econophysics.

1.1
A Brief History of Economics from the Physicist's Perspective

When physics started to develop, around the time of Galileo Galilei (1564–1642), there were hardly any fully matured fields in science from which to get help or inspiration. The only science that was somewhat advanced was mathematics, which is an analytical science (based on logic) and not empirical (based on observations/experiments carried out in controlled environments or laboratories). Yet, developments in mathematics, astronomical studies in particular, had a deep impact on the development of physics, of which the (classical) foundation was almost single-handedly laid down by Isaac Newton (1643–1727) in the seventeenth and early eighteenth century. Mathematics has remained at the core of physics since then. The rest of "main stream" sciences, like chemistry, biology, etc., have all tried to obtain inspiration from, utilize, and compare with physics since that time.

In contrast, development in the social sciences started much later. Even the earliest attempt to model an agricultural economy in a kingdom, the "physiocrats' model", named after the profession of its pioneer, the French royal physician Francois Quesnay (1694–1774), came only in the third quarter of the eighteenth century when physics was already put on firm ground by Newton. The physiocrats made the observation that an economy consists of the components like land and farmers, which are obvious. Additionally, they identified the other components as investment (in the form of seeds from previous savings) and protection (during harvest and collection, by the landlord or the king). The impact of the physical sciences in emphasizing these observations regarding components of an economy is clear. The analogy with human physiology then suggested that, like the healthy function of a body requiring proper performance of each of its components or organs, and the (blood) flow among them remaining uninterrupted, each component of the economy should be given proper care (suggesting rent for land and tax for protection!). Although the physiocrats' observations were appreciated later, the attempt to make conclusions based on the analogy with human physiology was not.

Soon, during their last phase, Mercantilists like Wilhelm von Hornick (1638–1712), James Stewart (1712–1780), and others, made some of the most profound and emphatic observations in economics, leading to the foundation of political economy. In particular, British merchants who traded in the colonies, including India, in their own set terms observed that instabilities arise as a result of growing unemployment in their home country. They also observed that whenever there is a net trade deficit and outflow of gold (export being less than import), this led to the formulation of the problem of effective demand: even though the merchants, or traders were independently trading (exporting or importing goods) with success, the country's economy as a whole did not do well due to lack of overall demand when there was a net flow of gold (the international exchange medium) to balance the trade deficit! This still remains as a major problem in macroeconomics. The only solution in those days was to introduce tax on imports: third party (in this case the government) intervention on the individual's choice of economic activity (trade). This immediately justified the involvement of the government in the economic activities of individuals.

In a somewhat isolated but powerful observation, Thomas Malthus (1766–1834) made a very precise model of the conflict between agricultural production and population growth. He assumed that the agricultural production can only grow (linearly) with the area of the cultivated land. With time t, in years, the area can only grow linearly ($\propto t$) or in arithmetic progression (AP). The consumption depends on the population which, on the other hand, grows exponentially ($\exp[t]$) or in geometric progression (GP). Hence, with time, or year $1, 2, 3, \ldots$, the agricultural production grows as $1, 2, 3, \ldots$, while the consumption demand or population grows in a series like $2, 4, 8, \ldots$. This means that it does not matter how large the area of cultivable land we start with, the population GP series soon overtakes the food production AP series and the population faces a disaster, resulting in famine, war or revolution.

They are inevitable, as an exponentially growing function will always win over a linearly growing function and such disasters will appear almost periodically in time.

Adam Smith (1723–1790) made the first attempt to formulate economic science. He painstakingly argued that a truly many-body system of selfish agents, each having no idea of benevolence or charity towards its fellow neighbors, or having no foresight (views very local in space and time), can indeed reach an equilibrium where the economy as a whole is most efficient; leading to the best acceptable price for each commodity. This "invisible hand" mechanism of the market to evolve towards the "most efficient" (beneficial to *all* participating agents) predates the demonstration of the "self-organization" mechanism in physics or chemistry of many-body systems, where each constituent cell or automata follows very local (in space and time) dynamical rules and yet the collective system evolves towards a globally "organized" pattern (cf. Ilya Prigogine (1917–), Per Bak (1947–2002) and others). This idea of "self-organizing" or "self-correcting economy" by Smith of course contradicted the prescription of the Mercantilists regarding government intervention in the economic activities of the individuals, and argued tampering by an external agency to be counterproductive.

Soon, the problem of price or value of any commodity in the market became a central issue. Following David Ricardo's (1772–1823) formulation of rent and labor theory of value, where the price depends only on the amount of labor put forth by the farmers or laborers, Karl Marx (1818–1883) formulated and advocated emphatically the surplus labor theory of value or wealth in any economy. However, neither could solve the price paradox: why diamonds are expensive, while coal is cheap. The amount of labor in mining is more or less the same for both diamonds and coal. Yet, the prices differ by an astronomical amount. This clearly demonstrates the failure of the labor theory of value. The alternative put forth was the utility theory of price: the more the utility of a commodity, the higher its price. But then, how does one explain why a bottle of water costs less than a bottle of wine? The argument could be made that water is more important for sustaining life and certainly has more utility! The solution identified was marginal utility. According to marginal utility theory, not the utility but rather its derivative with respect to the quantity determines the price: water is cheaper as its marginal utility at the present level of its availability is less than that for wine – this will surely change in a desert. This still does not solve the problem completely. Of course increasing marginal utility creates increasing demand for it, but its price must depend on its supply (and will be determined by equating the demand with the supply). If the offered (hypothetical) price p of a commodity increases, the supply will increase and the demand for that commodity will decrease. The price, for which supply S will be equal to demand D, will be the market price of the commodity: $S(p) = D(p)$ at the market (clearing) price. However, there are problems still. Which demand should be equated to which supply? It is not uncommon to see often that price as well as the demand for rice, for example in India, increases simultaneously. This can occur when the price of the other staple alternative, wheat, increases even more.

The solutions to these problems led ultimately to the formal development of economic science in the early twentieth century by Léon Walras (1834–1910), Alfred Marshal (1842–1924), and others: marginal utility theory of price and cooperative or coupled (in all commodities) demand and supply equations. These formulations went back to the self-organizing picture of any market, as suggested by Adam Smith, and incorporated this marginal utility concept, utilizing the following coupled demand-supply equations:

$$D_i(p_1, p_2, \ldots, p_i, \ldots, p_N, M) = S_i(p_1, p_2, \ldots, p_i, \ldots, p_N, M),$$

for N commodities and total money M in the market, each having relative prices p_i (determined by marginal utility rankings), and demand D_i and supply S_i, where $i = 1, 2, \ldots, N$ and the functions D or S are in general nonlinear in their arguments. These formal and abstract formulations of economic science were not appreciated in their early days and had a temporary setback. The lack of acceptance was due to the fact that neither utility nor marginal utility is measurable and the formal solutions of these coupled nonlinear equations in many (p_i) variables still remain elusive. The major reason for the lack of appreciation for these formal theories was a profound and intuitive observation by John Maynard Keynes (1883–1946) on the fall of aggregate (or macroeconomic) effective demand in the market (as pointed out earlier by the Mercantilists, this time due to "liquidity preference" of money by the market participants) during the great depression of the 1930s. His prescription was for government intervention (in direct contradiction with the laissez-faire ideas of leaving the market to its own devices for bringing back the equilibrium, as Smith, Walras, and others have proposed) to boost aggregate demand by fiscal measures. This prescription had immediate success in most cases. By the third quarter of the twentieth century, however, its limitations became apparent and the formal developments in microeconomics took the front seat again.

Several important, but isolated observations contributed later in significant ways. Vilfredo Pareto (1848–1923) observed that the number density $P(m)$ of wealthy individuals in any society decreases rather slowly with their wealth or income m: $P(m) \sim m^{-\alpha}$; for very large m (very rich people), $2 < \alpha < 3$ (*Cours d'Economic Politique*, Lausanne, 1897). It must be mentioned, at almost the same time, Joshiah Willard Gibbs (1839–1903) put forth that the number density $P(\epsilon)$ of particles (or microstates) with energy ϵ in a thermodynamic ensemble in equilibrium at temperature T falls off much faster: $P(\epsilon) \sim \exp[-\epsilon/T]$ (*Elementary Principles of Statistical Mechanics*, 1902). This was by then rigorously established in physics. The other important observation was by Louis Bachelier (1870–1946) who modeled the speculative price fluctuations (σ), over time τ, using Gaussian statistics (for a random walk): $P(\sigma) \sim \exp[-\sigma^2/\tau]$ (*Thesis: Théorie de la Spéculation*, Paris, 1900). This actually predated Albert Einstein's (1879–1955) random walk theory (1905) by five years. In another isolated development, mathematician John von Neumann (1903–1957) started developing game theories for microeconomic behavior of partners in oligopolistic competitions (to take care of the strategy changes by agents, based on earlier performance).

In mainstream economics, Paul Samuelson (1915–) investigated the dynamic stabilities of the demand-supply equilibrium by formulating, following Newton's equations of motion in mechanics, dynamical equations $\frac{dD_i}{dt} = \sum_j J_{ij} D_j(p_1, p_2,$ $\ldots, p_N, M)$ and $\frac{dS_i}{dt} = \sum_j K_{ij} S_i(p_1, p_2, \ldots, p_N, M)$, with the demand and supply (overlap) matrices $\underline{J}$ and $\underline{K}$, respectively for N commodities, and by looking for the equilibrium state(s) where $dS/dt = 0 = dD/dt$ at the market clearing prices $\{p_i^*\}$. Note that, in the absence of coupling, for each commodity the equilibrium price is obtained when the demand equals supply, that is $D_i(\{p\}, M) = S_i(\{p\}, M)$. Jan Tinbergen (1903–1994), a statistical physicist (student of Paul Ehrenfest of Leiden University) analyzed the business cycle statistics and initiated the formulation of econometrics. By this time, these formal developments in economics, with clear influence from other developed sciences (physics in particular), were becoming recognized. In fact, Tinbergen was the first recipient of the newly instituted Nobel Memorial Prize in Economics in 1969. The next year, the prize went to Samuelson. Soon after that, the formal development of certain economic concepts were made, like the axiomatic foundations of utility (ranking) theory, and the solution of general equilibrium theory by Kenneth Arrow (1921–), the ideas of George Stigler (1911–1991), who first performed Monte Carlo simulations of markets (similar to those of thermodynamic systems in physics), or that of John Nash (1928–), giving the proof of the existence of equilibrium solutions in strategic games, etc. All were awarded the Nobel Prize in Economics in 1972, 1982, and 1994, respectively. Although the impact of developments in physics has had a clear impact on economics, it has become more explicit in the last fifteen years.

The latest developments leading to econophysics had their seed in several earlier observations. Important among them was the observation by Benoit Mandelbrot (1924–) in 1963 that speculative fluctuations (in the cotton market for example) have a much slower rate of decay, compared to that suggested by the Gaussian statistics of Bachelier, and decreases following power-law statistics: $P(\sigma) \sim \sigma^{-a}$ with some robust exponent value (a) depending on the time scale of observations. With the enormous amount of stock market data now available on the internet, Eugene Stanley, Rosario Mantegna and coworkers established firmly the above mentioned (power law) form of the stock price fluctuation statistics in the late 1990s. Simultaneously, two important modeling efforts began, inspired directly by physics: the minority game models, for considering contiguous behavior (in contrast to perfect rational behavior) of agents in the market, and learning from the past performance of strategies, were developed by Brian Arthur, Damien Challet, Yi-Cheng Zhang and others, starting in 1994. The other modeling effort was to capture the income or wealth distribution in society, similar to energy distributions in (ideal) gases. These models intend to capture both the initial gamma/log-normal distribution for the income distributions of poor and middle-income groups and also the Pareto tail of the distribution for the rich. It turned out, as shown by the Kolkata group from 1990 to 2000, a random saving gas model can easily capture these features of the distribution function. However, the model had several well-documented previous, somewhat incomplete, versions available for quite some time. Meghnad Saha (1893–1956), the founder of the Saha Institute of Nuclear Physics, in Kolkata, and collaborators had already discussed at length in their text book in the 1950s, the possibility of using a Maxwell–Boltzmann velocity distribution (a gamma distribution) in an ideal gas to represent the income distribution in societies: "suppose in a country, the assessing department is required to find out the average income per head of the population. They will proceed somewhat in the similar way ... (the income distribution) curve will have this shape because the number of absolute beggars is very small, and the number of millionaires is also small, while the majority of the population have average income." ("Distribution of velocities" in *A Treatise on Heat*, M.N. Saha and B.N. Srivastava, Indian Press, Allahabad, 1950; pp. 132–134). This modeling had the obvious drawback that the distribution could not capture the Pareto tail. However, the accuracy of this Gibbs distribution for fitting the income data available now from the Internet has been pointed out recently by Victor Yakovenko and collaborators in a series of papers since 2000. The "savings" ingredient in the ideal-gas model, required for obtaining the gamma function form of the otherwise ideal gas (Gibbs) distribution, was also discovered more than a decade earlier by John Angle. He employed a different driver in his stochastic model of an inequality process. This inequality coming mainly from the stochasticity, together with the equivalent of saving introduced in the model. A proper Pareto tail of the gamma distribution comes naturally in this class of models when the saving propensity of the agents are distributed, as noted and analyzed first by the Kolkata group and by the Dublin group led by Peter Richmond.

Apart from the intensive involvements of physicists together with a few economists in this new phase of development, a happy outcome has been that econophysics has nearly established itself as a popular research discipline in statistical physics. Many physics journals have started publishing papers on such interdisciplinary fields. Courses on econophysics are also now being offered in several universities, mostly in their physics departments.

1.2
Outline of the Book

Here we shall give a brief outline of the book. We begin (in Chapter 2) with a discussion of the random walk, a versatile model of several natural phenomena, which shows how cumulative random effects can give rise to a well-understood distribution. The financial market, in particular, is thought by some, to exhibit a random walk; however, the deviation of the observed stock price (or market index) movements from that expected for a pure random walk, alerts us to the possibility of effects other than independent and uncorrelated random events playing a role. In the following chapter (Chapter 3), we look at these deviations in detail, focusing on the property of multifractality. Fractal or self-similar properties is often seen in many economic and financial systems, and multifractality is a generalization of the basic fractal concept. We also look at several types of cyclic temporal behavior. As the deviations from a pure random walk can also be a result of correlations between the different components of a system, in Chapter 4 we discuss methods of analyzing the cross-correlation between stock prices in a financial market. Using this knowledge one can build a picture of the network of interactions between the different players. It also throws interesting light on the difference between emerging and developed markets.

However, correlations by themselves do not explain other observed features of price fluctuations, such as, the existence of power-law tails in their distribution. Thus, in Chapter 5, we look in detail at power laws (i.e., scale-free distributions) and discuss them in the context of the financial market. While physicists have, for various reasons, been particularly interested in power laws, economic phenomena show several other kinds of distributions. One very commonly observed form is the log-normal distribution, which is discussed in Chapter 6. As limited data sets can often cause scientists to erroneously identify a log-normal as a power law, we believe that this often neglected distribution (in physics) should be much more widely discussed than it has been thus far. It is also possible, that an empirical distribution may not be properly fit by any single distribution. An example is the distribution of wealth (as well as income) in society. In these cases, the fitting of different parts of the data by various distribution functional forms can also suggest that multiple dynamical processes may be at play. In the next chapter (Chapter 8), we follow this up by discussing several models which reproduce these kinds of distributions.

Physicists have often been accused of simplifying reality too much in their efforts to study it. In the context of socioeconomic phenomena, it is often asked whether the basic constituents of physical models, which are simple particles, can capture the behavior arising from interactions between rational individuals, the complexity of whose decision-making behavior is beyond the power of advanced computers to mimic. In order to look at some aspects of how including strategy-based decision making in the dynamics of individuals can change the behavior of a system, in Chapter 9 we discuss several agent-based models, including the minority game.

Another simplification by physicists that often draws the ire of social scientists is the assumption of homogeneity or well-mixedness in the contacts between individuals. It goes without saying that if all kinds of interactions are possible, the physical theory becomes more tractable but it may not be capturing reality. With this aim in mind, in Chapter 10, we look at the emerging field of complex networks, in the context of economics. Examples of such networks occur widely in economics and finance, including the world trade web and the hierarchical organization structure within a company. Such analysis also alerts us to the possibly destabilizing effects of complex systems. We reflect on this point in the concluding chapter (Chapter 11), where we discuss how econophysics can bring a fresh perspective to the problem of how to achieve sustainable economic growth. The following appendices discuss all the physics concepts that have been used frequently in the econophysics literature.

References

1 Lux, T. and Westerhoff, F. (2009) Economic crisis. *Nature Physics*, **5**, 2–3.

2 Colander, D., Föllmer, H., Haas, A., Goldberg, M., Juselius, K., Kirman, A., Lux, T., and Sloth, B. (2009) The finan-

cial crisis and the systemic failure of academic economics. *Critical Review*, **21**, 249–267.

3 Bouchaud, J.-P. (2009) Economics needs a scientific revolution. *Nature*, **455**, 1181.

4 Mirowski, P. (1989) *More Heat Than Light: Economics as Social Physics, Physics as Nature's Economics*, Cambridge University Press, Cambridge.

5 http://airminded.org

6 Anderson, P.W., Arrow, K. and Pines D. (1988) *The Economy As An Evolving Complex System*, Perseus Books, Cambridge, Mass.

2
The Random Walk

"All truly great thoughts are conceived while walking."

– Friedrich Nietzsche

2.1
What is a Random Walk?

2.1.1
Definition of Random Walk

We begin the chapter by defining the random walk. Mathematically the trajectory consisting of successive "random" steps (e.g., decided by the flips of an unbiased coin), is known as a random walk. A particularly simple random walk would be that on the integers, which starts at time zero ($t = 0$), $S_0 = 0$ and at each step moves by 1 or -1 with equal probability. To define this walk more formally, we take independent random variables x_i, each of which is 1 with probability $1/2$ and -1 with probability $1/2$, and compose the set $S_t = \sum_{i=1}^{t} x_i$. This sequence S_t is then called the simple random walk on integers.

This walk can be illustrated (see Figure 2.1), for example, by repeatedly flipping an unbiased coin. At each time t, if it lands on heads H, you move one to the right on the number line, and if it lands on tails T, then you move one to the left. Now, suppose you have flipped the coin five times successively, then you have the chances of landing on any one of the values $1, -1, 3, -3, 5, -5$. You may land on 1 by flipping three heads and two tails in any order. Then, analyzing carefully, we find there are 10 different possible ways of landing on 1. Similarly, there are 10 different ways of landing on -1 (by flipping three tails and two heads), five different ways of landing on 3 (by flipping four heads and one tail), five different ways of landing on -3 (by flipping four tails and one head), only one way of landing on 5 (by flipping five heads), and only one way of landing on -5 (by flipping five tails).

These results are directly related to the properties of *Pascal's triangle*. It is easy to understand that the number of different walks of n steps where each step is of value $+1$ or -1 can be 2^n. For the simple random walk, each of these walks are

Econophysics. Sitabhra Sinha, Arnab Chatterjee, Anirban Chakraborti, and Bikas K. Chakrabarti
Copyright © 2011 WILEY-VCH Verlag GmbH & Co. KGaA, Weinheim
ISBN: 978-3-527-40815-3

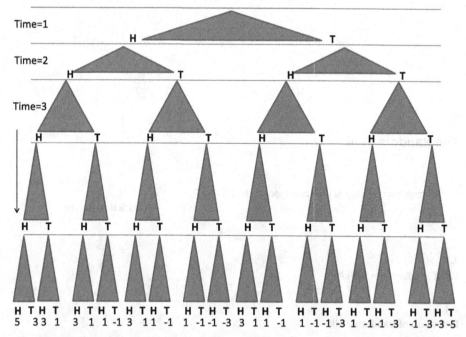

Figure 2.1 Random coin flips. If there is a head H we move right on the number line (add +1), and if there is a tail T we move left on the number line (add −1).

equally likely, so that in order for S_n to be equal to a number k, it is necessary and sufficient that the difference between the number of +1s and −1s in the walk exceeds k. Hence, the number of walks which satisfy $S_n = k$ is exactly the total number of ways we can choose $(n + k)/2$ elements from an n element set, which is nothing but an entry in *Pascal's triangle* denoted by $^nC_{(n+k)/2}$. Note that for this to be non-zero, it is necessary that $n + k$ is an even number. Thus, the probability that $S_n = k$ is equal to $2^{-n} {}^nC_{(n+k)/2}$.

The relation of the simple random walk of integers and the Pascal's triangle (see Figure 2.2) is easily demonstrated for small values of n. At time zero, the only possibility will be to remain at zero. However, at time step unity, one can move either to the left or the right of zero, meaning there is *one* chance of landing on −1 or *one* chance of landing on 1. At time step two, we will have to examine the moves from the time step before. If one had been at 1, one could move to 2 or back to zero. If one had been at −1, one could move to −2 or back to zero. So there is *one* chance of landing on −2, *two* chances of landing on zero (highlighted in Figure 2.2), and *one* chance of landing on 2. We shall study more interesting aspects of the random walk later in this chapter.

The results of random walk analysis is central in physics, chemistry, economics and a number of other fields as a fundamental model for random (stochastic) processes in time. There are many systems for which at smaller scales, the interactions

n	-5	-4	-3	-2	-1	0	1	2	3	4	5
$P[S_0 = k]$						1					
$2P[S_1 = k]$					1		1				
$2^2P[S_2 = k]$				1		2		1			
$2^3P[S_3 = k]$			1		3		3		1		
$2^4P[S_4 = k]$		1		4		6		4		1	
$2^5P[S_5 = k]$	1		5		10		10		5		1

Figure 2.2 Pascal's triangle. The construction of Pascal's triangle is as follows: On the first row S_0, we write only the number 1. To write the elements of the following rows, we add the number directly above and to the left, with the number directly above and to the right, such that if either the number to the right or left is absent, we substitute it with a zero. For example, to construct the value 2 (highlighted in the third row), we have added the 1s highlighted in the second row.

with the environment and their influence are in the form of random fluctuations, as in the case of "Brownian motion"[1]. If the motion of a pollen grain in a fluid like water is observed under a microscope, it would look somewhat like what is shown in the Figure 2.3.

It is interesting to note that the path traced by the pollen grain as it travels in a liquid (observed by R. Brown and analyzed physically first by A. Einstein), and the price of a fluctuating stock (studied first by L. Bachelier), can both be modeled as random walks (theory of stochastic processes). It is noteworthy that the formulation of the random walk model – as well as of a stochastic process – was first done in the framework of the economic study by L. Bachelier [1, 2], even five years prior to the work of A. Einstein!

1) The motion of a particle is called Brownian motion, in honor of the botanist Robert Brown who observed it for the first time in his studies of pollen (see original paper: R. Brown, Phil. Mag. 4, 161 (1828)). In 1828 he wrote "the pollen become dispersed in water in a great number of small particles which were perceived to have an irregular swarming motion". The physical theory of such motion, however, was derived by A. Einstein in 1905 (see original paper in German: Einstein, A. (1905) Ann. Phys 17, 549) when he wrote: "In this paper it will be shown that ... bodies of microscopically visible size suspended in a liquid perform movements of such magnitude that they can be easily observed in a microscope on account of the molecular motions of heat ..."

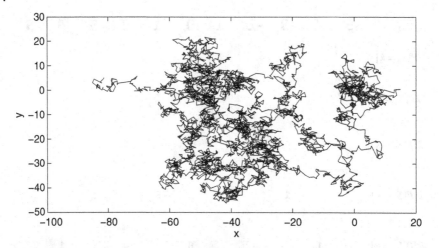

Figure 2.3 Simulated Brownian motion (5000 time steps).

There are, of course, other systems that present unpredictable "chaotic" behavior, this time due to dynamically generated internal "noise". Noisy processes in general, either truly stochastic or chaotic in nature, represent the rule rather than the exception. In this chapter, we will concentrate only on the theory of random or stochastic processes.

Louis Bachelier

Louis Jean-Baptiste Alphonse Bachelier (1870–1946) was a French mathematician. In his Ph.D. thesis, "The Theory of Speculation", which was published in 1900, he discussed the use of random walk theory to evaluate prices of government bonds. It was a pioneering work in financial mathematics; in fact, the first to use advanced mathematics in the study of finance. It is noteworthy that Bachelier's work on random walks was published five years prior to Einstein's celebrated study of Brownian motion. His instructor, Henri Poincare (another notable French mathematician and physicist), is known to have given some positive comments. His thesis had received a note of honorable mention, and was published in the prestigious Annales Scientifiques de l'Ecole Normale Superieure. Bachelier further developed the theory of diffusion processes. Bachelier became a "free professor" at the Sorbonne in 1909. He published the book, Le Jeu, la Chance, et le Hasard (Games, Chance, and Risk) in 1914. Later, Bachelier was given a permanent professorship at the Sorbonne. Bachelier was drafted into the French army during World War I, after which he found two temporary positions in Besancon and Dijon, before moving to Rennes in 1925. He was finally awarded a permanent professorship in 1927 at Besancon, where he worked for another 10 years.

2.1.2
The Random Walk Formalism and Derivation of the Gaussian Distribution

The original statement of the random walk problem was posed by Karl Pearson in 1905, who was intrigued by it. He appealed to the general readers of Nature 72, 294 (1905) to come up with a solution to the problem. Amongst the respondents was Lord Rayleigh, with whose help Pearson concluded that "the most probable place to find a drunken man who is at all capable of keeping on his feet is somewhere near his starting point!"

Let us consider that a drunkard begins at a lamp post and takes N steps of equal lengths in random directions, how far will the drunkard be from the lamp post? We will consider an idealized example of a random walk for which the steps of the walker are restricted to a line (a one-dimensional random walk). Each step is of equal length a, and at each interval of time, the walker either takes a step to the right with probability p or a step to the left with probability $q = 1 - p$. The direction of each step is *independent* of the preceding one. Let n be the number of steps to the right, and m the number of steps to the left. The total number of steps is $N = n + m$. The question we now ask: What is the probability $P_N(n)$ that a random walker in one dimension has taken, for example, n steps to the right out of N total steps?

Instead of the above example with a random walker, had we considered the physics (and probability distributions) of non interacting magnetic moments, or as in the earlier subsection, the flips of a coin, we would arrive at identical results (and hence we will use the terms interchangebly). All these examples have two characteristics in common. First, in each trial there are only two outcomes, for example, up or down, heads or tails, and right or left. Second, the result of each trial is *independent* of all previous trials, for example, the drunkard has no memory of his previous steps. This type of process is generally called a *Bernoulli process* (after the mathematician Jacob Bernoulli, 1654–1705). We will cast our discussion of Bernoulli processes in terms of the random walk. The main quantity of interest is the probability $P_N(n)$, which we now calculate for arbitrary N and n. We know that a particular outcome with n right steps and m left steps occurs with probability $p^n q^m$. We write the probability $P_N(n)$ as

$$P_N(n) = W_N(n, m) p^n q^m , \tag{2.1}$$

where $m = N - n$ and $W_N(n, m)$ is the number of distinct configurations of N steps with n right steps and m left steps.

From our earlier discussion of random coin flips, we will be able to deduce easily the first several values of $W_N(n, m)$. We can determine the general form of $W_N(n, m)$ by obtaining a recursion relation between W_N and W_{N-1}. A total of n right steps and m left steps out of N total steps can be found by adding one step to $N - 1$ steps. The additional step is either (a) right if there are $(n - 1)$ right steps and m left steps, or (b) left if there are n right steps and $(m - 1)$ left steps. Because there are $W_{N-1}(n-1, m)$ ways of reaching the first case and $W_{N-1}(n, m-1)$ ways for the second case, we obtain the recursion relation

$$W_N(n, m) = W_{N-1}(n-1, m) + W_{N-1}(n, m-1) .$$ (2.2)

If we begin with the known values $W_0(0,0) = 1$, $W_1(1,0) = W_1(0,1) = 1$, we can use the recursion relation to construct $W_N(n, m)$ for any desired N. For example,

$$W_2(2, 0) = W_1(1, 0) + W_1(2, -1) = 1 + 0 = 1 ,$$
$$W_2(1, 1) = W_1(0, 1) + W_1(1, 0) = 1 + 1 = 2 ,$$
$$W_2(0, 2) = W_1(-1, 2) + W_1(0, 1) = 0 + 1 = 1 .$$

Thus, we identify that $W_N(n, m)$ forms the Pascal's triangle (see Figure 2.2). It is straightforward to show by induction that the expression

$$W_N(n, m) = \frac{N!}{n!m!} = \frac{N!}{n!(N-n)!}$$ (2.3)

satisfies the relation (2.2), since by definition $0! = 1$. We can combine (2.1) and (2.3) to find the desired result

$$P_N(n) = \frac{N!}{n!(N-n)!} p^n q^{N-n} .$$ (2.4)

The form of (2.4) is called the "binomial distribution". Note that for $p = q = 1/2$, such as in the case of an unbiased coin, $P_N(n)$ reduces to

$$P_N(n) = \frac{N!}{n!(N-n)!} 2^{-N} .$$ (2.5)

The reason that (2.4) is called the binomial distribution is that its form represents a typical term in the expansion of $(p + q)^N$. By the binomial theorem we have

$$(p + q)^N = \sum_{n=0}^{N} \frac{N!}{n!(N-n)!} p^n q^{N-n} .$$ (2.6)

We use (2.4) and write

$$\sum_{n=0}^{N} P_N(n) = \sum_{n=0}^{N} \frac{N!}{n!(N-n)!} p^n q^{N-n} = (p + q)^N = 1^N = 1 ,$$ (2.7)

where we have used (2.6) and the fact that $p + q = 1$.

We note some properties of the binomial distribution. Suppose first that we have exactly one Bernoulli trial. We have two possible outcomes, 1 and 0, with the first having probability p and the second having probability $q = 1 - p$. Then, we obtain mean $\mu = p.1 + q.0 = p$ and variance $\sigma^2 = (1 - p)^2 p + (0 - p)^2 q = pq$.

Now, for N such *independent* trials, we have the results

1. mean $\mu = Np$
2. variance $\sigma^2 = Npq$,

which we will utilize soon.

We also note that for large N, the binomial distribution has a well-defined maximum at pN and can be approximated by a smooth, continuous function even though only integer values of n are physically possible. We now find the form of this continuous function of n, for very large N. The first step is to realize that for $N \gg 1$, $P_N(n)$ is a rapidly varying function of n near the maximum pN, and for this reason we do not want to approximate $P_N(n)$ directly. Because the logarithm of $P_N(n)$ is a slowly varying function, we expect the power series expansion of $\ln P_N(n)$ to converge more quickly. Hence, we expand $\ln P_N(n)$ in a Taylor series about the value of $n = \tilde{n}$ at which $\ln P_N(n)$ reaches its maximum value. We will write $p(n)$ instead of $P_N(n)$ because we will treat n as a continuous variable and hence $p(n)$ would be a probability density. We find that

$$\ln p(n) = \ln p(n = \tilde{n}) + (n - \tilde{n}) \frac{d \ln p(n)}{dn}\bigg|_{n=\tilde{n}}$$

$$+ \frac{1}{2}(n - \tilde{n})^2 \frac{d^2 \ln p(n)}{dn^2}\bigg|_{n=\tilde{n}} + \dots \tag{2.8}$$

Because we have assumed that the expansion (2.8) is about the maximum $n = \tilde{n}$, the first derivative $d \ln p(n)/dn|_{n=\tilde{n}}$ must be zero, and for the same reason the second derivative $d^2 \ln p(n)/d^2 n|_{n=\tilde{n}}$ must be negative. We also assume that the higher terms in (2.8) may be neglected, and introduce two parameters A and B, such that

$$\ln A = \ln p(n = \tilde{n}) \,,$$

and

$$B = -\frac{d^2 \ln p(n)}{dn^2}\bigg|_{n=\tilde{n}} \,.$$

The approximation (2.8) and the definitions above, allow us to write

$$\ln p(n) \approx \ln A - \frac{1}{2} B(n - \tilde{n})^2 \,, \tag{2.9}$$

or

$$p(n) \approx A \exp\left(-\frac{1}{2} B(n - \tilde{n})^2\right) \,. \tag{2.10}$$

Frequently we need to evaluate $\ln N!$ for $N \gg 1$. An approximation for the value of $\ln N!$ is known as the Stirling's approximation; the simple form is given by

$$\ln N! \approx N \ln N - N \,, \tag{2.11}$$

and a more accurate form of the approximation is given by

$$\ln N! \approx N \ln N - N + \frac{1}{2} \ln(2\pi N) \,.$$

We next use the above Stirling's approximation to evaluate the first two derivatives of $\ln p(n)$ and the value of $\ln p(n)$ at its maximum to find the parameters A, B, and $\tilde{n}$. We write

$$\ln p(n) = \ln N! - \ln n! - \ln(N - n)! + n \ln p + (N - n) \ln q . \qquad (2.12)$$

It is straightforward to obtain

$$\frac{d(\ln p(n))}{dn} = -\ln n + \ln(N - n) + \ln p - \ln q . \qquad (2.13)$$

The most probable value of n is found by finding the value of n that satisfies the condition $(d \ln p)/dn = 0$. We find

$$\frac{N - \tilde{n}}{\tilde{n}} = \frac{q}{p} , \qquad (2.14)$$

or $(N - \tilde{n})p = \tilde{n}q$. If we use the relation $p + q = 1$, we obtain

$$\tilde{n} = pN . \qquad (2.15)$$

Note that $\tilde{n} = \bar{n}$; that is, the value of n for which $p(n)$ is a maximum is also the mean value of n. The second derivative can be found from (2.13). We have

$$\frac{d^2(\ln p(n))}{dn^2} = -\frac{1}{n} - \frac{1}{N - n} . \qquad (2.16)$$

Hence, the coefficient B defined earlier is given by

$$B = -\frac{d^2 \ln p(n)}{dn^2}\bigg|_{n = \tilde{n}} = \frac{1}{\tilde{n}} + \frac{1}{N - \tilde{n}} = \frac{1}{Npq} . \qquad (2.17)$$

From the properties of the binomial distribution which we wrote earlier, we see that

$$B = \frac{1}{\sigma^2}$$

where σ^2 is the variance.

If we use the simple form of Stirling's approximation to find the normalization constant A from the relation $\ln A = \ln p(n = \tilde{n})$, we would find that $\ln A = 0$. Instead, we have to use the more accurate form of Stirling's approximation. The result is

$$A = \frac{1}{(2\pi Npq)^{1/2}} = \frac{1}{\sqrt{2\pi\sigma^2}} .$$

If we substitute our results for $\tilde{n}$, B, and A into (2.10), we find the distribution

$$p(n) = \frac{1}{\sqrt{2\pi\sigma^2}} \exp\left(-(n - \bar{n})^2 / 2\sigma^2\right) , \qquad (2.18)$$

which is called the "Gaussian distribution".

From our derivation we see that (2.18) is valid for large values of N and for values of n near $\bar{n}$, and hence occur very frequently when dealing with large numbers in statistical theories. However, even for relatively small values of N, the Gaussian approximation is a good approximation for most values of n. Another important feature of the Gaussian distribution is that its relative width, which may be defined as $\sigma_n/\bar{n}$, decreases as $N^{-1/2}$, which is, of course, true for the binomial distribution as well.

Friedrich Gauss

Johann Carl Friedrich Gauss (1777–1855) was a German mathematician and physicist who made significant contributions to several fields like number theory, statistics, differential geometry, and geophysics, in addition to mainstream physics and astronomy. Sometimes known as the *Princeps mathematicorum* (the Prince of Mathematicians), Gauss had a remarkable influence in many fields of mathematics and science. Gauss attended the Collegium Carolinum (now Technische Universität Braunschweig) from 1792 to 1795, and subsequently he moved to the University of Göttingen from 1795 to 1798. A major breakthrough in his research occurred in 1796 when he was able to show that any regular polygon with a number of sides which is a Fermat prime can be constructed by compass and straightedge. This was a major discovery in construction problems, which had preoccupied mathematicians since the times of the Ancient Greeks. On the personal front, the discovery led Gauss to choose mathematics instead of philology as a career. The Gaussian distribution is one of many things named after Gauss, who used it to analyze astronomical data, and determined the formula for its probability density function. However, Gauss was not the first to study this distribution or the formula for its density function. The important method of least squares was first published by Legendre in 1805, a few years before Gauss actually justified the method rigorously and published it in 1809 (though he is credited for using it since 1794). It is difficult to actually list all the numerous contributions of Gauss, simply because they are too many. He died in 1855 at Göttingen.

2.1.3
The Gaussian or Normal Distribution

The Gaussian distribution, also called the normal distribution, is perhaps the most important family of continuous probability distributions, applicable in many fields including physics and economics. Carl Friedrich Gauss became associated with this set of distributions when he used them to analyze astronomical data and rationalized the method of least squares; he defined the equation of its probability density function. It is often called the "bell curve" because the graph of its probability density resembles a bell.[2]

The importance of the normal distribution as a model of quantitative phenomena in the natural and behavioral sciences is due in part to the "central limit theorem". Under certain conditions (such as being independent and identically distributed with finite variance), the sum of a large number of random variables is approximately normally distributed – this is the *central limit theorem*. The practical importance of the central limit theorem is that the normal cumulative distribution function can be used as an approximation to some other well-known cumulative distribution functions; for example, a binomial distribution with parameters N and p is approximately normal for large N, and p not too close to 1 or 0. The approximating normal distribution has parameters $\mu = Np$, $\sigma^2 = Np(1 - p) = Npq$. It is noteworthy that a binomial distribution with parameter $\lambda = Np$ for large n and $p \to 0$ such that $\lambda = Np$ is constant, gives another well-known distribution, known as the "Poisson distribution", with parameters $\mu = \sigma^2 = \lambda$.

It was Karl Pearson who popularized the name "normal distribution" and expressed in the modern form the continuous probability density function as:

$$P(x) = \frac{1}{\sqrt{2\pi}\sigma} \exp\left(-(x - \mu)^2/2\sigma^2\right) ,$$

where $\sigma > 0$ is the standard deviation, the real parameter μ is the mean or expected value. Each member of the Gaussian PDF family (see Figure 2.4) may be defined by two important parameters, *location* and *scale*: the mean μ and variance (standard deviation squared) σ^2, respectively. This probability density function has the notable property that it has symmetry about its mean μ, and the mode and median both equal the mean μ.

The *standard* normal distribution is the normal distribution with a mean $\mu = 0$ and a variance $\sigma^2 = 1$:

$$P(x) = \frac{1}{\sqrt{2\pi}} \exp\left(-x^2/2\right) .$$

We note that as a Gaussian function with the denominator of the exponent equal to 2, the standard normal density function is an eigenfunction of the Fourier transform.

2) The normal distribution was first introduced by Abraham de Moivre in an article in 1733, which was later reprinted in the second edition of his *The Doctrine of Chances*, *(1738)* in the context of studying binomial distributions. The result was extended by Laplace in his book *Analytical Theory of Probabilities (1812)*, where he used the normal distribution in the analysis of errors of experiments. In 1809, Carl Friedrich Gauss assumed in his analyses a normal distribution of the errors. The name "bell curve" goes back to E. Jouffret who first used the term "bell surface" in 1872 for a "bivariate normal" with independent components. It is also known that the name "normal distribution" was coined independently by Charles S. Peirce, Francis Galton and Wilhelm Lexis around 1875.

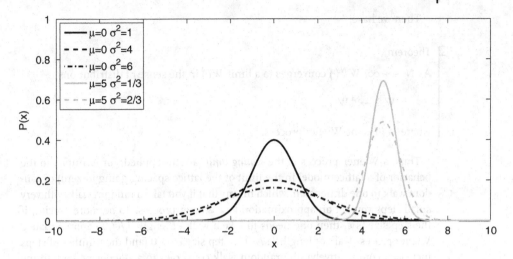

Figure 2.4 Gaussian Probability Distribution Functions with different means and standard deviations.

2.1.4
Wiener Process

The Bernoulli processes, which we mentioned earlier, can be naturally extended from the discrete versions to their continuous-time counterparts, which are called the "Poisson processes". The Poisson process is one of the best known Levy processes (stochastic processes with stationary independent increments). Another famous Levy process is the Wiener process.

The Wiener process in mathematics is defined as a continuous-time stochastic process, and is named in honor of Norbert Wiener. Norbert Wiener (1923) had ultimately proved the existence of Brownian motion and made significant contributions to related mathematical theories, including the study of the continuous-time martingales, which has become very important in financial applications.

Let $\{x_t\}_{t\in\mathbb{N}}$ be a sequence of independent and identically distributed (iid) random variables such that the mean is zero and the variance is unity, and define $S_0 = 0$ and $S_t = \sum_{i=1}^{t} x_i$, so that viewed as a function of discrete time t, S_t gives the instantaneous position of a random walker on integers. We wish to rescale both time and space so as to define a random function defined on $t \in [0, 1]$ and having value in $\mathbb{R}$. Recall that the central limit theorem asserts that

$$S_N/\sqrt{N} \to \mathcal{N}(0, 1)$$

in distribution as $N \longrightarrow \infty$, where $\mathcal{N}(0, 1)$ denotes the standard normal distribution. This suggests that we rescale S_t and define a piecewise constant random function $W^N(t)$ on $t \in [0, 1]$ by letting

$$W^N(t) = S_{\lfloor Nt \rfloor}/\sqrt{N}.$$

Then we have:

☐ Theorem

As $N \longrightarrow \infty$, $W^N(\cdot)$ converges to a limit $W(\cdot)$ in the sense of distributions

$$W^N \to {}^d W ,$$

where $W(\cdot)$ is the Wiener process.

Thus, a Wiener process is the *scaling limit* (a term applied, for example, to the behavior of a lattice model in the limit of the lattice spacing going to zero) of random walk in one dimension, which means that if you take a random walk with very small steps you get an approximation to a Wiener process. To be more precise, if the step size is ϵ, then one needs to take a walk of length L/ϵ^2 to approximate a Wiener process walk of length L. As the step size $\epsilon \to 0$ (and the number of steps increased comparatively), the random walk converges to a Wiener process in the appropriate sense.

A Wiener process occurs frequently in mathematics (the study of continuous-time martingales, stochastic calculus, diffusion processes), economics (the mathematical theory of finance, in particular the Black–Scholes option pricing model) and physics (the study of Brownian motion, the diffusion of minute particles suspended in fluid, and other types of diffusion via the Fokker–Planck and Langevin equations; see next subsection).

2.1.5
Langevin Equation and Brownian Motion

In this subsection, we shall study the basics of the Langevin equation in the language of colloidal suspensions (Brownian motion). Consider a sufficiently small colloidal particle of mass m suspended in a liquid at absolute temperature T. On its path through the liquid it will continuously collide with the liquid molecules and follow a random path exhibiting Brownian motion. In physics, this can serve as a prototype problem whose solutions provide deep insight into the mechanisms responsible for the existence of statistical fluctuations in a system at thermal equilibrium and "dissipation of energy". Moreover, such fluctuations constitute a background noise, which imposes certain limitations on the possible accuracy of delicate physical measurements.

Again as in previous cases, for simplicity, we consider the motion restricted to a single dimension. We consider a sufficiently small particle of mass m whose position is described by the variable coordinate $x(t)$ at any time t, and $v(t) = dx/dt$ is the corresponding velocity.

It would be very complex (and useless) to describe in details the interaction of the small particle with the motion of the molecules in the surrounding liquid (other degrees of freedom). However, all such degrees of freedom can be regarded as constituting a heat reservoir at the temperature T, and their interaction described by some net force $F_{res}(t)$. In addition, it is possible that the particle may also interact with an external force, such as gravity or electromagnetic fields, denoted by $F_{ext}(t)$. The velocity v of the particle may be appreciably different from the equilibrium mean value. Newton's equation of motion then reads

$$m\frac{dv}{dt} = F_{res}(t) + F_{ext}(t) . \tag{2.19}$$

We know very little about the nature of the force $F_{res}(t)$, except that it is some rapidly fluctuating function of the time t and varies in a highly irregular or random fashion. To advance, one has to cast the problem in statistical terms and, therefore, must consider a statistical ensemble of systems, each of them consisting of a particle and the surrounding liquid. For each member of the ensemble, the force $F_{res}(t)$ is some random function of time t. We also assume that the correlation time characterizing the force $F_{res}(t)$ is small on a macroscopic time scale, and there is no preferred direction in space (if the particle is imagined to be clamped to be stationary). Then the ensemble average $\bar{F}_{res}(t)$ vanishes.

Since $F_{res}(t)$ is a rapidly fluctuating function of time, v must be also fluctuating in time, and, as mentioned above, can be appreciably different from the equilibrium mean value. However, superimposed upon these fluctuations, the time dependence of v may also exhibit a more slowly varying drift, and we thus assume:

$$v = \bar{v} + \tilde{v} , \tag{2.20}$$

where $\tilde{v}$ denotes the rapidly fluctuating part of v, and whose mean value vanishes. The slowly varying part $\bar{v}$ is very significant (even though it is small) because it determines the behavior of the particle over a time-length.

We must also consider the fact that the interaction force $F_{res}(t)$ is actually affected by the motion of the particle in such a way that F_{res} itself must also contain a slowly varying part $\bar{F}$, which tends to restore the particle to equilibrium. Hence, similar to above equation for v, we must also write that:

$$F_{res} = \bar{F} + \tilde{F} , \tag{2.21}$$

where $\tilde{F}$ denotes the part of rapidly fluctuating part of F_{res}, and whose mean value vanishes. The slowly varying part $\bar{F}$ must be some function of $\bar{v}$, such that $\bar{F}(\bar{v}) = 0$ at equilibrium when $\bar{v} = 0$.

We note that if $\bar{v}$ is not very large, $\bar{F}(\bar{v})$ may be expanded as a power series in $\bar{v}$, whose first non vanishing term must be linear in $\bar{v}$. Therefore, $\bar{F}$ will of the general form

$$\bar{F} = -\gamma\bar{v} , \tag{2.22}$$

where γ is some positive "frictional" constant, and where the negative sign indicates that the force $\bar{F}$ acts in the opposite direction such that it tends to reduce the $\bar{v}$ to zero with time. Thus, we have the slowly varying part

$$m\frac{d\bar{v}}{dt} = \bar{F} + F_{ext} = -\gamma\bar{v} + F_{ext} , \tag{2.23}$$

and if we include the rapidly fluctuating parts $\tilde{v}$ and $\tilde{F}$, then we have

$$m\frac{dv}{dt} = -\gamma v + F_{\text{ext}} + \tilde{F}(t) , \qquad (2.24)$$

assuming $\gamma \tilde{v} \approx \gamma v$ (since the rapidly fluctuating part $\gamma \tilde{v}$ can be neglected compared to the predominantly fluctuating part $\tilde{F}(t)$). This is the *Langevin equation*, which describes the behavior of the colloidal particle at all later times, if the initial conditions are known. We note that since the Langevin equation contains the frictional force $-\gamma v$, it implies that there exist processes whereby the energy associated with the particle is dissipated in due course by the other degrees of freedom (molecules of the liquid surrounding the colloidal particle).

Now, while describing Brownian motion in the absence of external forces F_{ext}, we have

$$m\frac{dv}{dt} = -\gamma v + \tilde{F}(t) . \qquad (2.25)$$

To estimate the frictional force, we use the Stokes's law of hydrodynamics:

$$\gamma = 6\pi\eta r , \qquad (2.26)$$

where η is the viscosity of the liquid and r is the radius of the colloidal particle (assumed to be spherical).

Let the system be in thermal equilibrium. Then the mean displacement $\bar{x}$ of the particle must vanish by symmetry, as there is no preferred direction in space. Our aim is to calculate the mean-square displacement $\langle x^2 \rangle = \overline{x^2}$ of the particle in a time interval t.

We have $v = \dot{x}$ and $dv/dt = d\dot{x}/dt$, so that multiplying the Langevin equation by x throughout, we get

$$mx\frac{d\dot{x}}{dt} = m\left\{\frac{d}{dt}(x\dot{x}) - \dot{x}^2\right\} = -\gamma x\dot{x} + x\tilde{F}(t) . \qquad (2.27)$$

Now we take the ensemble average of the above equation and use the fact that, irrespective of the values of either x or v, we have $\langle x\tilde{F}\rangle = \langle x\rangle\langle\tilde{F}\rangle = 0$. Using the "equipartition theorem" of classical statistical mechanics, we have $(1/2)m\langle\dot{x}^2\rangle = (1/2)k_B T$, such that

$$m\left\langle\frac{d}{dt}(x\dot{x})\right\rangle = m\frac{d}{dt}\langle x\dot{x}\rangle = k_B T - \gamma\langle x\dot{x}\rangle . \qquad (2.28)$$

This is an ordinary differential equation which can be solved to get the value of the quantity $\langle x\dot{x}\rangle$. Hence, one gets

$$\langle x\dot{x}\rangle = C\exp(-\alpha t) + \frac{k_B T}{\gamma} , \qquad (2.29)$$

where C is a constant of integration and $\alpha = \gamma/m$ is the inverse of the characteristic time constant for the system. Assuming that each colloidal particle of the ensemble starts out at time $t = 0$ and the position $x = 0$, so that x is the displacement from the initial position, the constant C satisfies $0 = C + (k_B T)/\gamma$. Hence, we have

$$\langle x \dot{x} \rangle = \frac{1}{2} \frac{d}{dt} \langle x^2 \rangle = \frac{k_B T}{\gamma} \left(1 - \exp(-\alpha t)\right) , \tag{2.30}$$

or

$$\langle x^2 \rangle = \frac{2 k_B T}{\gamma} \left(t - \alpha^{-1}(1 - \exp(-\alpha t))\right) . \tag{2.31}$$

For us the case $t \gg \alpha^{-1}$ when $\exp(-\alpha t) \to 0$, is relevant and gives rise to the interesting equation

$$\langle x^2 \rangle = \frac{2 k_B T}{\gamma} t . \tag{2.32}$$

This signifies that the particle then behaves as a diffusing particle executing a random walk so that $\langle x^2 \rangle \propto t$. Indeed, the diffusion equation in physics for random walks gives a relation $\langle x^2 \rangle = 2 D t$, where D is the *diffusion constant*, and comparing these two we get

$$D = \frac{k_B T}{\gamma} , \tag{2.33}$$

which is known as the "Einstein relation". Using Stokes's law, we also have

$$\langle x^2 \rangle = \frac{k_B T}{3 \pi \eta r} t , \tag{2.34}$$

or

$$D = \frac{k_B T}{6 \pi \eta r} , \tag{2.35}$$

known as the "Einstein–Stokes equation".

2.2
Do Markets Follow a Random Walk?

Prices of assets in a financial market produce what is called a "financial time-series". Different kinds of financial time-series have been recorded and studied for decades, all over the world. Nowadays, all transactions on a financial market are recorded leading to huge amount of data available either commercially or for free in the Internet. Financial time-series analysis has been of great interest to not only the practitioners (an empirical discipline), but also the theoreticians for making inferences and predictions. The inherent uncertainty in the financial time-series and its theory makes it specially interesting to economists, statisticians, and physicists [3]. It is a formidable task to make an exhaustive review on this topic, but we try to give a flavor of some of the aspects here.

2.2.1
What if the Time-Series Were Similar to a Random Walk?

The answer is simple: It would not be possible to predict future price movements using the past price movements or trends. Louis Bachelier, who was the first one to investigate such studies in 1900 [1], had come to the conclusion that "The mathematical expectation of the speculator is zero" and he described this condition as "fair game." Let us discuss this issue in a bit more detail.

In economics, if $P(t)$ is the price of a stock or commodity at time t, then the "log-return" is defined as: $r_\tau(t) = \ln P(t + \tau) - \ln P(t)$, where τ is the interval of time. The definition of daily log-return is illustrated in Figure 2.5, using the price time-series for General Electric.

We generate a random time-series using random numbers from a normal distribution with zero mean and unit standard deviation, in order to compare with the real empirical returns, and plot them in Figure 2.6.

If we divide the time-interval τ into N sub-intervals (of width Δt), the total log-return $r_\tau(t)$ is by definition the sum of the log-returns in each sub-interval. If the

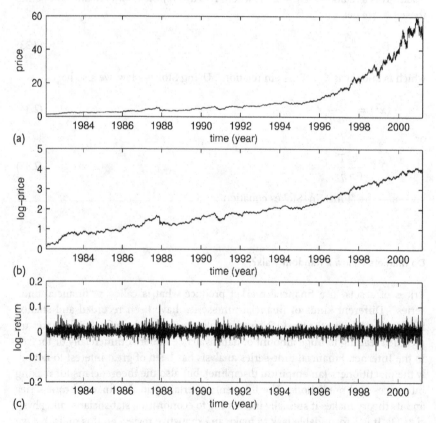

(a)

(b)

(c)

Figure 2.5 Price (USD), log-price and log-return plotted with time for General Electric during the period 1982–2000.

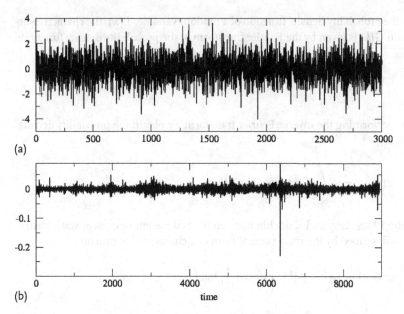

Figure 2.6 Time-series. (a) Random time-series (3000 time steps), (b) return time-series of the S&P 500 stock index (8938 time steps).

price changes in each sub-interval are independent (for the data shown in Figure 2.6) and identically distributed with a finite variance, then according to the central limit theorem the cumulative distribution function $F(r_\tau)$ would converge to a Gaussian (normal) distribution for large τ. As we have seen earlier, the Gaussian (normal) distribution has the properties: (a) when the average is taken, the most probable change is zero, (b) the probability of large fluctuations is very low, since the curve falls rapidly at extreme values, and (c) it is a *stable* distribution (discussed in the next chapter). The distribution of returns was first modeled for "bonds" by Bachelier [1], as a Normal distribution,

$$P(r) = \frac{1}{\sqrt{2\pi}\sigma} \exp(-r^2/2\sigma^2) ,$$

where σ^2 is the variance (second moment) of the distribution.

Note that the classical economical and financial theories had always assumed this normality, until Mandelbrot [4] and Fama [5] pointed out that the empirical return distributions are in fact fundamentally different – they are "fat-tailed" and more peaked compared to the normal distribution. Based on daily prices in different markets, Mandelbrot and Fama in the 1960's found that $F(r_\tau)$ was a stable Levy distribution, whose tail decays with an exponent $\alpha \simeq 1.7$, a result that suggested that short-term price changes were certainly not well-behaved (as the normal distribution) since most statistical properties are not defined when the variance does not exist. This changed the scenario dramatically and led to several studies later in the 1990s [6].

For the probability density function of a random variable, $P(x)$, the characteristic function $G(k)$ is given by the Fourier transform of the probability density function:

$$G(k) = \int_{-\infty}^{+\infty} P(x) \exp(ikx)\, dx \, ,$$

and by performing the inverse Fourier transform we obtain the probability density function:

$$P(x) = \frac{1}{2\pi} \int_{-\infty}^{+\infty} G(k) \exp(-ikx)\, dk \, .$$

In the 1930s, Levy and Khinchin had determined the entire class of stable distributions described by the most general form of a characteristic function:

$$\ln G(k) = i\mu k - \gamma \, |k|^{\alpha} \left[1 - i\beta \frac{k}{|k|} \tan\left(\frac{\pi}{2}\alpha\right) \right] \quad [\alpha \neq 1] \, ,$$

and

$$\ln G(k) = i\mu k - \gamma \, |k| \left[1 + i\beta \frac{k}{|k|} \frac{2}{\pi} \ln|k| \right] \quad [\alpha = 1] \, ,$$

where $0 < \alpha \leq 2$, γ is a positive scaling factor, μ is any real number and β is an asymmetry parameter between -1 and 1. Note that the analytical forms of the Levy stable distributions are known only for a few cases of α and β. For symmetric stable distributions $\beta = 0$, and if the distributions have zero mean (first moment), $\mu = 0$. The characteristic function for the Gaussian distribution, a special case of Levy stable distribution with $\alpha = 2$, $\beta = 0$ and $\mu = 0$, is thus

$$G(k) = \exp(-\gamma \, |k|^{2}) \, ,$$

where $\gamma \equiv \sigma^{2}/2$ is the positive scale factor. The symmetric stable Levy distribution with zero mean, of index α and scale factor γ is the inverse Fourier transform:

$$P_{\text{Levy}}(x) = \frac{1}{\pi} \int_{0}^{\infty} \exp(-\gamma \, |k|^{\alpha}) \cos(kx)\, dk \, .$$

If we assume further that $\gamma = 1$, and look at the asymptotic behavior of $|x|$:

$$P_{\text{Levy}}(|x|) \sim \frac{\Gamma(1+\alpha) \sin(\pi\alpha/2)}{\pi \, |x|^{1+\alpha}} \, ,$$

or

$$P_{\text{Levy}}(|x|) \sim |x|^{-(1+\alpha)} \, ,$$

which means that we find that it has a power law behavior. We also find that $\langle |x|^q \rangle$ diverge for $q \geq \alpha$ when $\alpha < 2$. It follows, in particular, that all Levy stable processes with $\alpha < 2$ have infinite variance, as mentioned earlier.

Slowly, several other interesting features of the financial data were unearthed. A point worth mentioning is that physicists have been analyzing financial data with the motive of finding common or "universal" regularities in the complex time-series, which is very different from those of the economists who are traditionally experts in statistical analysis of financial data. The results of the empirical studies on asset price series by the physicists show that the apparently random variations of asset prices share some statistical properties which are interesting, non trivial, and common for various assets, markets, and time periods. These are called "stylized empirical facts". This brings us to our next question.

2.2.2
What are the "Stylized" Facts?

Stylized facts have been usually formulated using general *qualitative* properties of asset returns and hence distinctive characteristics of the individual assets are not taken into account. Below we quote just a few from the paper by Cont [7], which reviews several empirical studies of the returns and other relevant issues.

1. **Fat tails**: large returns asymptotically follow a power law $F(r_\tau) \sim |r|^{-\alpha}$, with $\alpha > 2$ (with $\alpha = 3.01 \pm 0.03$ for the positive tail and $\alpha = 2.84 \pm 0.12$ for the negative tail [8]). With $\alpha > 2$, the second moment (the variance) is well-defined, excluding stable laws with infinite variance. There have been various suggestions for the form of the distribution: Student's t, hyperbolic, normal inverse Gaussian, exponentially truncated stable, and so on, but no general consensus exists on the exact form of the distribution describing the tails (see Figure 2.7).
2. **Aggregational Normality**: as one increases the time scale over which the returns are calculated, their distribution closes to normality. The shape is different at different time scales. The fact that the shape of the distribution changes with τ makes it clear that the random process underlying prices must have non-trivial temporal structure.[3]
3. **Absence of linear autocorrelations**: the autocorrelation of log-returns, $\rho(T) \sim \langle r_\tau(t + T)r_\tau(t)\rangle$ is illustrated in Figure 2.8. It normally rapidly decays to zero within a few minutes: for $\tau \geq 15$ min, it is practically zero [9]. This supports, in a way, the "efficient market hypothesis". When τ is increased, weekly and monthly returns exhibit some autocorrelation, but the statistical evidence varies from sample to sample.
4. **Volatility clustering**: price fluctuations are not identically distributed and the properties of the distribution, such as the absolute return or variance, change with time and this is called time-dependent or "clustered volatility" (see Fig-

---

3) Any non-Gaussian iid with finite variance has this property! What is special, however, is slow convergence.

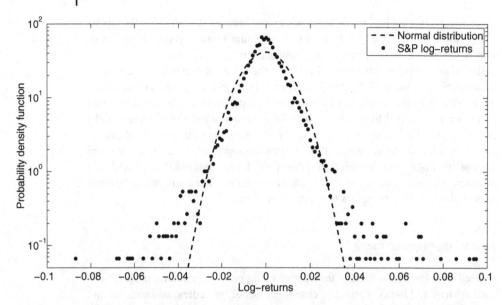

Figure 2.7 S&P 500 daily log-return distribution and normal kernel density estimate. For calculating log-returns, we have used the daily closure prices from 3 January 1950 to 29 October 2009, a total of 15 054 days. The mean is $-2.76 \cdot 10^{-4}$ and variance is $9.34 \cdot 10^{-5}$.

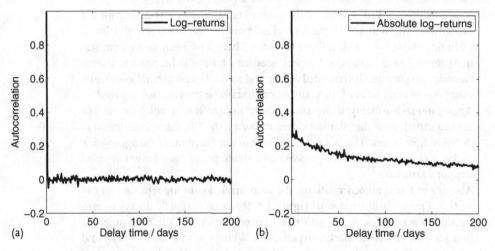

Figure 2.8 Autocorrelation functions. For calculating log-returns we have used the daily closure prices from 3 January 1950 to 29 October 2009, a total of 15 054 days.

ure 2.9). The volatility measure of absolute returns show a positive autocorrelation over a long period of time (see Figure 2.8) – decays roughly as a power law with exponent between 0.1 and 0.3 [9–11]. Therefore, high volatility events tend

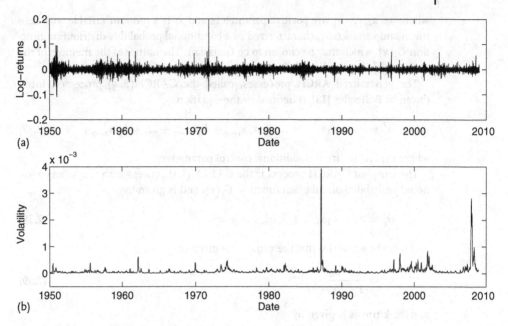

Figure 2.9 Returns and Volatility. For calculating log-returns we have used the daily closure prices from 3 January 1950 to 29 October 2009, a total of 15 054 days. For volatility calculations we have used the moving time window of 20 days.

to cluster in time and large changes tend to be followed by large changes and so also for small changes.

Some of these features have been studied very well by the class of economic models called ARCH and GARCH models.

2.2.3
Short Note on Multiplicative Stochastic Processes ARCH/GARCH

Considerable interest has been in the application of ARCH/GARCH models to financial time-series which exhibit periods of unusually large volatility followed by periods of relative tranquility or calmness. The assumption of constant variance or "homoscedasticity" is inappropriate or invalid in such circumstances. A stochastic process with auto-regressional conditional "heteroscedasticity" (**ARCH**) is actually a stochastic process with "non constant variances conditional on the past but constant unconditional variances" [12], which in simple terms mean locally non-stationary but asymptotically stationary. An ARCH(p) process is defined by the equation

$$\sigma_t^2 = a_0 + a_1 x_{t-1}^2 + \ldots + a_p x_{t-p}^2 , \tag{2.36}$$

where $a_0, a_1, \ldots, a_p$ are positive parameters and x_t is a random variable with zero mean and variance σ_t^2, characterized by a conditional probability distribution function $G_t(x)$, which may be chosen to be Gaussian. The nature of the memory of the variance σ_t^2 is controlled by the parameter p.

The generalized ARCH processes, called the GARCH(p, q) processes, introduced by Bollerslev [13], is defined by the equation

$$\sigma_t^2 = a_0 + a_1 x_{t-1}^2 + \ldots + a_q x_{t-q}^2 + b_1 \sigma_{t-1}^2 + \ldots + b_p \sigma_{t-p}^2 , \qquad (2.37)$$

where $b_1, \ldots, b_p$ are the additional control parameters.

The simplest GARCH process is the GARCH(1,1) process with Gaussian conditional probability distribution function $G_t(x)$, and is given by

$$\sigma_t^2 = a_0 + a_1 x_{t-1}^2 + b_1 \sigma_{t-1}^2 . \qquad (2.38)$$

It was shown in [14] that the variance is given by

$$\sigma^2 = \frac{a_0}{1 - a_1 - b_1} , \qquad (2.39)$$

and the kurtosis is given by

$$\kappa = 3 + \frac{6a_1^2}{1 - 3a_1^2 - 2a_1 b_1 - b_1^2} . \qquad (2.40)$$

The random variable x_t can be written in terms of σ_t by defining $x_t \equiv \epsilon_t \sigma_t$, where ϵ_t is a random Gaussian process with zero mean and unit variance.

One can also rewrite (2.38) as a random multiplicative process

$$\sigma_t^2 = a_0 + \left(a_1 \epsilon_{t-1}^2 + b_1 \right) \sigma_{t-1}^2 . \qquad (2.41)$$

2.2.4
Is the Market Efficient?

In financial econometrics, one of the most debatable issues is whether the market is "efficient" or not; the "efficient" asset market is one in which the information contained in past prices is instantly, fully, and continually reflected in the asset's current price. It was Eugene Fama who proposed the efficient market hypothesis (EMH) in his Ph.D. thesis work in the 1960s. He made the argument that in an active market which includes many well-informed and intelligent investors, securities would be fairly priced and reflect all the available information. In his own words:

> "An 'efficient' market is defined as a market where there are large numbers of ratio-
> nal profit-maximizers actively competing, with each trying to predict future market
> values of individual securities, and where important current information is almost
> freely available to all participants. In an efficient market, competition among the
> many intelligent participants leads to a situation where, at any point in time, actual
> prices of individual securities already reflect the effects of information based both

on events that have already occurred and on events which, as of now, the market expects to take place in the future. In other words, in an efficient market, at any point in time, the actual price of a security will be a good estimate of its intrinsic value."
– Eugene F. Fama, "Random Walks in Stock Market Prices," Financial Analysts Journal, September/October 1965 (reprinted January–February 1995).

Additionally, there continues to be disagreement on the degree of market efficiency. The three widely accepted forms of the efficient market hypothesis are:

"Weak" form: all past market prices and data are fully reflected in securities prices and hence technical analysis is of no use.

"Semistrong" form: all publicly available information is fully reflected in securities prices and hence fundamental analysis is of no use.

"Strong" form: all information is fully reflected in securities prices and hence even insider information is of no use.

The efficient market hypothesis has provided the basis for much of the financial market research in the last several decades. In the early 1970s, a lot of evidence seemed to have been supportive and consistent with the efficient market hypothesis: the prices followed a random walk and the predictable variations in returns, if any at all, came out to be statistically insignificant. While most of the studies in the 1970s concentrated or focussed mainly on predicting prices from past prices, the new studies in the 1980s dealt with the possibility of forecasting based on "variables" such as dividend yield (e.g., Fama & French (1988)[15]). Several later studies handled things like the reaction of the stock market to the announcement of various events such as takeovers, stock splits, and so on. In general, results from event studies typically showed that prices seemed to adjust to "new information" within a day of the announcement of the particular event, an inference that is *consistent* with the efficient market hypothesis. Again, studies beginning in the 1990s started looking at the deficiencies of asset pricing models. The accumulating evidence was suggesting more and more that stock prices could be predicted with a fair degree of reliability. To answer the question of whether the predictability of returns represented "rational" variations in expected returns or if it simply sprung as "irrational" speculative deviations from theoretical values, many further studies have been conducted in recent years. Researchers had discovered several other stock market "anomalies" that seem to *contradict* the efficient market hypothesis. Once an anomaly was discovered, investors attempting to gain or profit by exploiting such an inefficiency resulted in the disappearance of the anomaly! It is true that several anomalies which had been discovered via back-testing, have subsequently disappeared or proved to be impossible to exploit because of the cost of huge transactions. There is evidence in many cases that strong performers in one period frequently turned around and underperformed in subsequent periods. It was noted in numerous studies that there was very little or no correlation between strong performers from one period to the following period. This lack of consistent outperformance among active managers could be presented as further evidence in support of the efficient market hypothesis:

"Market efficiency is a description of how prices in competitive markets respond to new information. The arrival of new information to a competitive market can be likened to the arrival of a lamb chop to a school of flesh-eating piranha, where in-

vestors are – plausibly enough – the piranha. The instant the lamb chop hits the water, there is turmoil as the fish devour the meat. Very soon the meat is gone, leaving only the worthless bone behind, and the water returns to normal. Similarly, when new information reaches a competitive market there is much turmoil as investors buy and sell securities in response to the news, causing prices to change. Once prices adjust, all that is left of the information is the worthless bone. No amount of gnawing on the bone will yield any more meat, and no further study of old information will yield any more valuable intelligence."
        – Robert C. Higgins, Analysis for Financial Management (3rd edition 1992)

Thus, there have been many *twists* and *turns* in the research establishing the efficiency of the market. Before ending the discussion, we must mention that the nature of efficient markets is certainly paradoxical in the sense that if every practitioner truly believed that a market was "efficient", then the market would not have been efficient since no one would have then analyzed the behavior of the asset prices. Effectively, *efficient* markets depend on market participants who believe the market is actually *inefficient* and trade assets in order to profit the most from the market inefficiency.

2.3
Are there any Long-Time Correlations?

The random walk theory of prices assumes that the returns are uncorrelated. But are they truly uncorrelated or are there long-time correlations in the financial time-series? This question has been studied especially since it may lead to deeper insights about the underlying processes that generate the time-series [6, 16].

Now we study two measures to quantify the long-time correlations and study the strength of trends.

2.3.1
Detrended Fluctuation Analysis (DFA)

In the DFA method [17] the time-series ξ_t of length T is first divided into N non-overlapping periods of length τ, such that $N\tau = T$. In each period $i = 1, 2, \ldots, N$ the time-series is first fitted through a linear function $z_t = at + b$, called the local trend. Then it is detrended by subtracting the local trend in order to compute the fluctuation function,

$$F(\tau) = \left[\frac{1}{\tau} \sum_{t=(i-1)\tau+1}^{i\tau} (\xi_t - z_t)^2 \right]^{\frac{1}{2}}. \tag{2.42}$$

The function $F(\tau)$ is re-computed for different box sizes τ (different scales) to obtain the relationship between $F(\tau)$ and τ. A power law relation between $F(\tau)$ and the box size τ, $F(\tau) \sim \tau^\alpha$, indicates the presence of scaling. The scaling, or "correlation exponent" α, quantifies the correlation properties of the signal: if $\alpha =$

0.5 the signal is uncorrelated (white noise); if $\alpha > 0.5$ the signal is anti-correlated; if $\alpha < 0.5$, there are positive correlations in the signal.

2.3.2
Power Spectral Density Analysis

Power spectral density (PSD), or energy spectral density (ESD), is used in physics to identify the periodicities and to quantify the frequency contents of *stationary* stochastic processes, or *deterministic* functions of time such as waves, and respectively possess the dimensions of power per Hz and energy per Hz.

If $\Psi(t)$ is a finite-energy (square integrable) wave, then the spectral density $\Phi(\omega)$ of the wave is the square of the absolute value of the continuous Fourier transform of the wave:

$$\Phi(\omega) = \left| \frac{1}{\sqrt{\pi}} \int_{-\infty}^{\infty} \Psi(t) \exp(-i\omega t) dt \right|^2 = \frac{G(\omega) G^*(\omega)}{2\pi}$$

where ω is the angular frequency and $G(\omega)$ is the continuous Fourier transform of $\Psi(t)$, and $G^*(\omega)$ is the complex conjugate. The multiplicative factor of $1/(2\pi)$ is a normalizing constant used in the definition of the Fourier transforms.

Instead of a continuous wave, if the signal is discrete with values Ψ_n, over an *infinite* number of elements n, then energy spectral density is defined as:

$$\Phi(\omega) = \left| \frac{1}{\sqrt{\pi}} \sum_{n=-\infty}^{\infty} \Psi_n \exp(-i\omega n) \right|^2 = \frac{G(\omega) G^*(\omega)}{2\pi}$$

where $G(\omega)$ (and $G^*(\omega)$, the complex conjugate) is now replaced by the discrete-time Fourier transform of Ψ_n.

Caution has to be maintained when the number of defined values is finite, because strictly speaking the sequence does not actually have an energy spectral density. However, the sequence can be treated as periodic, using a discrete Fourier transform to make a discrete spectrum, or it can be extended with zeros and a spectral density can be computed as in the infinite-sequence case.

Moreover, we have seen that the above definitions of energy spectral density require that the Fourier transforms of the signals *exist*. The power spectral density (PSD) is an alternative quantity that describes the distribution of the power of a signal or time series with frequency. With the help of the Wiener–Khinchin theorem, we can define:

> *The power spectral density is the Fourier transform of the autocorrelation function $C(\tau)$ of the signal, if the signal is a stationary random process.*

This gives the formula for power spectral density:

$$S(\omega) = \int_{-\infty}^{\infty} C(\tau) \exp(-i\omega\tau) d\tau,$$

or

$$S(f) = \int_{-\infty}^{\infty} C(\tau) \exp(-2\pi i f \tau) d\tau , \tag{2.43}$$

which exists *if and only if* the signal is a *stationary* process. If the signal is non-stationary, then the autocorrelation function must be a function of two variables, so truly no power spectral density exists. However, it is still possible to generalize and construct a time-varying spectral density.

2.3.3
DFA and PSD Analyses Of the Autocorrelation Function Of Absolute Returns

The analysis of financial correlations was done in 1997 by the group of H.E. Stanley [10]. The correlation function of the financial indices of the New York stock exchange and the S&P 500 between January 1984 and December 1996 were analyzed at one minute intervals. The study confirmed that the autocorrelation function of the returns fell off exponentially, but the absolute value of the returns did not. Correlations of the absolute values of the index returns could be described through two different power laws, with crossover time $t_\times \approx 600$ min, corresponding to 1.5 trading days. Results from power spectrum analysis and DFA analysis were found to be consistent. The power spectrum analysis of Figure 2.10 yielded exponents $\beta_1 = 0.31$ and $\beta_2 = 0.90$ for $f > f_\times$ and $f < f_\times$, respectively. This is consistent with the result that $\alpha = (1 + \beta)/2$ and $t_\times \approx 1/f_\times$, as obtained from detrended fluctuation analysis with exponents $\alpha_1 = 0.66$ and $\alpha_2 = 0.93$ for $t < t_\times$ and $t > t_\times$, respectively.

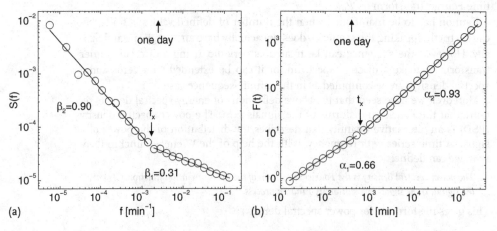

Figure 2.10 Power spectrum analysis (a) and detrended fluctuation analysis (b) of the autocorrelation function of absolute returns, from Liu *et al.* [10].

References

1 Bachelier, L. (1900) *Annales Scientifiques de l'Ecole Normale Superieure*, **III**-7, 21.

2 Bouchaud, J.P. (2005) *Chaos*, **15**, 026104.

3 Tsay, R.S. (2002) *Analysis of Financial Time Series*, John Wiley & Sons, Inc., Hoboken.

4 Mandelbrot, B.B. (1963) *J. Business*, **36**, 394.

5 Fama, E. (1965) *J. Business*, **38**, 34.

6 Mantegna, R.N. and Stanley, H.E. (2000) *An Introduction to Econophysics*, Cambridge University Press, Cambridge.

7 Cont, R. (2001) *Quant. Fin.*, **1**, 223–236.

8 Gopikrishnan, P., Meyer, M., Amaral, L.A.N., and Stanley, H.E. (1998) *Eur. Phys. J. B* (Rapid Note), **3**, 139.

9 Cont, R., Potters, M., and Bouchaud, J.-P. (1997) in *Scale Invariance and Beyond (Proc. CNRS Workshop on Scale Invariance. Les Houches, 1997)* (eds Dubrulle, Graner, Sornette), Springer, Berlin.

10 Liu, Y., Cizeau, P., Meyer, M., Peng, C.-K., and Stanley, H.E. (1997) *Physica A*, **245**, 437.

11 Cizeau, P., Liu, Y., Meyer, M., Peng, C.-K. and Stanley, H.E. (1997) *Physica A*, **245**, 441.

12 Engle, R.F. (1982) *Econometrica*, **50**, 987.

13 Bollerslev, T. (1986) *J. Econometrics*, **31**, 307.

14 Baillie, R.T. and Bollerslev, T. (1992) *J. Econometrics*, **52**, 91.

15 Fama, E. and French, K. (1988) *J. Financial Economics*, **22**, 3–27.

16 Chakraborti, A., Muni Toke, I., Patriarca M., and Abergel, A. (2009) *"Econophysics: Empirical facts and agent-based models"*, available at *arXiv:0909.1974*.

17 Vandewalle, N. and Ausloos, M. (1997) *Physica A*, **246**, 454.

3
Beyond the Simple Random Walk

"A cauliflower shows how an object can be made of many parts, each of which is like a whole, but smaller. Many plants are like that. A cloud is made of billows upon billows upon billows that look like clouds. As you come closer to a cloud you don't get something smooth but irregularities at a smaller scale."

– Benoit B. Mandelbrot, *A Theory of Roughness* (2004)
http://www.edge.org/3rd_culture/mandelbrot04/mandelbrot04_index.html

The use of probability distributions to study natural phenomena had initially started with models having finite means (the first moment) and variances (related to the second moment). The most notable example is of course the Gaussian law of errors, where the most probable value from a large number of observations is the measure of central tendency. The dispersion of these values around the average provides a measure of the associated uncertainty, and is related to the variance. The Gaussian (μ, σ^2) distribution (discussed in the preceding chapter) with mean μ and variance σ^2:

$$P(x) = \frac{1}{\sqrt{2\pi\sigma^2}} \exp\left(-\frac{(x-\mu)^2}{2\sigma^2}\right), \quad (-\infty < x < \infty), \qquad (3.1)$$

is widely seen in different contexts, and is the earliest known example of a stable distribution.

A distribution is said to be *stable*, when a linear combination of random variables independently chosen from the distribution has the same distribution, up to a translation in the mean and a scale factor in the variance. Thus, a sum of independent, identically distributed random variables always converge to a stable distribution. In terms of symbols, if x_1 and x_2 are random variables chosen from a stable distribution $P_{\text{stable}}(x)$, then for any pair of positive constants a, b, the composite variable $ax_1 + bx_2$ has the same distribution, but possibly with a different mean and variance. If the mean is identical to the original distribution, then it is said to be *strictly stable* (or *stable in the narrow sense*) [1]. This is a generalization of the classical *Central Limit Theorem*, according to which adding together a large number of random terms produces a variable which eventually is seen to follow a Gaussian distribution. However, a crucial requirement here is that the random variables have finite variance. In the absence of this condition, stable distributions

Econophysics. Sitabhra Sinha, Arnab Chatterjee, Anirban Chakraborti, and Bikas K. Chakrabarti
Copyright © 2011 WILEY-VCH Verlag GmbH & Co. KGaA, Weinheim
ISBN: 978-3-527-40815-3

with unbounded second (and possibly, first) moment can be observed: Examples are the Cauchy (γ, δ) distribution

$$P(x) = \frac{1}{\pi} \frac{\gamma}{\gamma^2 + (x - \delta)^2} , \quad (-\infty < x < \infty) , \tag{3.2}$$

and the Levy (γ, δ) distribution

$$P(x) = \sqrt{\frac{\gamma}{2\pi}} \frac{1}{(x - \delta)^{3/2}} \exp\left(-\frac{\gamma}{2(x - \delta)}\right), \quad (\delta < x < \infty) . \tag{3.3}$$

With the exception of the normal distribution, stable distributions are leptokurtotic and have long tails. Kurtosis measures how peaked a distribution is, and leptokurtosis indicates that a distribution is more peaked than the Gaussian around the mean.

The class of stable distributions was characterized in the 1920s by Paul Levy; however, the lack of closed-form formulas for the distribution functions for all but a few (viz. Gaussian, Cauchy and Levy) has limited their wider use. Possibly the first observation of a non-Gaussian distribution in economics is the claim by Vilfredo Pareto that income follows a distribution having a tail decaying as a power law. However, in recent times, the use of non-Gaussian distributions in economics has undergone a resurgence ever since the observation by Benoit Mandelbrot that cotton price changes seem to follow a Levy-stable distribution.

The non-Gaussian long-tailed stable distributions are associated with random walks having fractal or self-similar scaling characteristics. These walks are often termed as Levy flights, and are a generalization of the simpler Brownian motion associated with Gaussian distribution. Unlike Brownian motion, where the mean-square displacement grows linearly with time, in Levy walks it can be faster than linear.

Consider a random walk of N steps in one dimension. Furthermore, let the length of each step x be chosen from the same distribution $P(x)$. Levy had asked the question as to what should be the distribution such that the net displacement after N steps $(= x_1 + x_2 + \ldots + x_N)$ has the same distribution $P(x)$, up to a scale factor. In 1853, Augustine Cauchy had proposed a form for the probability in Fourier k-space

$$\tilde{P}_N(k) = \exp(-N|k|^\beta) , \tag{3.4}$$

which yields the Cauchy distribution (given above) for $\beta = 1$. Note that the distribution of N-step displacement $P_N(x) = \frac{1}{N} P(x/N)$, where $P(\ )$ is the distribution for the length of each step. Levy answered his question by showing that if $P(\ )$ is to be non-negative, an essential condition for a probability function, then β has to lie between 0 and 2. At the tails of the distribution, it is approximated by $|x|^{-1-\beta}$. When $\beta = 2$, the mean and variance are finite, and the distribution converges to a Gaussian. However, when $\beta < 2$, the variance of the distribution is unbounded, implying that there is no characteristic length for the steps in the random walk. In other words, the trajectories produced by Levy flights are scale-invariant fractals.

Example 3.1

Example ([2]): Let us consider a random walk along one dimension, where the discrete probability distribution for step length is given by

$$P(x) = \frac{\lambda - 1}{2\lambda} \sum_{j=0}^{\infty} \lambda^{-j} \left[\delta(x, +b^j) + \delta(x, -b^j) \right]. \qquad (3.5)$$

The distribution parameter λ lies between b and 1. The function $\delta(x, y)$ is the Kronecker delta ($= 1$, when $x = y$, and $= 0$, otherwise). The distribution gives the probabilities for steps of length $1, b, b^2, \ldots$. However, each time the step length is increased b times, the corresponding probability decreases by λ. Thus, on average, we will see a sequence of λ successive steps of length 1 before there is an instance of a step of length b. Similarly, there will be typically λ steps of length b before a step of length b^2 is observed, and so on. The Fourier transform of the distribution is

$$\tilde{P}(k) = \frac{\lambda - 1}{2\lambda} \sum_{j=0}^{\infty} \lambda^{-j} \cos(b^j k), \qquad (3.6)$$

which is the well-known Weierstrass function. The self-similarity of this function can be shown explicitly from the equation $\tilde{P}(k) = \frac{1}{\lambda} \tilde{P}(bk) + \frac{\lambda-1}{\lambda} \cos(k)$, which for small k yields, $\tilde{P}(k) = \exp(-|k|^\beta)$ where $\beta = \log(\lambda)/\log(b)$.

As already mentioned above, the mean-square displacement for Levy flight diverges. The mean-square displacement can increase as a quadratic function of time or even faster if acceleration is allowed. Let the spatiotemporal probability density for a random walker to have a displacement of r at time t be $\Psi(r, t) = \psi(t|r) P(r)$, where $P(r)$ is the probability distribution of the length of a single step. The probability that a step of length r takes time t is given by $\psi(t|r)$. For simplicity, $\psi(t|r)$ can be taken as the Dirac δ-function $\delta(t - |r|/v(r))$ ensuring that $r = vt$, where v is the velocity of the walk. The velocity need not be a constant, but can be a function of the step length. Considering constant velocity, the probability for time spent in a flight segment is $\psi(t) = 1/(1 + t)^{1+\beta}$. Different values of β result in different time dependences of the mean-square displacement. For example, it goes as t^2 for $0 < \beta < 1$, $t^2/\ln t$ for $\beta = 1$, $t^{3-\beta}$ for $1 < \beta < 2$, $t \ln t$ for $\beta = 2$ and as t for $\beta > 2$.

3.1
Deviations from Brownian Motion

In the previous chapter we looked at random walk (and its most commonly known physical manifestation, Brownian motion) as a model for market movements. Despite the fact that Bachelier's work had remained relatively unknown for almost half a century afterwards, the dominant paradigm had been to assume the market to behave essentially as a random walk and that price changes are uncorrelated. Eugene Fama's work on the efficient market hypothesis in the 1960s sought to put this idea on a more rigorous footing [3]. However, in the 1960s, starting from the work of Mandelbrot among others, it started to appear that there was structure present in the price or index movements that were not expected in a purely random motion. When compared to Brownian motion, price time-series showed anomalies such as fat tails in the distribution of returns (a measure of price fluctuations) and long memory in the volatility (for instance, measured as either the standard deviation over a specified window or simply the absolute value of the return). As understanding the origin of these deviations from the simple random walk is essential to developing a complete theory of market dynamics, there have been no dearth of efforts at applying physics models to this problem. In particular, the appearance of self-similarity in the statistics of volatility clustering (which has some analogy to the persistent regions of dynamical disorder in fluids) has meant that tools for studying turbulence in physical systems have been sought to be used in this context. This has led to some physicists questioning the neoclassical economist's position of markets being informationally efficient (that the prices at any given time tend to reflect the entire pertinent information available at that time). It may be that, in contrast, markets are systematically biased towards arriving at prices that do not accurately reflect all information.

It is worth pointing out here that Bachelier's theory of random walk for market prices does match at least one important empirical observation: successive changes in price are uncorrelated, at least in the first approximation. Indeed, this is the fundamental reason behind the unpredictability of markets. However, to explain other features such as fat tails and the long-range correlation of volatility, we clearly need to go beyond the random walk. As recent empirical studies have shown that financial data share many statistical properties with turbulent flows, where the velocity reveals strong, intermittent fluctuations, much as the volatility of financial markets, a more sophisticated model for market movements may come from this field. However, the mechanism that converts a rather predictable human behavior into a sequence of nearly unpredictable price changes, has not been investigated in details until recently. The availability of high frequency, trade by trade data, and the shift of paradigm from efficient markets to models with agents having bounded rationality have prompted recent studies on this aspect.

Bachelier's hypothesis that markets follow a random walk imply that all financial time-series should have the property that the dispersion of the price fluctuations about the mean grows linearly with the time-scale of observation. Thus, if $\langle r_t \rangle_\tau$ is the average return over the time-scale τ

$$\langle (r_t - \langle r_t \rangle_\tau)^2 \rangle \simeq \sigma^2 \tau , \tag{3.7}$$

where the averaging is done over many windows of size τ and the root mean square per unit time σ is the volatility. This is valid for τ varying from a few minutes to a few years. Thus, the returns appear to be uncorrelated over the entire range of observational time-scales. This is usually taken to be the evidence of information efficiency of markets where one cannot predict future prices because any trend or systematic biases that may appear and which can be used to make such forecasting has already been exploited and thus, accounted for in the price. As the mean return itself varies slowly over time, looking at scales beyond several years is obviously going to bring other factors into consideration, although some studies suggest that over a very long scale there is a tendency to revert towards the mean. Note that, the volatility is used as a quantifier of the inherent risk in the market. However, apart from this single observation the Bachelier model of uncorrelated stochastic processes cannot account for any of the other stylized features.

Among these "unexplained" features are (i) the occurrence of a nontrivial multifractal scaling such that higher moments of the price returns scale anomalously with time

$$R_q(\tau) = \langle (r_t - \langle r_t \rangle_\tau)^q \rangle \simeq A_q \tau^{\zeta_q} , \tag{3.8}$$

where $\zeta_q \neq q/2$ (which is the case for the Brownian random walk). A related observation is (ii) the intermittent nature of the volatility time-series, measured as the absolute value of the price returns, $|r_t|$. The occurrence of high-volatility events appear to be strongly correlated, a feature that is often referred to as *volatility clustering*. Autocorrelation of volatility shows that it decays very slowly as a power law of the time-lag $(\Delta t)^{-\nu}$, with ν in the range 0.1–0.3 [4].

Paul Levy

Paul Levy (1886–1971) was a French mathematician who did seminal work on the theory of statistical distributions. Levy attended the Ecole Polytechnique and while still an undergraduate published his first paper on semiconvergent series in 1905. After graduating and a year in military service, he studied at the Ecole de Mines in 1907. In 1910 he began research on functional analysis, for which he received his doctorate in 1912. Levy became a professor at the Ecole des Mines in Paris in 1913, before joining Ecole Polytechnique in Paris in 1920 as a professor of analysis (eventually retiring in 1959). During the First World War, he was involved in solving mathematical problems on defending against airborne attacks. In 1919 Levy gave lectures at the Ecole Polytechnique on the calculus of probabilities and the role of Gaussian distribution in the theory of errors, eventually resulting in his own investigations in the theory of stable distributions. Besides his contributions to probability and functional analysis, he also worked on partial differential equations [4].

3.2
Multifractal Random Walk

Benoit Mandelbrot had first proposed a generalization of the random walk that has the multifractal property mentioned above. Not surprisingly, he had also tried applying it to explain financial time-series. However, such Mandelbrot cascades have certain theoretical problems associated with them, in particular, the violation of causality. By contrast, Bacry, Muzy, and Delour (BMD) [5, 6] have proposed a model of generalized random walks with multifractal properties that does not have these problems.

We have already mentioned that the correlation function of the square returns (which serves as a proxy for the true volatility) decays as a function of the time-lag Δt like a power law with exponent ν. Thus, the fourth moment for return can be written as

$$R_4(\tau) \sim \sigma^4 \tau^2 (1 + A\tau - \nu) , \qquad (3.9)$$

where A is the amplitude of the long-range part of the square volatility correlation. The fourth moment of the price difference therefore behaves as the sum of two power laws, not as an unique power law as for a multifractal process, for which by definition $R_q(\tau) \sim \tau^{\zeta_q}$ exactly. However, when ν is small and in a restricted range of τ, this sum of two power laws is indistinguishable from a unique power law with an effective exponent ζ_4^{eff} with a value lying between 2 and $2 - \nu$, so that $\zeta_4^{\text{eff}} < 2\zeta_2 = 2$.

In the BMD model, the key ingredient is the volatility correlation shape which mimics that arising in cascade models. Indeed, the treelike structure underlying a Mandelbrot cascade implies that the volatility logarithm covariance decreases very slowly as a logarithm function,

$$\langle \ln(\sigma_t) \ln(\sigma_{t+\Delta t}) \rangle - \langle \ln(\sigma_t) \rangle^2 = C_0 - \lambda^2 \ln(\Delta t + \tau) . \qquad (3.10)$$

The BMD model involves the above equation within the continuous-time limit of a discrete stochastic volatility model. One first discretizes time in units of an elementary time step τ_0 and sets $t = i\tau_0$. The volatility σ_i at time i is a log-normal random variable such that $\sigma_i = \sigma_0 \exp(\xi_i)$, where the Gaussian process ξ_i has the same covariance as in the above equation

$$\langle \xi_i \rangle = -\lambda^2 \ln(T/\tau_0) = \mu_0 , \qquad (3.11)$$

$$\langle \xi_i \xi_j \rangle - \mu_0^2 = -\lambda^2 \ln(T/\tau_0) - \lambda^2 \ln(|i - j| + 1) , \qquad (3.12)$$

for $|i - j|\tau_0 \leq T$. Here T is a large cut-off time-scale beyond which the volatility correlation vanishes. In the above equation, the angular brackets stands for the mathematical expectation. The choice of the mean value μ_0 is such that $\langle \sigma^2 \rangle = \sigma_0^2$. As before, the parameter λ^2 measures the intensity of volatility fluctuations (called in the finance jargon the "vol of the vol") and corresponds to the intermittency parameter. The price returns are constructed as

$$x((i + 1)\tau_0) - x(i\tau_0) = r_{\tau_0}(i) = \sigma_i \epsilon_i = \sigma_0 e^{\xi_i} \epsilon_i \, , \tag{3.13}$$

where the ϵ_i are a set of independent, identically distributed random variables of zero mean and variance equal to τ_0. One also assumes that the ϵ_i and ξ_i are independent. In the original BMD model, ϵ_i are Gaussian, and the continuous-time limit $\tau_0 = dt \to 0$ is taken. Since x is the logarithm of the price p, the exponential of a sample path of the BMD model can be compared to real price charts.

The multifractal scaling properties of this model can be computed explicitly. Moreover, using the properties of multivariate Gaussian variables, one can get closed expressions for all even moments $R_q(\tau)$ where $q = 2k$. For $q = 2$

$$R_2(\tau = l\tau_0) = \sigma_0^2 l\tau_0 = \sigma_0^2 \tau \, , \tag{3.14}$$

independent of λ^2. For $q \neq 2$, we must distinguish between the cases $q\lambda^2 < 1$ and $q\lambda^2 > 1$. For the former, the corresponding moments are finite, and one finds, in the scaling region $\tau_0 \ll \tau \ll T$, a true multifractal behavior

$$R_q(\tau) = A_q \tau^{\zeta_q} \, , \tag{3.15}$$

where $\zeta_q = q(1/2 + \lambda^2) - q^2 \lambda^2 / 2$, and A_q is a prefactor that can be exactly calculated.

For $q\lambda^2 > 1$, the moments diverge, suggesting that the unconditional distribution of $x(t + \tau) - x(t)$ has power-law tails with an exponent $\mu = 1/\lambda^2$. Since volatility correlations are absent for $\tau \gg T$, the scaling becomes that of a standard random walk, for which $\zeta_q = q/2$. The corresponding distribution of returns thus becomes progressively Gaussian.

3.3
Rescaled Range (R/S) Analysis and the Hurst Exponent

A frequently used method for studying multifractal economic time-series is the R/S or rescaled range technique. Studying the Nile river and the problems related to water storage, the British hydrologist Harold Edwin Hurst had created the rescaled range analysis technique to study the power-law scaling of flood statistics.

Let us assume that water is flowing into a reservoir from a long distance away. If the inflow at a particular time t be $\xi(t)$, then the average inflow over time interval τ is $\langle \xi \rangle_\tau = (1/\tau) \sum_{t=1}^{\tau} \xi(t)$. The flow fluctuates from time to time, and this is measured by the deviation from the mean, $\xi(t) - \langle \xi \rangle_\tau$. The cumulative deviation from the mean over the period τ is $X(t, \tau) = \sum_{u=1}^{t} (\xi(u) - \langle \xi \rangle_\tau)$. The difference between the maximum and minimum of this quantity is defined as the range, $R(\tau) = \max(X(t, \tau)) - \min(X(t, \tau))$ for the range $1 \leq t \leq \tau$.

The rescaled range is calculated by dividing the range $R(\tau)$ by the standard deviation $S(\tau) = \sqrt{\frac{1}{\tau} \sum_{t=1}^{\tau} (\xi(t) - \langle \xi \rangle_\tau)^2}$. The Hurst exponent is estimated by calculating the average rescaled range over multiple regions of the data. The expected value of R/S, calculated over a set of regions (starting with a region size of 8 or 10) converges on the Hurst exponent power function as the time interval becomes larger $(\tau \to \infty)$

$$E(R(\tau)/S(\tau)) = C\tau^H .\tag{3.16}$$

By comparison, for the short-range dependence of the classical random walk, $H = 1/2$.

A linear regression line through a set of points, composed of the logarithm of τ (the interval over which the average rescaled range is calculated) and the logarithm of the average rescaled range over a set of intervals of size τ, is calculated. The slope of regression line is the estimate of the Hurst exponent. This method for estimating the Hurst exponent was developed and analyzed by Benoit Mandelbrot and co-authors during the 1970s.

The Hurst exponent applies to datasets that are statistically self-similar. Statistically self-similar means that the statistical properties for the entire dataset are the same for subsections of the dataset. For example, the two halves of the dataset have the same statistical properties as the entire dataset. This is applied to estimating the Hurst exponent, where the rescaled range is estimated over sections of different size.

First, the rescaled range is calculated for the entire dataset. Then the rescaled range is calculated for the two halves of the dataset, resulting in RS_0 and RS_1. These two values are averaged, resulting in $\langle RS \rangle_1$. In this way the process continues by dividing each of the previous sections in half and calculating the rescaled range for each new section. The rescaled range values for each section are then averaged. At some point the subdivision stops, since the regions get too small. Usually regions will have at least 8 data points. To estimate the Hurst exponent using the rescaled range algorithm, a vector of points is created, where x_i is the logarithm (to base 2) of the size of the interval used to calculate $\langle RS \rangle_i$ and y_i is the logarithm to base 2 of the $\langle RS \rangle_i$ value. The Hurst exponent is estimated by the slope of the linear fit through these points. The algorithm outlined here uses nonoverlapping data regions where the size of the dataset is a power of two. Each subregion is a component power of two. Other versions of the rescaled range algorithm use overlapping regions and are not limited to data sizes that are a power of two.

3.4
Is there Long-Range Memory in the Market?

As mentioned earlier, according to the classical random walk or Brownian motion view, markets have no memory. The effect of any perturbation decays rapidly, and it will be impossible to make any predictions on how the market will behave in the future. While this indeed seems to be true for the price returns, the slow decay of the volatility auto-correlation suggests that there are at least a few economic variables that may exhibit long-range memory. While many technical traders believe that there are in fact long periodic cycles associated with market movements that occur over the space of a few years to decades, there is as yet no clear-cut evidence to support the existence of such extremely long market movements with clearly defined periodicities.

3.4.1
Mandelbrot and the Joseph Effect

Referring to the Old Testament prophet who had predicted seven years of plenty followed by seven years of famine in Egypt, Mandelbrot suggested the name "Joseph effect" to describe the existence of unusually long periods of relatively high or low returns. In contrast to this, he coined the term Noah effect to suggest that extreme events can result in unusually large deviations.

3.4.2
Cycles in Economics

An extensive literature exists for each of three different modes of periodic fluctuations in the economy – the business cycle, the Kuznets cycle and the Kondratieff wave. The business cycle is the well-known short-term fluctuation of business activity. It appears as varying production rates and employment with peaks of activity separated by some 3–7 years. The business cycle lies within the experience of most people and is the focus of attention in the press and in government policy debates.

The Kuznets cycle is much less generally recognized. It exists as a statistical observation that many time-series in the economy seem to fluctuate with a periodicity of some 15–25 years. The cause of the Kuznets cycle has been a subject of debate. Other cyclic modes in the economy are of sufficient magnitude to mask the Kuznets cycle from popular awareness.

The Kondratieff cycle (also known as the "long-wave") was forcefully presented in the literature by Nikolai Kondratieff in the 1920s. Kondratieff was a Russian economist who made extensive studies of the long-term behavior of the Western capitalist economies. His statistical analyses of economic activity showed that many variables in the Western economies had fluctuated with peaks about 45 to 60 years apart. Such peaks of economic activity have been placed around 1810, 1860, and 1920. Kondratieff believed that the 50-year cycle was caused by internal structural dynamics of the economic system, but he did not propose a sharply defined set of mechanisms. Most other economists took the position that the long-term fluctuation had occurred but that it was caused by events external to the economy, such as gold discoveries, wars, major technical innovations, and fluctuations in population growth.

Simulation studies with the Jay Forrester's Systems Dynamics National Model of the economy have shown that realistically modeled physical and policy relationships in the production of consumer durables and capital equipment can generate simultaneously all three major periodicities. The short-term business cycle can result from interaction between backlogs, inventories, production and employment without requiring involvement of capital investment. The Kuznets cycle is consistent with policies governing production and acquisition of capital equipment. The Kondratieff cycle can arise from the structural setting of the capital equipment sector, which supplies capital to the consumer goods sector but also at the same time must procure its own input capital equipment from its own output [7–9].

3.4.3
Log-Normal Oscillations

Recently, a particular kind of oscillation pattern, referred to as log-normal oscillations, have been seen in financial market time-series data [10]. Log-periodic oscillations, that is when an order parameter follows a power law with a complex valued exponent, is not unique to economics. Such a phenomenon can be seen in physics, for example in the dynamics of Ising spins that are located on the nodes of a hierarchical self-similar structure referred to as the diamond fractal [11]. It is constructed by starting with a single bond connecting two spins, and then at the next step replacing the bond by two links each of which connect the two existing nodes by traveling via two new nodes. This new system has four nodes and four bonds. If the procedure is repeated at every bond in all subsequent steps up to n, we eventually generate 4^n links connecting $(2/3)(2 + 4^p)$ spins. When the temperature of the system is below its critical value, so that normally we would expect spontaneous magnetization, the magnetization approaches zero as any existing external field is reduced. Log-periodic oscillations are observed instead, which become more prominent as the strength of coupling between spins, J, is increased. These oscillations are not a critical phenomenon but arise from the hierarchical organization of the spins. The majority of spins in the system have only 2 neighbors, while the set of those that have higher number of neighbors reduce rapidly as a function of the number of neighboring spins. Thus, when the system is subjected to an external field, the highly connected spins will be polarized while the less connected ones will be disordered. As the field is increased, even more weakly connected spins get ordered by the field, and a jump will be seen in the order parameter. At very low temperatures, a series of such jumps will be observed as more and more weakly connected spins get successively polarized. At the critical point, the effect of the field will be minimal as only the very few spins with the highest number of connections will show any response. However, the recent attention given to log-periodic oscillations is because of its implied association with catastrophic events at the critical point. As of now it is still unclear whether the apparently log-periodic oscillations that we see in economic phenomena are related to any critical event in the underlying system.

References

1 Nolan, J.P. (2010) *Stable distributions: Models for Heavy Tailed Data*, Birkhauser, Boston.

2 Klafter, J., Shlesinger, M.F., and Zumofen, G. (1996) Beyond Brownian motion. *Physics Today*, 33–39.

3 Fama, E. (1965) The behavior of stock market prices. *J. Buisness*, **38**, 34–105.

4 Liu, Y., Gopikrishnan, P., Cizeau, P., Meyer, M., Peng, C.-K., and Stanley, H.E. (1999) *Phys. Rev. E*, **60**(2), 1390–1400.

5 Bacry, E., Delour, J., and Muzy, J.F. (2000) Modelling fluctuations of financial time series: from cascade process to stochastic volatility model. *Eur. Phys. J. B*, **17**, 537–548.

6 Bacry, E., Delour, J., and Muzy, J.F. (2001) Multifractal random walk. *Phys. Rev. E*, **64**, 026103.

7 Forrester, J.W. (1978) Innovation and the economic long wave. MIT Systems Dynamics Group working paper.

8 Forrester, J.W. (1981) The Kondratieff cycle and changing economic conditions. MIT Systems Dynamics Group working paper.

9 Forrester, J.W. (1981) *Futurist*, **19**(3), 16–20.

10 Sornette, D. (2002) *Why Stock Markets Crash: Critical Events in Complex Financial Systems*, Princeton University Press, Princeton.

11 Plischke, M. and Bergersen, B. (2006) *Equilibrium Statistical Physics*, 3rd edition, World Scientific, Singapore.

4
Understanding Interactions through Cross-Correlations

"Because nothing is completely certain but subject to fluctuations, it is dangerous
for people to allocate their capital to a single or a small number of securities. [...]
No one has reason to expect that all securities ... will cease to pay off at the same
time, and the entire capital be lost."
– *From the 1776 prospectus of an early mutual fund in the Netherlands* [1]

As the above quote makes evident, those interested in the study of financial market
movements have long been concerned with the correlation between price movements of different commodities or stocks. Recently, with the emergence of the understanding that markets are examples of complex systems with many interacting
components, cross-correlations have been useful for inferring the existence of collective modes in the underlying dynamics of financial markets. While it may not
be possible to directly observe the interactions among traders buying and selling
various stocks, we can measure the resulting effect on the market by looking at
how their actions introduce correlations into the price behavior of different stocks.
Thus, the dynamics of the network of agents is being inferred from the observation of the interaction between stock price movements. Note that, the interactions
between stocks are indirect in the sense that they are mediated through the actions
of the buyers and sellers of those stocks. This is analogous to using the process of
Brownian motion to infer the motion of air molecules by observing pollen grains
with which the air molecules are colliding.

It is natural to expect that stocks which interact strongly between themselves will
have correlated price movements. Such interactions may arise because the corresponding companies belong to the same business sector (so that they compete for
the same set of customers and share similar market environments), or they may
belong to related sectors (e.g., automobile and energy sector stocks would show
similar response to a sudden rise in gasoline prices), or they may be owned by the
same business house and therefore perceived by investors to be linked. In addition, all stocks may respond similarly to news breaks that affect the entire market
(e.g., the outbreak of a war) and this induces market-wide correlations. On the other hand, information that is related only to a particular company will only affect the
price movement of a single stock, effectively decreasing its correlation with other
stocks.

Econophysics. Sitabhra Sinha, Arnab Chatterjee, Anirban Chakraborti, and Bikas K. Chakrabarti
Copyright © 2011 WILEY-VCH Verlag GmbH & Co. KGaA, Weinheim
ISBN: 978-3-527-40815-3

The study of empirical correlations between financial assets is extremely important from a practical point of view as it is vital for risk management of a stock portfolio. High correlation between the components of a given portfolio or option book increases the probability of large losses. For example, if a portfolio is dominated by stocks from a single sector, say petroleum, they may all be hard hit by a crisis in oil production and supply and as a result suffer a steep decline in value. Thus, consideration of correlations have always been important to theories for constructing optimal portfolios, such as that of Markowitz [2], which try to answer the following question: for a set of financial assets, for each of which the average return and risk (measured by the variance of fluctuations) is known, how shall we determine the optimal weight of each asset constituting the portfolio that gives the highest return at a fixed level of risk, or alternatively, that has the lowest risk for a given average rate of return?

The effects governing the cross-correlation behavior of stock price fluctuations can be classified into that of (i) market (i.e., common to all stocks), (ii) sector (i.e., related to a particular business sector) and (iii) idiosyncratic (i.e., limited to an individual stock). The empirically obtained correlation structure can then be analyzed to find out the relative importance of such effects in actual markets. Physicists investigating financial market structure have focussed on the spectral properties of the correlation matrix. Pioneering studies in this area have investigated the deviation of these properties from those of a random matrix, which would have been obtained had the price movements been uncorrelated. It has been found that the bulk of the empirical eigenvalue distribution matches fairly well with those expected from a random matrix, as does the distribution of eigenvalue spacings [3, 4]. Among the few large eigenvalues that deviate from the random matrix predictions, the largest represent the influence of the entire market common to all stocks, while the remaining eigenvalues correspond to different business sectors [5], as indicated by the composition of the corresponding eigenvectors [6]. However, although models, in which the market is assumed to be composed of several correlated groups of stocks, are found to reproduce many spectral features of the empirical correlation matrix [7], one needs to filter out the effects of the market-wide signal as well as noise in order to identify the group structure in an actual market. Recently, such filtered matrices have been used to reveal significant clustering among a large number of stocks from the New York Stock Exchange (NYSE) [8] and the National Stock Exchange of India (NSE) [9, 10].

4.1
The Return Cross-Correlation Matrix

To measure correlation between the price movements across different stocks, we first need to measure the price fluctuations such that the result is independent of the scale of measurement. For this, we calculate the logarithmic return of price. If $P_i(t)$ is the stock price of the ith stock at time t, then the (logarithmic) price return

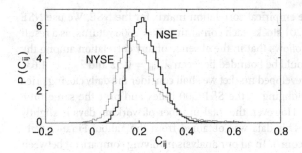

Figure 4.1 The probability density function of the elements of the correlation matrix **C** for 201 stocks in the NSE of India and NYSE for the period January 1996 to May 2006. The mean value of elements of **C** for the NSE and NYSE, $\langle C_{ij} \rangle$, are 0.22 and 0.20, respectively.

is defined as

$$R_i(t, \Delta t) \equiv \ln P_i(t + \Delta t) - \ln P_i(t) . \tag{4.1}$$

For daily return, $\Delta t = 1$ day. By subtracting the average return and dividing the result with the standard deviation of the returns (which is a measure of the volatility of the stock), $\sigma_i = \sqrt{\langle R_i^2 \rangle - \langle R_i \rangle^2}$, we obtain the normalized price return

$$r_i(t, \Delta t) \equiv \frac{R_i - \langle R_i \rangle}{\sigma_i} , \tag{4.2}$$

where $\langle \ldots \rangle$ represents time average. Once the return time series for N stocks over a period of T days are obtained, the cross-correlation matrix **C** is calculated, whose element $C_{ij} = \langle r_i r_j \rangle$, represents the correlation between returns for stocks i and j. By construction, **C** is symmetric with $C_{ii} = 1$ and C_{ij} has a value in the domain $[-1, 1]$. Figure 4.1 shows that, the correlation among stocks in NSE is larger on the average compared to that among the stocks in the NYSE. This supports the general belief that developing markets tend to be more correlated than developed ones. To understand the factors responsible for the correlations in more detail, we shall now perform an eigenvalue analysis of the correlation matrix.

4.1.1
Eigenvalue Spectrum of Correlation Matrix

If the time series are uncorrelated, then the resulting random correlation matrix is known as a Wishart matrix, whose statistical properties are well known. In the limit $N \to \infty$, $T \to \infty$, such that $Q \equiv T/N \geq 1$, the eigenvalues of this matrix are distributed according to the Sengupta–Mitra distribution [11]

$$P(\lambda) = \frac{Q}{2\pi} \frac{\sqrt{(\lambda_{\max} - \lambda)(\lambda - \lambda_{\min})}}{\lambda} , \tag{4.3}$$

for $\lambda_{\min} \leq \lambda \leq \lambda_{\max}$ and, 0 otherwise. The bounds of the distribution are given by $\lambda_{\max} = [1 + (1/\sqrt{Q})]^2$ and $\lambda_{\min} = [1 - (1/\sqrt{Q})]^2$. We now compare this with the statistical properties of the empirical correlation matrix for the NSE. We use NSE data where there are $N = 201$ stocks each containing $T = 2606$ returns; as a result $Q = 12.97$. Therefore, it follows that in the absence of any correlation among the stocks, the distribution should be bounded between $\lambda_{min} = 0.52$ and $\lambda_{max} = 1.63$.

For comparison with a developed market we shall consider the daily closing price of 434 stocks of NYSE belonging to the S&P 500 index and over the same time period as the Indian data. However, the total number of working days is slightly different, viz., 2622 days. This data was obtained from the Yahoo! Finance website (http://finance.yahoo.com/). In all our analysis involving comparison between NYSE and NSE, to make the two data sets of the same size we have used multiple random samples of 201 stocks each from the set of 434 NYSE stocks. The results obtained were independent of the particular sample of 201 stocks chosen.

As observed in developed markets [3, 4, 6, 12], the bulk of the eigenvalue spectrum $P(\lambda)$ for the empirical correlation matrix is in agreement with the properties of a random correlation matrix spectrum $P_{rm}(\lambda)$, but a few of the largest eigenvalues deviate significantly from the RMT bound (Figure 4.2). However, the number of these deviating eigenvalues are relatively few for the NSE compared to NYSE. To verify that these outliers are not an artifact of the finite length of the observa-

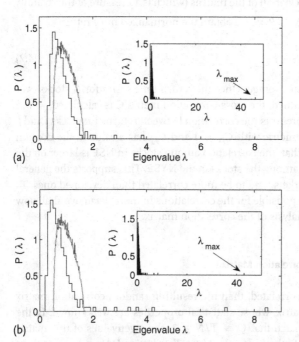

Figure 4.2 The probability density function of the eigenvalues of the correlation matrix **C** for the NSE of India (a) and NYSE (b). For comparison, the theoretical distribution predicted by (4.3) is shown using broken curves, which overlaps with the distribution obtained from the surrogate correlation matrix generated by randomly shuffling each time series. In both figures, the inset shows the largest eigenvalue.

tion period, we have randomly shuffled the return time series for each stock, and then recalculated the resulting correlation matrix. The eigenvalue distribution for this surrogate matrix matches exactly with the random matrix spectrum $P_{rm}(\lambda)$, indicating that the outliers are not due to "measurement noise" but are genuine indicators of correlated movement among the stocks. Therefore, by analyzing the deviating eigenvalues, we may be able to obtain an understanding of the structure of interactions between the stocks in the market.

The random nature of the smaller eigenvalues is also indicated by an observation of the distribution of the corresponding eigenvector components. Note that, these components are normalized for each eigenvalue λ_j such that, $\sum_{i=1}^{N}[u_{ji}]^2 = N$, where u_{ji} is the ith component of the jth eigenvector. For random matrices generated from uncorrelated time series, the distribution of the eigenvector components is given by the Porter–Thomas distribution:

$$P(u) = \frac{1}{\sqrt{2\pi}} \exp -\frac{u^2}{2} . \tag{4.4}$$

As shown in Figure 4.3, this distribution fits the empirical histogram of the eigenvector components for the eigenvalues belonging to the bulk. However, the eigen-

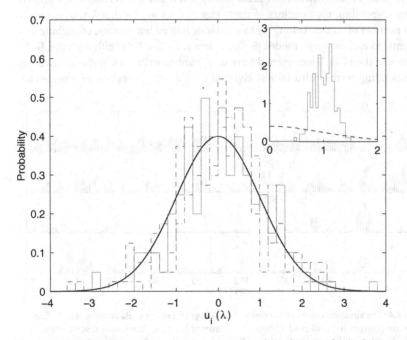

Figure 4.3 The distribution of eigenvector components corresponding to three eigenvalues belonging to the bulk predicted by RMT and (inset) corresponding to the largest eigenvalue of the cross-correlation matrix **C** for 201 stocks in the NSE of India for the period January 1996 to May 2006. In both cases, the Gaussian distribution expected from RMT is shown for comparison.

vectors of the few largest eigenvalues (e.g., the largest eigenvalue λ_{max}, as shown in the inset) deviate quite significantly, indicating their nonrandom nature.

4.1.2
Properties of the "Deviating" Eigenvalues

The largest eigenvalue λ_0 for the NSE cross-correlation matrix is more than 28 times greater than the maximum predicted by RMT. This is comparable to the NYSE, where λ_0 is about 26 times greater than the random matrix upper bound. Upon testing with synthetic US data containing the same number of missing data points as in the Indian market, we observe that λ_0 remains almost unchanged compared to the value obtained from the original US data. The corresponding eigenvector shows a relatively uniform composition, with all stocks contributing to it and all elements having the same sign (Figure 4.4a). As this is indicative of a common factor that affects all the stocks with the same bias, the largest eigenvalue is associated with the *market mode*, that is the collective response of the entire market to external information [3, 6]. Of more interest for understanding the market structure are the intermediate eigenvalues, that is those occurring between the largest eigenvalue and the bulk of the distribution predicted by RMT. For the NYSE, it was shown that corresponding eigenvectors of these eigenvalues are localized, that is only a small number of stocks, belonging to similar or related businesses, contribute significantly to each of these modes [5, 6]. However, for the NSE, although the Technology and the IT & Telecom stocks are dominant contributors to the eigenvector corresponding to the third largest eigenvalue, a direct inspection of eigenvector

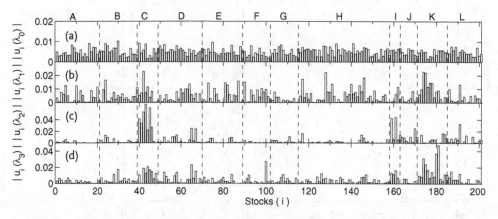

Figure 4.4 The absolute values of the eigenvector components $u_i(\lambda)$ of stock i corresponding to the four largest eigenvalues of **C** for the NSE. The stocks i are arranged by business sectors separated by broken lines. A: Automobile & transport, B: Financial, C: Technology D: Energy, E: Basic materials, F: Consumer goods, G: Consumer discretionary, H: Industrial, I: IT & Telecom, J: Services, K: Healthcare & Pharmaceutical, L: Miscellaneous.

composition does not yield a straightforward interpretation in terms of a related group of stocks corresponding to any particular eigenvalue (Figure 4.4).

To obtain a quantitative measure of the number of stocks contributing to a given eigenmode, we calculate the inverse participation ratio (IPR), defined for the kth eigenvector as $I_k \equiv \sum_{i=1}^{N}[u_{ki}]^4$, where u_{ki} are the components of eigenvector k. An eigenvector having components with equal value, that is $u_{ki} = 1/\sqrt{N}$ for all i, has $I_k = 1/N$. We find this to be approximately true for the eigenvector corresponding to the largest eigenvalue, which represents the market mode. To see how different stocks contribute to the remaining eigenvectors, we note that if a single stock had a dominant contribution in any eigenvector, for example $u_{k1} = 1$ and $u_{ki} = 0$ for $i \neq 1$, then $I_k = 1$ for that eigenvector. Thus, IPR gives the reciprocal of the number of eigenvector components (and therefore, stocks) with significant contribution. On the other hand, the average value of I_k, for eigenvectors of a random correlation matrix obtained by randomly shuffling the time series of each stock is $\langle I \rangle = 3/N \approx 1.49 \times 10^{-2}$. Figure 4.5 shows that the eigenvalues belonging to the bulk of the spectrum indeed have this value of IPR. But at the

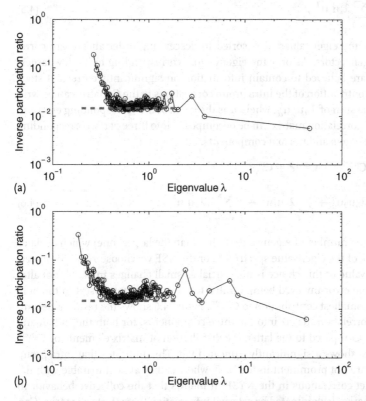

Figure 4.5 Inverse participation ratio as a function of eigenvalue for the correlation matrix **C** of the NSE (a) and NYSE (b). The broken line indicates the average value of $\langle I \rangle = 1.49 \times 10^{-2}$ for the eigenvectors of a matrix constructed by randomly shuffling each of the N time-series.

lower and higher end of eigenvalues, both the US and Indian markets show deviations, suggesting the existence of localized modes. However, these deviations are much less significant and fewer in number in the latter compared to the former. This implies that distinct groups, whose members are mutually correlated in their price movement, do exist in the NYSE, while their existence is far less clear in the NSE.

4.1.3
Filtering the Correlation Matrix

The above analysis suggests the existence of a market-induced correlation across all stocks, which makes it difficult to observe the correlations that might be due to interactions between stocks belonging to the same sector. Therefore, we now use a filtering method to remove market mode, as well as the random noise [8]. The correlation matrix is first decomposed as

$$C = \sum_{i=0}^{N} \lambda_i \mathbf{u}_i \mathbf{u}_i^T , \qquad (4.5)$$

where λ_i are the eigenvalues of C sorted in descending order and $\mathbf{u}_i$ are corresponding eigenvectors. As only the eigenvectors corresponding to the few largest eigenvalues are believed to contain information on significantly correlated stock groups, the contribution of the intra-group correlations to the C matrix can be written as a partial sum of $\lambda_\alpha \mathbf{u}_\alpha \mathbf{u}_\alpha^T$, where α is the index of the corresponding eigenvalue. Thus, the correlation matrix can be decomposed into three parts, corresponding to the *market, group* and *random* components:

$$C = C^{\text{market}} + C^{\text{group}} + C^{\text{random}}$$

$$= \lambda_0 \mathbf{u}_0 \mathbf{u}_0^T + \sum_{i=1}^{N_g} \lambda_i \mathbf{u}_i \mathbf{u}_i^T + \sum_{i=N_g+1}^{N-1} \lambda_i \mathbf{u}_i \mathbf{u}_i^T , \qquad (4.6)$$

where N_g is the number of eigenvalues (other than the largest one) which deviates from the bulk of the eigenvalue spectrum. For the NSE we choose $N_g = 5$. However, the exact value of this choice is not crucial as small changes in N_g do not alter the results, the error involved being limited to the eigenvalues closest to the bulk that have the smallest contribution to C^{group}. Figure 4.6 shows the result of decomposing the correlation matrix into the three components, for both the Indian and US markets. Compared to the latter, the distribution of matrix elements of C^{group} in the former shows a significantly truncated tail. This indicates that intra-group correlations are not prominent in the NSE, whereas they are comparable with the overall market correlations in the NYSE. It follows that the collective behavior in the Indian market is dominated by external information that affects all stocks. Correspondingly, correlations generated by interactions between stocks, as would be the case for stocks in a given business sector, are much weaker, and hence, such correlated sectors would be difficult to observe.

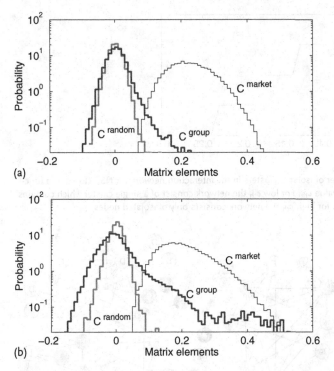

Figure 4.6 The distribution of elements of a correlation matrix corresponding to the market, $\mathbf{C}^{\text{market}}$, the group, $\mathbf{C}^{\text{group}}$, and the random interaction, $\mathbf{C}^{\text{random}}$. For the NSE (a) $N_g = 5$ whereas for the NYSE (b) $N_g = 10$. The short tail for the distribution of the $\mathbf{C}^{\text{group}}$ elements in the NSE indicates that the correlation generated by mutual interaction among stocks is relatively weak.

We indeed find this to be true when we use the information in the group correlation matrix to construct the network of interacting stocks [8]. The adjacency matrix **A** of this network is generated from the group correlation matrix $\mathbf{C}^{\text{group}}$ by using a threshold c_{th} such that $A_{ij} = 1$ if $C_{ij}^{\text{group}} > c_{\text{th}}$, and $A_{ij} = 0$ otherwise. Thus, a pair of stocks are connected if the group correlation coefficient C_{ij}^{group} is larger then a preassigned threshold value, c_{th}. To determine an appropriate choice of $c_{\text{th}} = c^*$ we observe the number of isolated clusters (a cluster being defined as a group of connected nodes) in the network for a given c_{th} (Figure 4.7). This number is much less in the NSE compared to that observed in the NYSE for any value of c_{th} [8]. Figure 4.8 shows the resultant network for $c^* = 0.09$, for which the largest number of isolated clusters of stocks are obtained. The network has 52 nodes and 298 links partitioned into 3 isolated clusters. From these clusters, only two business sectors can be properly identified, namely the Technology and the Pharmaceutical sectors. The fact that the majority of NSE stocks cannot be arranged into well-segregated groups reflecting business sectors illustrates our conclusion that intra-group interaction is much weaker than the market-wide correlation in the Indian market. An

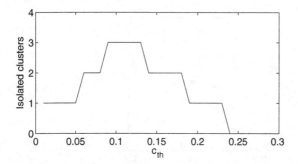

Figure 4.7 The number of isolated clusters in the interaction network for NSE stocks as a function of the threshold value c_{th}. For low c_{th} the network consist of a single cluster which contains all the nodes, whereas for high c_{th} the network consists only of isolated nodes.

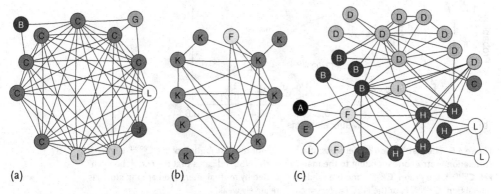

(a) (b) (c)

Figure 4.8 The structure of interaction network in the Indian financial market at threshold $c^* = 0.09$. The node label indicates the business sector of the stock, classified by the NSE. (a) The top left cluster comprises of mostly Technology stocks, while the middle cluster (b) is composed almost entirely of Healthcare & Pharmaceutical stocks. By contrast, the cluster on the right (c) is not dominated by any particular sector. The node labels indicate the business sector to which a stock belongs and are as specified in the caption to Figure 4.4.

alternative possibility is to construct the minimum spanning tree from the data in C^{group} using the technique described in Section 4.5.

4.2
Time-Evolution of the Correlation Structure

In this section, we study the temporal properties of the correlation matrix. We note here that if the deviations from the random matrix predictions are indicators of genuine correlations, then the eigenvectors corresponding to the deviating eigenvalues should be stable in time, over the period used to calculate the correlation matrix. We choose the eigenvectors corresponding to the 10 largest eigenvalues for

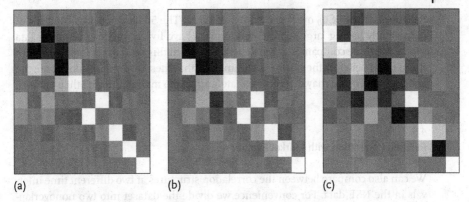

(a)   (b)   (c)

Figure 4.9 Grayscale pixel representation of the overlap matrix as a function of time for daily data during the period 1996–2001 taken as the reference. Here, the grayscale coding is such that white corresponds to $O_{ij} = 1$ and black corresponds to $O_{ij} = 0$. The length of the time window used to compute C is $T = 1250$ days (5 years) and the separa-tions used to calculate O_{ij} are $\tau = 6$ months (a), 1 year (b) and 2 years (c). The diagonal represents the overlap between the components of the corresponding eigenvectors for the 10 largest eigenvalues of the original and shifted windows. The bottom right corner corresponds to the largest eigenvalue.

the correlation matrix over a period $A = [t, t + T]$ to construct a 10×201 matrix $\mathbf{D}_A$. A similar matrix $\mathbf{D}_B$ can be generated by using a different time period $B = [t + \tau, t + \tau + T]$ having the same duration but a time-lag τ compared to the other. These are then used to generate the 10×10 overlap matrix $\mathbf{O}(t, \tau) = \mathbf{D}_A \mathbf{D}_B^T$. In the ideal case, when the 10 eigenvectors are absolutely stable in time, $\mathbf{O}$ would be an identity matrix. For the NSE data we have used time lags of $\tau = 6$ months, 1 year and 2 years, for a time window of 5 years and the reference period beginning in January 1996. As shown in Figure 4.9 the eigenvectors show different degrees of stability, with the one corresponding to the largest eigenvalue being the most stable. The remaining eigenvectors show decreasing stability with an increase in the lag period.

Next, we focus on the temporal evolution of the composition of the eigenvector corresponding to the *largest* eigenvalue. Our purpose is to find the set of stocks that have consistently high contributions to this eigenvector, and they can be identified as the ones whose behavior is dominating the market mode. We can study the time-development by dividing the return time-series data into M overlapping sets of length T. Two consecutive sets are displaced relative to each other by a time-lag δt. For example, T can be taken as six months (125 trading days), while δt can be taken to be one month (21 trading days). The resulting correlation matrices, $\mathbf{C}_{T,\delta t}$, can then be analyzed to obtain further understanding of the time-evolution of correlated movements among the different stocks.

The largest eigenvalue of $\mathbf{C}_{T,\delta t}$ follows closely the time variation of the average correlation coefficient. This indicates that the largest eigenvalue λ_0 captures the behavior of the entire market. However, the relative contribution to its eigenvector $\mathbf{u}_0$ by the different stocks may change over time. We assume that if a company is a really important player in the market, then it will have a significant contribution in the composition of u_0 over many time windows. The 50 largest stocks in terms of consistently having large representation in u_0 have five companies from the Tata group and three companies of the Reliance group in this set. This is consistent with the general belief in the business community that these two groups dominate the Indian market, and may disproportionately affect the market through their actions.

4.3
Relating Correlation with Market Evolution

We can also compare between the correlation structures at two different time intervals in the NSE data. For convenience we divide the dataset into two nonoverlapping parts corresponding to the periods between January 1996 to December 2000 (Period I) and between January 2001 to May 2006 (Period II). The corresponding correlation matrices C are generated following the same set of steps as for the entire dataset. The average value for the elements of the correlation matrix is slightly lower for the later period, suggesting a greater homogeneity between the stocks at the earlier period (Figure 4.10).

Next, we look at the eigenvalue distribution of C for the two periods. The Q value for Period I is 6.21, while for Period II it is 6.77. Thus the bounds for the random distribution is almost the same in the two cases. In contrast, the largest deviating eigenvalues, λ_0, are different: 48.56 for Period I and 45.88 for Period II. This implies the relative dominance of the market mode in the earlier period, again suggesting that with time the market has become less homogeneous. The number of deviating eigenvalues remains the same for the two periods.

When the interaction networks between stocks are generated for the two periods, they show less distinction into clearly defined sectors than was obtained with the data for the entire period (Figure 4.12). This is possibly because the shorter datasets create larger fluctuations in the correlation values, thereby making it difficult to segregate the existing market sectors. However, we do observe that, using

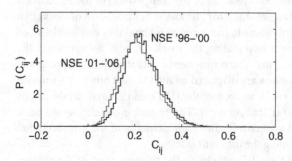

Figure 4.10 The probability density functions of the elements in the correlation matrix C for the NSE during Period I: the period January 1996 to December 2000 and Period II: January 2001 to May 2006. The mean value of the elements of C for the two periods are 0.23 and 0.21, respectively.

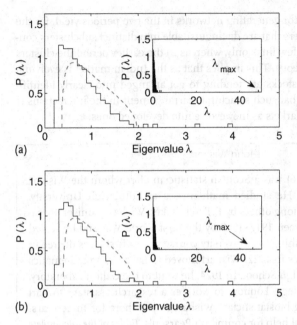

Figure 4.11 The probability density function of the eigenvalues of the NSE correlation matrix **C** for the periods (a) January 1996 to December 2000 and (b) January 2001 to May 2006. For comparison, the theoretical distribution predicted by (4.3) is shown using broken curves. In both figures, the inset shows the largest eigenvalue.

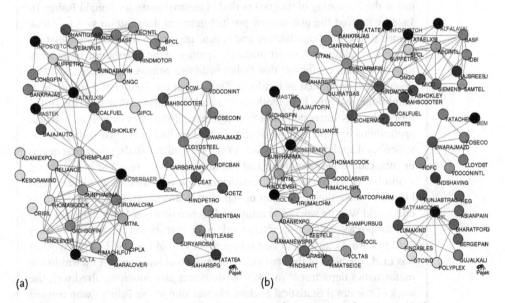

Figure 4.12 The structure of interaction network in the Indian financial market for a threshold of $c^* = 0.09$ for the period January 1996 to December 2000 (a) and January 2001 to May 2006 (b). The node colors indicate the business sector to which a stock belongs.

the same threshold value for generating networks in the two periods yield, for the later period, isolated clusters that are distinguishable into distinct subclusters connected to each other via a few links only, whereas in the earlier period the clusters are much more homogeneous. This implies that as the Indian market is evolving, the interactions between stocks are tending to get arranged into clearly identifiable groups. We propose that such structural rearrangement in the interactions is a hallmark of emerging markets as they evolve into developed ones.

John Wishart

John Wishart (1898–1956) was a Scottish statistician after whom the Wishart distribution is named. He studied mathematics at Edinburgh University, where he was taught among others by E.T. Whittaker. Wishart's university career was disrupted between 1917–1919 by the First World War and he served in France in 1918. Finishing his university course with a first class degree in mathematics and physics in 1922, Wishart moved to Leeds as a mathematics teacher at West Leeds High School. In 1924, he went to the Galton Laboratory, based at University College, London, to work as a research assistant to Karl Pearson, the pioneering biostatistician. Wishart stayed there for three years, providing computational help for compiling Pearson's *Tables of the Incomplete Beta Function* and in learning about statistical research. Subsequently, after a brief period as a mathematical demonstrator at Imperial College, he joined the staff of the Rothamsted Experimental Station in the statistical department just at the beginning of its great period of expansion under Ronald Fisher. In 1928 he derived the generalized product-moment distribution which is now named the Wishart distribution and is fundamental to multivariate statistical analysis. In addition, Wishart studied properties of the distribution of the multiple correlation coefficient that Fisher had been considering earlier. In 1931, Wishart was appointed as Reader in Statistics at the Faculty of Agriculture at Cambridge, teaching both mathematics and agriculture students. This suited him as he was interested in both mathematical statistics as well as practical applications of experimental design. Indeed he had published widely on the agricultural applications of statistics such as fertilizer trials, sugar beet experiments, crop experimentation and pig nutrition. As a result of his teaching, a number of Cambridge mathematicians took up statistics as a career and helped advance the application of statistics to diverse areas. During the Second World War, he worked in army intelligence between 1940–1942 and then as a temporary assistant secretary working on statistics at the Admiralty from 1942 to 1946. Returning to Cambridge at the end of the war, in 1953 Wishart was made the head of the statistical laboratory that had been set up within the mathematics department in 1949. Wishart was also much involved with the work of the Royal Statistical Society. He was one of the Fellows who formed the organizing committee of the Agriculture research section in 1933. In 1945 he became chairman of the Royal Statistical Society's Research Section. He also traveled extensively abroad to teach agricultural statistics, including China (1934), Spain (1947), US (1949), India (1954) and Mexico (1956). Indeed it was in Acapulco, Mexico, that Wishart died in a bathing accident while he was visiting as a representative of the United Nations Food and Agriculture Organization to arrange setting up a research center to apply statistical techniques in agricultural research.

4.4
Eigenvalue Spacing Distributions

Our observation that the bulk of the eigenvalues agrees with the theoretical distribution for that of correlation matrices obtained from random time-series is, in general, not sufficient to conclusively prove the randomness of the bulk. In principle, it is possible that matrices having the same distribution of eigenvalues can have very different correlation structures in the eigenvalue correlations. Thus, we need to explicitly test the correlations among the eigenvalues of the empirical correlation matrix C to make a judgement about their randomness. As C is a symmetric matrix containing only real numbers, we need to focus on the properties of an ensemble of such matrices whose off-diagonal elements are chosen from a random distribution with zero mean. If the order of the matrix, $N \rightarrow \infty$, then the eigenvalue correlations follow an universal distribution – that of the Gaussian orthogonal ensemble – independent of the distribution from which the matrix elements are chosen [13]. We will test the empirical matrix C for the following well-known properties of matrices belonging to the Gaussian orthogonal ensemble: (i) the distribution of eigenvalue spacings $P(\Delta\lambda)$ (where $\Delta\lambda = \lambda_{i+1} - \lambda_i$ is the interval between two adjacent eigenvalues), (ii) the distribution of the sum of the gaps that an eigenvalue has with its two adjacent eigenvalues, $P(\sum_{adj} \Delta\lambda)$ (where $\sum_{adj} \Delta\lambda = (\lambda_{i+1} - \lambda_i) + (\lambda_i - \lambda_{i-1}) = \lambda_{i+1} - \lambda_{i-1}$) and (iii) the "number variance", $\sum^2$, the variance of the number of eigenvalues in a specific interval around each eigenvalue. Before proceeding further we therefore make a short detour into the general theory of such matrices.

The distribution of eigenvalues of random matrices became a subject of interest to physicists in the 1950s. Around this time, Eugene Wigner realized that the distribution of spacing between the energy levels of highly excited nuclei resembled that for the eigenvalues of a random Hermitian matrix[1].

The symmetry of a random matrix is an important factor governing the probability distribution of its eigenvalues. For example, real symmetric matrices (A, say) are invariant with respect to an orthogonal transformation: $A' = OAO^T$, where

---

1) A Hermitian or self-adjoint matrix H is defined as a square matrix whose entries are complex conjugates of the corresponding elements of the matrix transpose, that is $a_{ij} = a_{ji}^*$. If the entries are all real, then H is simply a real symmetric matrix.

O, O^T is an orthogonal matrix and its transpose (i.e., $OO^T = 1$). The corresponding statistical ensemble is often referred to as the Gaussian orthogonal ensemble. The $N \times N$ random matrix for such an ensemble have matrix elements that are independent, identically distributed random numbers. It can be constructed by choosing the diagonal entries from the Gaussian distribution with zero mean and unit variance, $N(0, 1)$, and the off-diagonal entries for either the lower or upper triangular matrix chosen from $N(0, \frac{1}{2})$ (the entries of the other triangular matrix being obtained through the symmetry, $a_{ij} = a_{ji}$). The eigenvalues of such a matrix follows a distribution derived by Wigner,

$$P(\lambda) = \frac{1}{\pi}\sqrt{2N - \lambda^2}, \qquad \text{if} \quad |\lambda| < \sqrt{2N},$$

$$= 0, \qquad\qquad\qquad \text{otherwise},\qquad\qquad\qquad\qquad (4.7)$$

known as the Wigner semicircle law [14].

Wigner's original observation of how the spacing between energy levels in a complex nuclei varies about an average value relates to the problem of determining the distribution of intervals between successive eigenvalues arranged in ascending or descending values [15]. For this, we need to assume that a characteristic or average interval can be defined for the entire eigenvalue spectrum. If the local average (calculated over a part of the entire spectrum) varies as a function of the magnitude of the eigenvalues, then we need to split the contribution to the variation due to this nonuniformity of the average from that of the local fluctuations that we are interested in. This can be done by the process of *unfolding*, provided the scale at which the average varies is much longer than that of the difference between successive eigenvalues. Let $\rho(\lambda)$ denote the average density of the eigenvalues (i.e., the reciprocal of the intervals) as a function of the eigenvalues. Then the function

$$\eta(\lambda) = \int_{-\infty}^{\lambda} \rho(\lambda')d\lambda', \qquad\qquad\qquad\qquad\qquad (4.8)$$

indicates the number of eigenvalues that are less than the value λ. The unfolding then gives the scaled eigenvalues $e_i = \eta(\lambda_i)$ ($i = 1, \ldots, N$), which have a uniform average interval of unity. Each eigenvalue spacing now fluctuates about this constant quantity, and can be numerically computed from the histogram of successive differences of e_i. For the Gaussian orthogonal ensemble, the Wigner semicircle law can be integrated to obtain the corresponding function for unfolding, as:

$$\eta(\lambda) = \frac{1}{2\pi}\left(\lambda\sqrt{(2N - \lambda^2)} + 2N\arcsin\left(\frac{\lambda}{\sqrt{2N}} + \pi N\right)\right). \qquad (4.9)$$

The theoretical distribution for the interval distribution of 2×2 random matrices was derived by Wigner, and this is empirically found to be very similar to the distribution obtained as N becomes large. For the Gaussian orthogonal ensemble, the interval distribution for the unfolded eigenvalues is

$$P(\Delta\lambda) = \frac{\pi}{2}\Delta\lambda \exp\left[-\frac{\pi}{4}(\Delta\lambda)^2\right]. \qquad\qquad\qquad (4.10)$$

As the interval decreases, the distribution decreases linearly to zero. This is often referred to as linear level repulsion between the eigenvalues of the Gaussian orthogonal ensemble, indicating that the eigenvalues of the random matrix are correlated (as absence of any correlation would have implied a Poisson distribution).

4.4.1
Unfolding of Eigenvalues for the Market Correlation Matrix

Consider again the cumulative distribution function, $\eta(\lambda)$, which counts the number of eigenvalues $\lambda_i \leq \lambda$. It can be decomposed into two parts [6], one corresponding to an average and the other to fluctuations around this mean:

$$\eta(\lambda) = \eta_{\text{av}}(\lambda) + \eta_{\text{fluc}}(\lambda) . \tag{4.11}$$

As on average the probability density of the fluctuating component $\rho_{\text{fluc}} = d\eta_{\text{fluc}}(\lambda)/d\lambda = 0$, the averaged eigenvalue density is given by $d\eta_{\text{av}}(\lambda)/d\lambda$. Thus, the dimensionless, unfolded eigenvalues are $e_i = \eta_{\text{av}}(\lambda_i)$.

To obtain the mean component $\eta_{\text{av}}(\lambda)$, one can first approximate it by using a series of Gaussian functions in a phenomenological procedure known as *Gaussian broadening* [16], followed by fitting the cumulative distribution $\eta(\lambda)$ with the analytical expression for it (given by the corresponding expression of eigenvalue distribution for Wishart matrices). Consider the eigenvalue distribution of the correlation matrix, which can be expressed as a superposition of δ-functions about each eigenvalue: $\rho(\lambda) = (1/N) \sum_{i=1}^{N} \delta(\lambda - \lambda_i)$. Each δ-functions can be approximated by a Gaussian distribution centered around the corresponding eigenvalue and with a standard deviation $\sigma = (\lambda_{i+a} - \lambda_{i-a})/2$, with $2a$ being the broadening window size. Integrating $\rho(\lambda)$ then gives an approximation of $\eta_{\text{av}}(\lambda)$ as a series of error functions, which can then be used to obtain the unfolded eigenvalues, e_i. To avoid the "overfitting" of the averaged component, we can fit the cumulative distribution $\eta(\lambda)$ with its analytical expression obtained by averaging the theoretical eigenvalue density distribution for Wishart matrices from $-\infty$ to λ. The fitted function is an estimate of $\eta_{\text{av}}(\lambda)$, from which the unfolded eigenvalues are obtained.

4.4.2
Distribution of Eigenvalue Spacings

The distribution of the gaps between successive (unfolded) eigenvalues, $\Delta\lambda$, expresses the pairwise correlations between the eigenvalues. The distribution for the Gaussian orthogonal ensemble (GOE) has been obtained previously (4.10) and we shall compare that for the empirical matrix against this. The spacings for the unfolded eigenvalues obtained from the Gaussian broadening procedure shows that the distribution constructed from 30 minute returns agrees quite well with the GOE results. Using alternative unfolding results give identical results.

Plerou *et al.* [6] have fitted the distribution to a one-parameter Brody functional form:

$$P_{\text{Brody}}(\Delta\lambda) = \left[\Gamma\left(\frac{\beta+2}{\beta+1}\right)\right]^{1+\beta} (1+\beta)(\Delta\lambda)^{\beta}$$

$$\times \exp\left(-\left[\Gamma\left(\frac{\beta+2}{\beta+1}\right)\right]^{1+\beta} (\Delta\lambda)^{1+\beta}\right). \tag{4.12}$$

The GOE corresponds to the parameter $\beta = 1$, while $\beta = 0$ corresponds to uncorrelated eigenvalues, that is when the eigenvalue spacings have a Poisson distribution. For the empirical matrix, constructed from correlations of 30 minute returns, $\beta = 0.99 \pm 0.02$, which is essentially completely in agreement with the GOE value. A Kolmogorov–Smirnov test[2] shows that at 80% confidence level, the hypothesis that the GOE correctly describes the empirical distribution of $\Delta\lambda$ cannot be rejected.

Similar results are obtained for the empirical correlation matrix for daily returns. Thus, the eigenvalue spacings for the market correlations are distributed as per the theoretically expected distribution for real symmetric random matrices.

4.4.3
Distribution of Next Nearest Spacings between Eigenvalues

Another independent verification for the random nature of the bulk of the eigenvalue distribution is to inspect the distribution of the sum of the gaps that an eigenvalue has with its two adjacent eigenvalues, $P(\sum_{\text{adj}} \Delta\lambda)$ (where $\sum_{\text{adj}} \Delta\lambda = (\lambda_{i+1} - \lambda_i) + (\lambda_i - \lambda_{i-1}) = \lambda_{i+1} - \lambda_{i-1}$). According to Mehta and Dyson [17], this distribution will follow the statistics of the eigenvalue spacings of a Gaussian symplectic ensemble (GSE), that is corresponding to real quaternion Hamiltonians,

$$P_{\text{GSE}}(\delta\lambda) = \frac{2^{18}(\Delta\lambda)^4}{3^6 \pi^3} \exp\left(-\frac{64(\Delta\lambda)^2}{9\pi}\right). \tag{4.13}$$

At 40% confidence level, a Kolmogorov–Smirnov test cannot reject the hypothesis that the distribution for the empirical matrix constructed from 30 min returns is identical to the GSE eigenvalue spacing distribution.

4.4.4
The Number Variance Statistic

Finally, we use the "number variance" defined as the variance of the number of unfolded eigenvalues in an interval of length l around each eigenvalue,

$$\sum{}^2(l) = \langle[n(\eta, l) - l]^2\rangle, \tag{4.14}$$

---

2) The Kolmogorov–Smirnov non-parametric test for comparing between two probability distributions estimates the minimum distance between a sample and a reference distribution or between two sample distributions. The null hypothesis is that the two distributions are the same.

where $n(\eta, l)$ is the number of (unfolded) eigenvalues in the interval $[\eta - l/2, \eta + l/2]$. The averaging is performed over all eigenvalues. If the eigenvalues are uncorrelated, then the number variance scales linearly with the interval length l. For a strongly correlated system on the other hand, such as in the case of the simple harmonic oscillator, it is a constant. In general,

$$\sum^2(l) = l - 2 \int_0^l (l - x) Y(x) dx , \tag{4.15}$$

where $Y(x)$ is the two-level cluster function. For the Gaussian orthogonal ensemble, it is

$$Y(x) = \left(\frac{\sin(\pi x)}{\pi x} \right)^2 + \frac{ds}{dx} \int_x^\infty \frac{\sin(\pi x')}{\pi x'} dx' . \tag{4.16}$$

When l is large, $\sum^2 \sim \ln l$. This agrees with the results obtained for the correlation matrices constructed from both the 30 minute as well as the daily returns.

Therefore, we conclude that the bulk of the eigenvalues of the cross-correlation matrix for financial matrices is consistent with those for matrices generated from random time-series.

Eugene Paul Wigner

Eugene Paul Wigner (1902–1995) was a Hungarian born American physicist and mathematician who contributed significantly to the spectral theory of random matrices, among other fields. Wigner had been educated at the Technische Hichschule in Berlin (the present Technical University) where he studied chemical engineering and received his Ph.D. in 1925. In the 1920s Wigner started working on quantum mechanics. Serving as a lecturer first at Berlin and then at Gottingen, Wigner pioneered the use of group theory and the study of symmetries in quantum mechanics. Here he formulated his law for the conservation of parity. From the 1930s, Wigner became interested in the atomic nuclei, as part of which he became interested in the eigenvalue statistics of random matrices. In 1930, Princeton University offered a position to Wigner, subsequent to which he came to the United States. Apart from two years (1936–1938) as Professor of Physics at the University of Wisconsin, he spent his academic life at Princeton University, serving as a professor of mathematical physics from 1938 until his retirement in 1971. He became a naturalized US citizen in 1937. In 1960, Wigner wrote a now famous article on the philosophy of mathematics and physics that became his best-known work outside of physics: "The Unreasonable Effectiveness of Mathematics in the Natural Sciences". He was awarded the Nobel Prize in Physics in 1963 "for his contributions to the theory of the atomic nucleus and the elementary particles, particularly through the discovery and application of fundamental symmetry principles". After his retirement from Princeton in 1971, Wigner concentrated on exploring the philosophical consequences of quantum mechanics. In

1939, Wigner helped Leo Szilard persuade Albert Einstein to write the historic letter to President Franklin D. Roosevelt that set in motion the US atomic bomb project. During World War II he worked at the Metallurgical Laboratory at the University of Chicago, where he helped Enrico Fermi construct the first atomic pile.

4.5
Visualizing the Network Obtained from Cross-Correlations

Mantegna (1999) pioneered the use of graph theoretic visualization of the topological relation of stocks obtained from the cross-correlation information that is associated with a meaningful economic taxonomy [18]. To convert the correlation between two stocks i, j into a distance in an abstract space, the simplest choice is to define a metric $d(i, j) = 1 - \rho_{ij}^2$. The measure satisfies the three criteria for a metric: $d(i, j) = 0$ if and only if $i = j$; $d(i, j) = d(j, i)$; and $d(i, j) \leq d(i, k) + d(k, j)$. The distance matrix $\mathbf{D} = \{d(i, j)\}$ can now be used to determine the minimal spanning tree [19] for the stocks comprising a portfolio. The spanning tree of a connected graph (i.e., where it is possible to find a path between any pair of nodes) having N nodes, is any subgraph connecting all the nodes with $N - 1$ links (so that there are no closed paths or cycles). If the links are weighted, then the minimum spanning tree is the spanning tree whose total weight over all links is the least for all spanning trees. This provides an arrangement of stocks selecting the most relevant connections between them; moreover Mantegna [18] was able to show that the spanning tree representation isolates groups of stocks which belong to the same industry or subindustry sector.

Constructing the minimum spanning tree (MST) for the stocks used for computing the S & P 500 index (443 stocks were present for the entire period investigated) reveals a hierarchical structure. For example, concentrating on a particular section of the MST which is well-connected shows four groups: the first three consisting of oil companies, communication and electrical utility companies and companies dealing in raw materials, respectively, while the fourth comprised financial services, capital goods, retailing, food, drink, tobacco and consumer nondurables companies. With a few exceptions the groups appear to be homogeneous with respect to industry (and often also subindustry) sectors. It suggests that price returns of companies in the same industry or subindustry respond similarly to market conditions. The tree classification can also be used to refine the conventional economic classification into sectors and subsectors. For example, while ores, aluminum and copper are usually classified together as nonferrous metals subindustry, the MST analysis suggests that they behave differently with ores companies forming a group that is distant from all the others groups of stocks in the tree while aluminum and copper companies are part of the group containing companies dealing in raw materials.

We have used the technique proposed by Mantegna, to graphically present the interaction structure of the stocks in the National Stock Exchange of India (Figure 4.13). The distance d_{ij} between two stocks i and j are calculated from the cross-correlation matrix $\mathbf{C}$, according to $d_{ij} = \sqrt{2(1 - C_{ij})}$. These are then used to construct a minimum spanning tree, which connects all the N nodes of a network with $N-1$ edges such that the total sum of the distance between every pair of nodes, $\sum_{i,j} d_{ij}$, is minimum. Unlike the NYSE however, for the NSE such a method fails

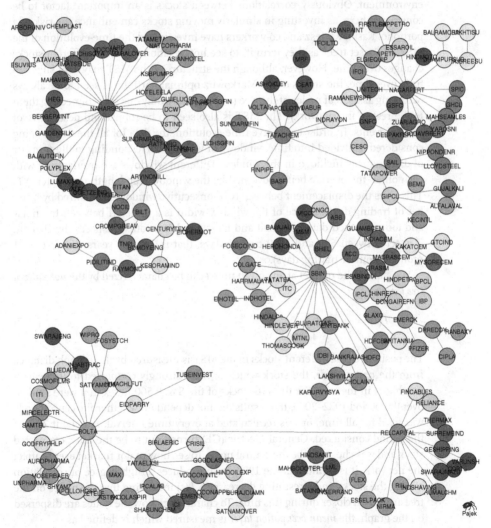

Figure 4.13 The minimum spanning tree connecting 201 stocks of the National Stock Exchange of India. The node colors indicate the business sector to which a stock belongs.

to clearly segregate any of the business sectors. Instead, stocks belonging to very different sectors are equally likely to be found within each cluster. This suggests that the market mode is dominating over all intra-sector interactions.

This graphical approach has been carried forward and suggested as a tool for portfolio optimization by the group of Kimmo Kaski and Janos Kertesz in a series of papers [20–22]. Optimizing a portfolio by suitable choice of different stocks and their quantities in order to maximize return and/or minimize risk is one of the central problems in financial economics. The theory of Markowitz [2] is one of the classical strategies to reduce risk from investments in a potentially hazardous financial environment. Obviously, correlations between stocks is an important factor to be considered here, as investing in similarly moving stocks can enhance the risk of a portfolio. Kaski, Kertesz and co-workers have investigated the time-evolution of the MST (or "asset tree" as they term it) to see how relations between different stocks changes over time. However, although the structure of the MST may alter substantially over time, the stocks of the Markowitz optimal portfolio is seen to be always occurring as the leaf nodes of the tree (i.e., they have only one link connecting them to the rest of the network). The MST is also seen to be robust and descriptive of market events. In order to observe time-evolution, the data for all the stocks being considered is divided into time windows of width T corresponding to the number of daily returns included in the window. Different windows are overlapping with each other, the overlap being governed by the window step length parameter δT. This gives the displacement between two consecutive windows, measured by number of trading days. Choice of the window width is a trade-off between too noisy and too smoothed data for small and large window widths, respectively. Usually, T is set between 500 and 1500 trading days, that is 2 and 6 years, and $\delta T = 1$ month (about 21 trading days).

The state of the market at any given time t can be characterized by the *normalized tree length*

$$L(t) = \frac{1}{N-1} \sum_{d_{ij} \in D^t} d_{ij} \,. \tag{4.17}$$

The position of the different stocks in the MST is measured by their graph distance from the *central node*, the stock which is most strongly connected to its nearest neighbors in the tree. For the 116 stocks of the S & P 500 index that were studied over the period 1982–2000, the results did not depend on whether the central node was fixed for all time, or was recalculated at every time interval. For about 70% of the period considered, General Electric (GE) turned out to be the most connected stock and was chosen to be the central node. To see how major market events affect the network, it is found that the 1987 crash produces a high degree of correlation among the stocks and a resulting decrease in $L(t)$ indicating that the nodes on the tree are drawn closer during this time. To characterize how the nodes are dispersed on the graph, the *mean occupation layer* is measured which is defined as

$$l(t) = \frac{1}{N} \sum_{i=1}^{N} \text{lev}(v_i^t) \,, \tag{4.18}$$

where $\text{lev}(v_i)$ denotes the level of vertex v_i in relation to the central node, whose level is taken to be zero. It is seen that $l(t)$ decreases to a very low value at the time of a market crisis.

We now consider a minimum risk Markowitz portfolio $P(t)$ with the asset weights $w_1, w_2, \ldots, w_N$. In the Markowitz portfolio optimization scheme financial assets are characterized by their average return and risk, both determined from historical price data, where risk is measured by the standard deviation of returns. The aim is to optimize the asset weights so that the overall portfolio risk is minimized for a given portfolio return. In the minimum spanning tree framework, the task is to determine how the assets are located with respect to the central node. Intuitively, we expect the weights to be distributed on the outskirts of the graph. In order to describe what happens, we define the *weighted portfolio layer*:

$$l_P(t) = \sum_{i \in P} w_i \text{lev}(v_i^t) , \qquad (4.19)$$

with the constraint $w_i \geq 0$ for all i, assuming that there is no short-selling. The portfolio layer is seen to be higher than the mean layer practically at all times. The difference in layers depends to a certain extent on the window width: for $T = 500$ it is about 0.76 and for $T = 1000$ about 0.97. As the stocks of the minimum risk portfolio are found on the outskirts of the graph, we expect larger graphs (higher L) to have greater *diversification potential*, that is the scope of the stock market to eliminate specific risk of the minimum risk portfolio. In order to look at this, one can calculate the mean-variance frontiers for the ensemble of stocks using, for example, $T = 500$ as the window width.

The robustness of the minimum spanning tree topology can be measured by the survival ratio of tree edges (fraction of edges is found common in both graphs) at time t, $\frac{1}{N-1}|E^t \cap E^{t-1}|$, where E^t refers to the set of edges of the graph at time t, $\cap$ is the intersection operator and $|\ldots|$ gives the number of elements in the set. Under normal circumstances, the graphs at two consecutive time windows t and $t+1$ (for small values of δT) should look very similar. Whereas some of the differences can reflect real changes in the asset taxonomy, others may simply be due to noise. As $\delta T \to 0$, $\sigma_t \to 1$, indicating that the graphs *are* stable in the limit.

Another possible graphical representation of the correlation between N stocks is through creating a graph of $N - 1$ links with smallest values of d ("asset graph"). The restriction on the number of edges to $N - 1$ is to make it comparable with the MST. In other words, the graph is constructed by inserting edges sequentially, according to the rank of the corresponding element of the $\mathbf{D}$ matrix such that we start with the smallest (i.e., with the highest correlation). For this graph there is no condition that cycles cannot be present. Applying this technique on the preceding data (stocks used for computing S & P 500 index) reveals strongly interconnected clusters. Indeed, some of the nodes form cliques. For example, the cluster formed by energy companies and the cluster of utilities companies are heavily intra-connected, but neither have strong connections to other clusters. In general, there is good agreement between the clusters in the graph and business sectors defined economically.

4.6
Application to Portfolio Optimization

The random character of the bulk of the eigenvalue distribution has significant implications for optimal portfolio construction (e.g., using the Markowitz theory) [6]. Consider a portfolio $\Pi(t)$, where w_i is the fraction composed of stock i with return G_i. Therefore, the return on $\Pi(t)$ is

$$\Phi = \sum_{i=1}^{N} w_i G_i , \tag{4.20}$$

where the fractions w_i are normalized such that $\sum_{i=1}^{N} w_i = 1$. The risk associated with the portfolio is expressed by the variance

$$\Omega^2 = \sum_{i=1}^{N}\sum_{j=1}^{N} w_i w_j C_{ij}\sigma_i\sigma_j , \tag{4.21}$$

where σ_i is the volatility (as measured by the standard deviation of the returns) of stock i, and C_{ij} are elements of the cross-correlation matrix. An optimal portfolio is constructed by minimizing the risk Ω^2 subject to holding the return on the portfolio constant at a certain Φ. Minimizing Ω^2 subject to this constraint (as well as the constraint that $\sum_{i=1}^{N} w_i = 1$) by using Lagrange multipliers yields a system of linear equations for w_i. Solving these equations yields the optimal portfolios whose return Φ can be shown as a function of the risk Ω^2.

In order to see how the randomness of the cross-correlation matrix $\mathbf{C}$ affects the selection of the optimal portfolio, the empirical data for a two-year time period, e.g., 1994–1995, can be divided into two 1-year periods. Using the cross-correlation matrix $\mathbf{C}_{1994}$ for 1994, and the returns G_i for 1995, a family of optimal portfolios can be constructed. Thus, the return Φ can be visualized as a function of the predicted risk Ω_p^2 for 1995. However, the actual or realized risk for 1995, Ω_r^2, can also be calculated as we have access to the cross-correlation matrix $\mathbf{C}_{1995}$ for 1995. It is seen that the predicted risk is significantly smaller when compared to the realized risk, $(\Omega_r^2 - \Omega_p^2)/\Omega_p^2 \simeq 170\%$. Almost the entire significant information in the cross-correlations is in the eigenvectors for the eigenvalues falling outside the bulk of the distribution predicted by random matrix theory. If a filtered correlation matrix $\mathbf{C}'$ is constructed that contains only the deviating eigenvectors, that should provide us with a better match to the eventual risk-return relation. Constructing a diagonal matrix which retains only the deviating eigenvalues (the remaining eigenvalues being set to 0) and transforming it to the basis of $\mathbf{C}$, we obtain $\mathbf{C}'$ after setting its diagonal elements equal to 1 (so that $\text{Tr}(\mathbf{C}') = \text{Tr}(\mathbf{C}) = N$). Repeating the above calculations for finding the optimal portfolio, but using the filtered correlation matrix shows that the realized risk is now much closer to the predicted risk, $(\Omega_r^2 - \Omega_p^2)/\Omega_p^2 \simeq 25\%$. Thus, taking the structure of the correlation matrix and filtering it to obtain the component corresponding to the deviating eigenvalues can help in constructing optimal portfolios that are more robust.

In addition, the difference between emerging and developed markets in terms of the global and sector correlations have obvious importance in choosing stocks to minimize risk. If the entire market is responding homogeneously to external shocks, the market may inherently be more risky than one in which different sectors respond differently.

4.7
Model of Market Dynamics

To understand the relation between the interaction structure among stocks and the eigenvalues of the correlation matrix, we perform a multivariate time-series analysis using a simple two-factor model of market dynamics. We assume that the normalized return at time t of the ith stock from the kth business sector can be decomposed into (i) a market factor, $r_m(t)$, that contains information or signal common to all stocks, (ii) a sector factor, $r_g^k(t)$, representing effects exclusive to stocks in the kth sector, and (iii) an idiosyncratic term, $\eta_i(t)$, which corresponds to random variations unique for that stock. Thus,

$$r_i^k(t) = \beta_i r_m(t) + \gamma_i^k r_g^k(t) + \sigma_i \eta_i(t) , \qquad (4.22)$$

where β_i, γ_i^k and σ_i represent relative strengths of the three terms mentioned above, respectively. For simplicity, these strengths are assumed to be time-independent. We choose $r_m(t)$, $r_g^k(t)$ and $\eta_i(t)$ from a zero mean and unit variance Gaussian distribution. We further assume that the normalized returns r_i, also follow Gaussian distribution with zero mean and unit variance. Although the empirically observed return distributions have power-law tails, as these distributions are not Levy stable, they will converge to Gaussian if the returns are calculated over sufficiently long intervals. The assumption of unit variance for the returns ensures that the relative strengths of the three terms will follow the relation:

$$\beta_i^2 + (\gamma_i^k)^2 + \sigma_i^2 = 1 . \qquad (4.23)$$

As a result, for each stock we can assign σ_i and γ_i independently, and obtain β_i from (4.23). We choose σ_i and γ_i from a uniform distribution having width δ and centered about the mean values σ and γ, respectively.

We now simulate an artificial market with N stocks belonging to K sectors by generating time-series of length T for returns r_i^k from the above model. These K sectors are composed of $n_1, n_2, \ldots, n_K$ stocks such that $n_1 + n_2 + \cdots + n_K = N$. The collective behavior is then analyzed by constructing the resultant correlation matrix $\mathbf{C}$ and obtaining its eigenvalues. Our aim is to relate the spectral properties of $\mathbf{C}$ with the underlying structure of the market given by the relative strength of the factors. We first consider the simple case, where the contribution due to market factor is neglected, that is $\beta_i = 0$ for all i, and the strength of sector factor is equal for all stocks within a sector, that is $\gamma_i^k = \gamma^k$, is independent of i. In this case, the spectrum of the correlation matrix is composed of K large eigenvalues,

$1 + (n_j - 1)(\gamma^j)^2$, where $j = 1 \ldots K$, and $N - K$ small eigenvalues, $1 - (\gamma^j)^2$, each with degeneracy $n_j - 1$, where $j = 1 \ldots K$ [23]. Now, we consider a nonzero market factor which is equal for all stocks, that is $\beta_i = \beta$ for all i, and the strength of sector factor is also the same for all stocks, that is $\gamma_i^k = \gamma$ (independent of i and k). In this case too, there are K large eigenvalues and $N - K$ small eigenvalues. Our numerical simulations suggest that the largest and the second largest eigenvalues are

$$\lambda_0 \sim N\beta^2 ,$$
$$\lambda_1 \sim n_l(1 - \beta)^2 , \tag{4.24}$$

respectively, where n_l is the size of the largest sector, while the $N - K$ small degenerate eigenvalues are $1 - \beta^2 - \gamma^2$. We now choose the strength γ_i^k and σ_i from a uniform distribution with mean γ and σ, respectively, and with width $\delta = 0.05$.

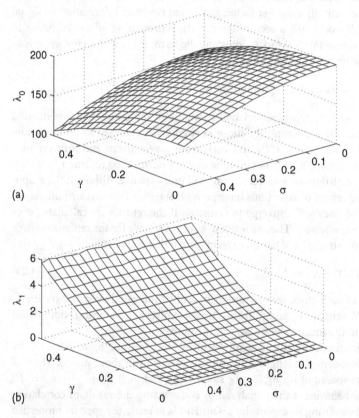

(a)

(b)

Figure 4.14 The variation of the largest (a) and second largest (b) eigenvalues of the correlation matrix of simulated return in the two-factor model (4.22) with the model parameters γ and σ (corresponding to strength of the sector and idiosyncratic effects, respectively). The matrix is constructed for $N = 200$ stocks each with return time-series of length $T = 2000$ days. We assume there to be 10 sectors, each having 20 stocks.

Figure 4.14 shows the variation of the largest and second largest eigenvalues with σ and γ. The strength of the market factor is determined from (4.23).

Note that, decreasing the strength of the sector factor relative to the market factor results in decreasing the second largest eigenvalue λ_1. As $Q = T/N$ is fixed, the RMT bounds for the bulk of the eigenvalue distribution, $[\lambda_{min}, \lambda_{max}]$, remain unchanged. Therefore, a decrease in λ_1 implies that the large intermediate eigenvalues occur closer to the bulk of the spectrum predicted by RMT, as is seen in the case of the NSE. The analysis of the model supports our hypothesis that the spectral properties of the correlation matrix for the NSE are consistent with a market in which the effect of information common for all stocks (i.e., the market mode) is dominant, resulting in all stocks exhibiting a significant degree of correlation.

4.8
So what did we Learn?

In conclusion, the analysis given above does imply that the stocks in emerging market are much more correlated than in developed markets. Although, the bulk of the eigenvalue spectrum of the correlation matrix of stocks **C** in emerging market is similar to that observed for developed markets, the number of eigenvalues deviating from the RMT upper bound are smaller in number. Further, most of the observed correlation among stocks is found to be due to effects common to the entire market, whereas correlation due to interaction between stocks belonging to the same business sector are weak. This dominance of the market mode relative to modes arising through interactions between stocks makes an emerging market appear more correlated than developed markets. Using a simple two-factor model we show that a dominant market factor, relative to the sector factor, results in spectral properties similar to that observed empirically for the emerging financial market of India. This study helps in understanding the evolution of markets as complex systems, suggesting that strong interactions may emerge within groups of stocks as a market evolves over time. How such self-organization occurs and its relation to other changes that a market undergoes during its development, for example large increase in transaction volume, is a question worth pursuing in the future with the tools available to econophysicists.

This chapter also makes a significant point regarding the physical understanding of markets as complex dynamical systems. In recent times, the role of the interaction structure within a market in governing its overall dynamical properties has come under increasing scrutiny. However, such intra-market interactions affect only very weakly certain market properties, which is underlined by the observation of identical fluctuation behavior in markets having very different interaction structures, viz., the NYSE and NSE [9, 10]. The system can be considered as a single homogeneous entity responding only to external signals in explaining these statistical features, for example the price fluctuation distribution. This suggests that the basic assumption behind the earlier approach of studying financial markets as essentially executing random walks in response to independent external shocks [24], which ignored the internal structure, may still be considered accurate for explaining market fluctuation phenomena. In other words, complex interacting systems like financial markets can have simple mean field-like description for some of their properties.

References

1 Rouwenhorst, K.G. (2005) The origins of mutual funds. in *The Origins of Value: The financial innovations that created modern capital markets* (eds W.N. Goetzmann and K.G. Rouwenhorst), Oxford Univ Press, New York.

2 Markowitz, H. (1959) *Portfolio Selection: Efficient diversification of investments*, John Wiley, New York.

3 Laloux, L., Cizeau, P., Bouchaud, J.-P., and Potters, M. (1999) Noise dressing of financial correlation matrices. *Phys. Rev. Lett.*, **83**, 1467–1470.

4 Plerou, V., Gopikrishnan, P., Rosenow, B., Amaral, L.A.N., and Stanley, H.E. (1999) Universal and nonuniversal properties of cross correlations in financial time series. *Phys. Rev. Lett.*, **83**, 1471–1474.

5 Gopikrishnan, P., Rosenow, B., Plerou, V., and Stanley, H.E. (2001) Quantifying and interpreting collective behavior in financial markets. *Phys. Rev. E*, **64**, 035106.

6 Plerou, V., Gopikrishnan, P., Rosenow, B., Amaral, L.A.N., Guhr, T., and Stanley, H.E. (2002) Random matrix approach to cross correlations in financial data. *Phys. Rev. E*, **65**, 066126.

7 Noh, J.D. (2000) Model for correlations in stock markets. *Phys. Rev. E*, **61**, 5981–5982.

8 Kim, D.-H. and Jeong, H. (2005) Systematic analysis of group identification in stock markets. *Phys. Rev. E*, **72**, 046133.

9 Pan, R.K. and Sinha, S. (2007) Collective behavior of stock price movements in an emerging market. *Phys. Rev. E*, **76**, 046116.

10 Sinha, S. and Pan, R.K. (2007) Uncovering the internal structure of the Indian financial market: Large cross-correlation behavior in the NSE (eds A. Chatterjee and B.K. Chakrabarti). *Econophysics of*

markets and business networks, Springer, Milan, pp. 3–19.

11 Sengupta, A.M. and Mitra, P.P. (1999) Distribution of singular values for some random matrices. *Phys. Rev. E*, **60**, 3389–3392.

12 Utsugi, A., Ino, K., and Oshikawa, M. (2004) Random matrix theory analysis of cross correlations in financial markets. *Phys. Rev. E*, **70**, 026110.

13 Mehta, M.L. (1991) *Random Matrices*, 3rd edition, Academic Press, Boston.

14 Wigner, E.P. (1957) Statistical properties of real symmetric matrices with many dimensions. *Canadian Mathematical Congress Proceedings*, University of Toronto Press, Toronto, pp. 174–184.

15 Timberlake, T. (2006) Random numbers and random matrices: Quantum chaos meets number theory. *Am. J. Phys*, **74**, 547–553.

16 Bruus, H. and Anglés d'Auriac, J.-C. (1996) The spectrum of the two-dimensional Hubbard model at low filling. *Europhys. Lett.*, **35**, 321–326.

17 Mehta, M.L. and Dyson, F.J. (1963) Statistical Theory of the Energy Levels of Complex Systems V. *J. Math. Phys.*, **4**, 713–719.

18 Mantegna, R.N. (1999) Hierarchical structure in financial markets. *Eur. Phys. J. B*, **11**, 193–197.

19 Papadimitriou, C.H. and Steiglitz, K. (1982) *Combinatorial Optimization*, Prentice-Hall, Englewood Cliffs.

20 Onnela, J.-P., Chakraborti, A., Kaski, K., and Kertesz, J. (2002) Dynamic asset trees and portfolio analysis. *Eur. Phys. J. B*, **30**, 285–288.

21 Onnela, J.-P., Chakraborti, A., and Kaski, K. (2003) Dynamics of market correlations: Taxonomy and portfolio analysis. *Phys. Rev. E*, **68**, 056110.

22 Onnela, J.-P., Kaski, K., and Kertesz, J. (2004) lustering and information in correlation based financial networks. *Eur. Phys. J. B*, 38, 353–362.

23 Lillo, F. and Mantegna, R.N. (2005) Spectral density of the correlation matrix of factor models: A random matrix theory approach. *Phys. Rev. E*, 72, 016219.

24 Bachelier, L. (1900) Théorie de la spéculation, *Annales Scientifiques de l'École Normale Supérieure, 3 série*, 17, 21–86.

5
Why Care about a Power Law?

"The reason we care about power law correlations is that we're conditioned to think
they're a sign of something interesting and complicated happening."
– Cosma R. Shalizi, *Power-law distributions, 1/f noise, long-memory time series.*
http://www.cs.umich.edu/~crshalizi/notebooks/power-laws.html

Social scientists have often remarked that physicists working in social or economic
phenomena seem to be obsessed with power laws. Although there are no a priori reasons why, say, log-normal distributions should be any less interesting, it
remains a fact that many papers in econophysics are about distributions exhibiting a power-law tail. The reason for this attraction to power laws is that it signals
the occurrence of scale-independent behavior, which for physicists has an intimate
connection to phase transitions and critical phenomena. In contrast, most other
distributions, such as the Gaussian or the Poisson, are characterized by a "typical
value" (or "scale") about which all the recorded data points occur. Thus, the absence
of such a characteristic scale in a process would indicate that the corresponding
variable may vary over a wide range of values, sometimes spanning several orders
of magnitude [1]. A large number of examples for power-law distributions occurring in reality have been put forward, such as the size distribution of cities, the
intensity of earthquakes and the frequency of word usage in human languages, to
name a few. However, none of these are beyond controversy. In fact, critics would
say that most of the so-called power laws reported in the scientific literature are an
artifact of limited datasets and crude estimation methods [2]. In econophysics, the
most reliable examples of possible power-law distributions occur in the financial
arena, and in the following sections we shall look at these in detail.

5.1
Power Laws in Finance

Financial markets can be viewed as complex systems with a large number of interacting components that are subject to external influences or information flow.
Physicists are being attracted in increasing numbers to the study of financial mar-

Econophysics. Sitabhra Sinha, Arnab Chatterjee, Anirban Chakraborti, and Bikas K. Chakrabarti
Copyright © 2011 WILEY-VCH Verlag GmbH & Co. KGaA, Weinheim
ISBN: 978-3-527-40815-3

kets by the prospect of discovering universalities in their statistical properties [3–5]. This has partly been driven by the availability of large amounts of electronically recorded data with very high temporal resolution, making it possible to study various indicators of market activity.

5.1.1
The Return Distribution

One of the principal statistical features characterizing the activity in financial markets is the distribution of fluctuations of individual stock prices and overall market indicators such as the index. To study these fluctuations such that the result is independent of the scale of measurement, we define the logarithmic return for a time-scale Δt as

$$R(t, \Delta t) \equiv \ln P(t + \Delta t) - \ln P(t) , \tag{5.1}$$

where $P(t)$ is the price (or market index) at time t and Δt is the time-scale over which the fluctuation is observed. This definition of price change has the added advantage that to obtain the return at a longer time-scale one simply needs to add together the changes occurring at the shorter time-scale.

Market indices, rather than individual stock prices, have been the focus of most previous studies, as the former are more easily available, and also gives overall information about the market. By contrast, individual stocks are susceptible to sector-specific, as well as, stock-specific influences, and may not be representative of the entire market. These two quantities, in fact, characterize the market from different perspectives, the microscopic description being based on individual stock price movements, while the macroscopic point of view focusses on the the collective market behavior as measured by the market index.

The importance of interactions among stocks, relative to external information, in governing market behavior has emerged only in recent times. The earliest theories of market activity, for example Bachelier's random walk model, assumed that price changes are the result of several independent external shocks, and therefore, predicted the resulting distribution to be Gaussian [6]. As an additive random walk may lead to negative stock prices, a better model would be a multiplicative random walk, where the price changes are measured by logarithmic returns [7]. While the return distribution calculated from empirical data is indeed seen to be Gaussian at long time-scales, at shorter times the data show much larger fluctuations than would be expected from this distribution [8]. Such deviations were also observed in commodity price returns, for example in Mandelbrot's analysis of cotton price, which was found to follow a Levy-stable distribution [9]. However, it contradicted the observation that the distribution converged to a Gaussian at longer time-scales. Later, it was discovered that while the bulk of the return distribution for the S & P 500 index appears to be fit well by a Levy distribution, the asymptotic behavior shows a much faster decay than expected. Hence, a truncated Levy distribution, which has exponentially decaying tails, was proposed as a model for the distribution of returns [10]. Subsequently, it was shown that the tails of the cumulative return distribution for this index actually follow a power law

$$P_c(r > x) \sim x^{-\alpha} , \tag{5.2}$$

with the exponent $\alpha \approx 3$ (the "inverse cubic law") [11], well outside the stable Levy regime $0 < \alpha < 2$. This is consistent with the fact that at longer time-scales the distribution converges to a Gaussian. This "inverse cubic law" had been reported initially for a small number of stocks from the S & P 100 list [12]. Later, it was established from statistical analysis of stock returns in the German stock exchange [13], as well as for three major US markets, including the New York Stock Exchange (NYSE) [14]. The distribution was shown to be quite robust, retaining the same functional form for time-scales of up to several days. Similar behavior has also been seen in the London Stock Exchange [15]. An identical power-law tail has also been observed for the fluctuation distribution of a number of market indices [16, 17]. This apparent universality of the distribution may indicate that different markets self-organize to an almost identical nonequilibrium steady state.

These observations are somewhat surprising, although not at odds with the "efficient market hypothesis" in economics, which assumes that the movements of financial prices are an immediate and unbiased reflection of incoming news and future earning prospects. To explain these observations various multiagent models of financial markets have been proposed, where the scaling laws seen in empirical data arise from interactions between agents [18]. Other microscopic models, where the agents (i.e., the traders comprising the market) are represented by mutually interacting spins and the arrival of information by external fields, have also been used to simulate the financial market [19–22]. Among nonmicroscopic approaches, multifractal processes have been used extensively for modeling such scale invariant properties [23, 24]. The multifractal random walk model has generalized the usual random walk model of financial price changes and accounts for many of the observed empirical properties [25].

However, on the empirical front, there is some controversy about the universality of the power-law nature for the tails of the index return distribution. In the case of developed markets, for example the All Ordinaries index of the Australian stock market, the negative tail has been reported to follow the inverse cubic law while the positive tail is closer to Gaussian [26]. Again, other studies of the Hang Seng and Nikkei indices report the return distribution to be exponential [27, 28]. For developing economies, the situation is even less clear. There have been several claims that emergent markets have a return distribution that is significantly different from developed markets. For example, a recent study contrasting the behavior of indices from seven developed markets with the KOSPI index of the Korean stock market found that while the former exhibit the inverse cubic law, the latter follows an exponential distribution [17]. Another study of the Korean stock market reported that the index distribution has changed to exponential from a power-law nature only in recent years [29]. On the other hand, the IBOVESPA index of the Sao Paulo stock market has been claimed to follow a truncated Levy distribution [30, 31]. However, there have also been reports of the inverse cubic law for emerging markets, for example for the Mexican stock market index IPC [32] and the WIG20 index of the

Polish stock market [33]. A comparative analysis of 27 indices from both mature and emerging markets found their tail behavior to be similar [34]. It is of course difficult to conclude about the nature of the fluctuation distribution for individual stock prices from the index data, as the latter is a weighted average of several stocks. Therefore, in principle, the index can show a distribution quite different from that of its constituent stocks if their price movements are not correlated.

Many of the studies reported above have only used graphical fitting to determine the nature of the observed return distribution. This has recently come under criticism as such methods often result in erroneous conclusions. Hence, a more accurate study using reliable statistical techniques needs to be carried out to decide whether emerging markets do behave similarly to developed markets in terms of fluctuations. Here, we will show how to carry out such an analysis using data from the Indian financial markets. The Indian data is very important for deciding whether emerging markets behave differently from developed markets, as it is one of the fastest growing financial markets in the world. A recent study of individual *stock prices* in the National Stock Exchange (NSE) of India has claimed that the corresponding return distribution is exponentially decaying at the tails [35], and not the inverse cubic law that is observed for developed markets [13, 14]. However, a more detailed study over a larger dataset has established the inverse cubic law for individual *stock prices* [36]. On the other hand, to get a sense of the nature of fluctuations for the entire market, one needs to look at the corresponding distribution for the market index. Although the individual stock prices and the market index are related, it is not obvious that they should have the same kind of distribution, as this relation is dependent on the degree of correlation between different stock price movements.

We will show in the following how to use both individual price as well as overall market index data to establish the nature of the distribution. Although we use the specific example of Indian markets, it is needless to say that the analysis applies to other markets as well.

5.1.2
Stock Price Return Distribution

Most early studies of stock price fluctuations in emerging markets were done on low-frequency daily data. Let us instead focus on high-frequency tick-by-tick data, which will be complemented by analysis of daily data over much longer periods. The dataset that we have chosen for this purpose is from the National Stock Exchange (NSE) of India, the largest among the 23 exchanges in India, with more than 85% of the total value of transactions for securities in all market segments of the entire Indian financial market in recent times [37]. This data set is of unique importance, as we have access to daily data right from the time the market commenced operations in the equities market in November 1994, up to the present when it has become the world's third largest stock exchange (after NASDAQ and NYSE) in terms of transactions [38]. Over this period, the market has grown rapidly, with the number of transactions having increased by more than three orders of magnitude. Therefore, if markets do show discernible transition in the return distribution during their evolution, the Indian market data is best placed to spot evidence for it, not least because of the rapid transformation of the Indian economy in the liberalized environment since the 1990s.

We focus on two important questions: (i) Does the market exhibit a price fluctuation different from the inverse cubic law form seen in developed markets, and (ii) if the market is indeed following the inverse cubic law at present, whether this has been converged, starting from an initially different distribution when the market had just begun operation. Both of these questions are answered negatively in the following analysis.

The two datasets having different temporal resolutions that we analyze are the following. (i) The high-frequency tick-by-tick data contains information about all transactions carried out in the NSE between January 2003 and March 2004. This information includes the date and time of trade, the price of the stock during transaction and the volume of shares traded. This database is available in the form of CDs published by the NSE. For calculating the price return, we have focused on 489 stocks that were part of the BSE 500 index (a comprehensive indicator for the Indian financial market) during this period. The number of transactions for each company in this set is $\sim 10^6$, on the average. The total number of transactions for the 489 stocks is on the order of 5×10^8 during the period under study. (ii) The daily closing price of all the stocks listed in NSE during its period of existence between November 1994 and May 2006. This was obtained from the NSE website [39] and manually corrected for stock splitting.[1] For comparison with US markets, in particular the NYSE, we consider the 500 stocks listed in S & P 500 during the period November 1994 to May 2006, the daily data being obtained from Yahoo! Finance [40].

To measure the price fluctuations such that the result is independent of the scale of measurement, we calculate the logarithmic return of price. If $P_i(t)$ is the stock price of the ith stock at time t, then the (logarithmic) price return is defined as

$$R_i(t, \Delta t) \equiv \ln P_i(t + \Delta t) - \ln P_i(t) . \tag{5.3}$$

However, the distribution of price returns of different stocks may have different widths, owing to differences in their volatility, defined (for the ith stock) as $\sigma_i^2 \equiv \langle R_i^2 \rangle - \langle R_i \rangle^2$. To compare the distribution of different stocks, we normalize the returns by dividing them with their volatility $\sigma_i(t)$, as in [4]. The resulting normalized

1) When the price of a stock increases to such levels that trading in it reduces significantly, the company often decides to split a single share into multiple parts in order to increase liquidity. In the stock price data we are analyzing here, this may be reflected as a sudden drop in the price. As it is not related to any change in the intrinsic value of the stock, this should be corrected in the data so as to avoid obtaining an erroneous return distribution.

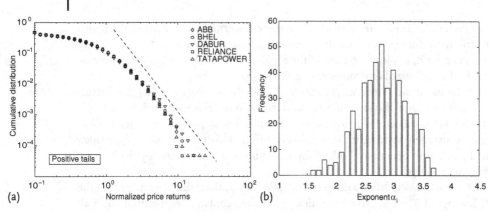

(a)

(b)

Figure 5.1 (a) Cumulative distribution of the positive tails of the normalized 5 min returns distribution of 5 stocks chosen arbitrarily from those listed in the NSE for the period January 2003 to March 2004. The broken line indicates a power law with exponent $\alpha = 3$.

(b) The histogram of the power-law exponents obtained by regression fit for the positive tail of individual cumulative return distributions of 489 stocks. The median of the exponent values is 2.84.

price return[2] is given by

$$r_i(t, \Delta t) \equiv \frac{R_i(t) - \langle R_i(t) \rangle}{\sigma_i(t)} , \tag{5.4}$$

where $\langle \cdots \rangle$ denotes the time average over the given period.

For analysis of the high-frequency data, we consider the aforementioned 489 stocks. Choosing an appropriate Δt, we obtain the corresponding return by taking the log ratio of consecutive average prices, averaged over a time window of length Δt. Figure 5.1a shows the cumulative distribution of the normalized returns r_i with $\Delta t = 5$ min for five stocks, arbitrarily chosen from the dataset. We observe that the distribution of normalized returns r_i for all the stocks have the same functional form with a long tail that follows a power-law asymptotic behavior. The distribution of the corresponding power-law exponent α_i for all the 489 stocks that we have considered is shown in Figure 5.1b.

As all the individual stocks follow very similar distributions, we can merge the data for different stocks to obtain a single distribution for normalized returns. The aggregated return dataset with $\Delta t = 5$ min has 6.5×10^6 data points. The corresponding cumulative distribution is shown in Figure 5.2a, with the exponents for the positive and negative tails estimated as

$$\alpha = \begin{cases} 2.87 \pm 0.08 & \text{(positive tail)} \\ 2.52 \pm 0.04 & \text{(negative tail)} \end{cases} . \tag{5.5}$$

From this figure we confirm that the distribution does indeed follow a power-law decay, albeit with different exponents for the positive and negative return tails.

---

2) The normalization of return $R_i(t)$ is performed by removing its own contribution from the volatility, that is $\sigma_i(t) = \sqrt{\frac{1}{N-1} \sum_{t' \neq t} \left[R_i(t') - \langle R_i(t) \rangle \right]^2}$.

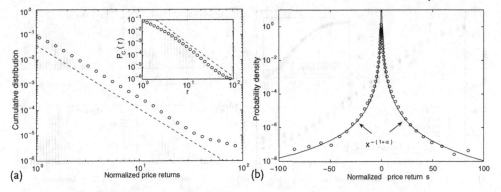

Figure 5.2 (a) Cumulative distribution of the negative and (inset) positive tails of the normalized returns for the aggregated data of 489 stocks in the NSE for the period January 2003 to March 2004. The broken line for visual guidance indicates the power-law asymptotic form. (b) Probability density function of the normalized returns. The solid curve is a power-law fit in the region 1–50. We find that the corresponding cumulative distribution exponent, $\alpha = 2.87$ for the positive tail and $\alpha = 2.52$ for the negative tail.

Such a difference between the positive and negative tails have also been observed in the case of stocks in the NYSE [14]. To further verify that the tails are indeed consistent with a power-law form, we perform an alternative measurement of α using the Hill estimator [43, 44]. We arrange the returns in decreasing order such that $r_1 > \cdots > r_n$ and obtain the Hill estimator (based on the largest $k + 1$ values) as $H_{k,n} = \frac{1}{k}\sum_{i=1}^{k} \log \frac{r_i}{r_{k+1}}$, for $k = 1, \cdots, n-1$. The estimator $H_{k,n} \to \alpha^{-1}$ when $k \to \infty$ and $k/n \to 0$. For our data, this procedure gives $\alpha = 2.86$ and 2.56 for the positive and the negative tail, respectively (when $k = 20\,000$), which are consistent with (5.5).

Next, we extend this analysis for longer time-scales, to observe how the nature of the distribution changes with increasing Δt. As has been previously reported for US markets, the distribution is found to decay faster as Δt becomes large. However, up to $\Delta t = 1$ day, that is the daily closing returns, the distribution clearly shows a power-law tail (Figure 5.3a). The deviation is because of the decreasing size of the dataset with increase in Δt. Note that, while for $\Delta t < 1$ day we have used the high-frequency data, for $\Delta t = 1$ day we have considered the longer dataset of closing price returns for all stocks traded in NSE between November 1994 to May 2006. In Figure 5.3b we have also shown the distributions of the power-law exponents for the individual stocks, for time-scales varying between $10\,\text{min} \le \Delta t \le 60\,\text{min}$. We observe that the bulk of the exponents fall between 2 and 4, consistent with the results from the merged datasets.

To compare the distribution of returns in this emerging market with that observed in mature markets, we have considered the daily return data for the 500 stocks from NYSE listed in S & P 500 over the same period. As seen in Figure 5.4, the distributions for NSE and NYSE are almost identical, implying that the price fluctuation distribution of emerging markets cannot be distinguished from that of developed markets, contrary to what has been claimed recently [35].

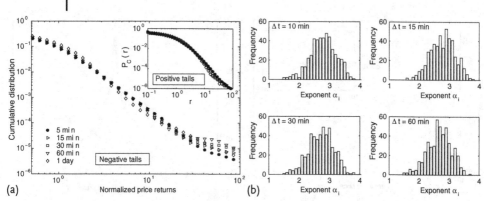

(a)

(b)

Figure 5.3 (a) Cumulative distribution of the negative and (inset) positive tails of the normalized returns distribution for different time-scales ($\Delta t \leq 1$ day). (b) Histograms of the power-law exponents for each of the 489 stocks, obtained by regression fit on the positive tail of cumulative return distributions, for different time-scales (10 min $\leq \Delta t \leq$ 60 min).

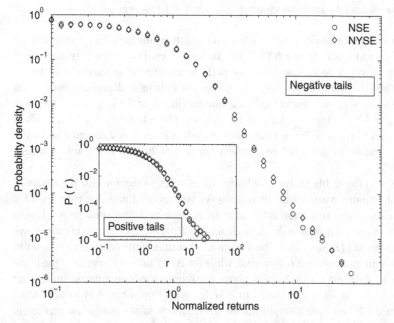

Figure 5.4 Comparison of the negative and (inset) positive tails of the normalized daily returns distribution for all stocks traded at NSE (○) and 500 stocks traded at NYSE (◇) during the period November 1994 to May 2006.

We now turn to the second question, and check whether it is possible to see any discernible change in the price fluctuation distribution as the stock market evolved over time. For this we focus on the daily return distribution for *all* stocks that were traded during the entire period of existence of the NSE. This period is divided into four intervals (a) 1994–1996, (b) 1997–1999, (c) 2000–2002, and (d) 2003–2006 [36],

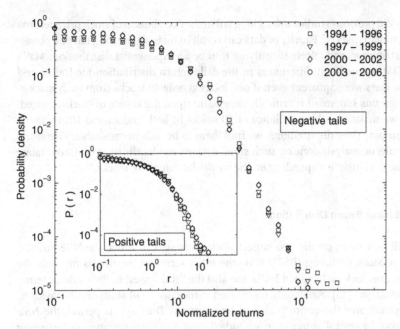

Figure 5.5 The negative and (inset) positive tails of the normalized daily returns distribution for all NSE stocks traded during the periods 1994–1996 (□), 1997–1999 (∇), 2000–2002 (◇) and 2003–2005 (○).

each corresponding to an increase in the number of transactions by an order of magnitude. Figure 5.5 shows that the return distribution at all four periods are similar, the negative tail even more so than the positive one. While the numerical value of the tail exponent may appear to have changed somewhat over the period that the NSE has operated, the power-law nature of the tail is apparent at even the earliest period of its existence. We therefore conclude that the convergence of the return distribution to a power-law functional form is extremely rapid, indicating that a market is effectively always at the nonequilibrium steady state characterized by the inverse cubic law.

We have also verified that stocks in the Bombay Stock Exchange (BSE), the second largest in India after NSE, follow a similar distribution [41]. Moreover, the return distribution of several Indian market indices (e.g., the NSE Nifty) also exhibit power-law decay, with exponents very close to 3 [42]. As the index is a composite of several stocks, this behavior can be understood as a consequence of the power-law decay for the tails of individual stock price returns, provided the movement of these stocks are correlated [14, 41]. Even though the Indian market microstructure has been refined and modernized significantly in the period under study as a result of the reforms and initiatives taken by the government, the nature of the return distribution has remained invariant, indicating that the nature of price fluctuations in financial markets is most probably independent of the level of economic development.

Why did previous studies miss a long tail in the distribution of stock price returns in the Indian market? Paucity of data can result in missing the long tail of a power-law distribution and falsely identifying it to be an exponential distribution. Matia *et al.* [35] claimed that differences in the daily return distribution for Indian and US ma rkets were apparent even if one looks at only 49 stocks from each market. However, this statement is critically dependent upon the choice of stocks. Indeed, when we make an arbitrary choice of 50 stocks in both Indian and US markets, and compare their distributions, we find them to be indistinguishable. Therefore, the results of analysis done on such small datasets can hardly be considered stable, with the conclusions depending on the particular sample of stocks.

5.1.3
Market Index Return Distribution

We will now focus on the two largest stock exchanges in India, the NSE and the Bombay Stock Exchange (BSE). NSE, the more recent of the two, is not only the most active stock exchange in India, but also the third largest in the world in terms of transactions [45]. As already mentioned earlier, we shall study the behavior of this market over the entire period of its existence. During this period, the NSE has grown by several orders of magnitude (Figure 5.6) demonstrating its emerging character. In contrast, BSE is the oldest stock exchange in Asia, and was the largest in India until the creation of NSE. However, over the past decade its share of the Indian financial market has fallen significantly. Therefore, we contrast two markets which have evolved very differently in the period under study.

In this subsection, we show that the Indian financial market, one of the largest emerging markets in the world, has index fluctuations similar to that seen for developed markets. Further, we find that the nature of the distribution is invariant with respect to different market indices, as well as the time-scale of observation. Taken together with the results of the previous subsection on the distribution of individual stock price returns in Indian markets, this strongly argues in favor of the universality of the nature of fluctuation distribution, regardless of the stage of development of the market or the economy underlying it.

The primary dataset that is used here is that of the Nifty index of the NSE which, along with the Sensex of BSE, is one of the primary indicators of the Indian market. It is composed of the top 50 highly liquid stocks which make up more than half of the market capitalization in India. We have used (i) high frequency data from January 2003 to March 2004, where the market index is recorded every time a trade takes place for an index component. The total number of records in this database is about 6.8×10^7. We have also looked at data over much longer periods by considering daily closing values of (ii) the Nifty index for the 16-year period July 1990 to May 2006, and (iii) the Sensex index of BSE for the 15-year period January 1991 to May 2006. In addition, we have also looked at the BSE 500 index for the much shorter period February 1999 to May 2006. Sensex consists of the 30 largest and most actively traded stocks, representative of various sectors of BSE, while the BSE 500 is calculated using 500 stocks representing all 20 major sectors of the economy.

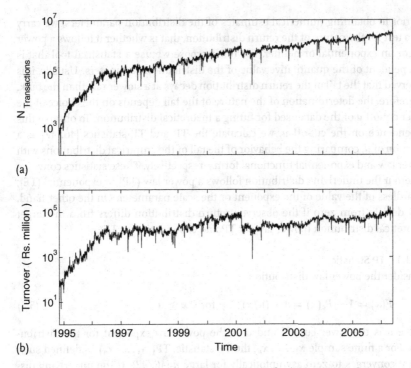

Figure 5.6 Time-evolution of the National Stock Exchange of India from 1994–2006 in terms of (a) the total number of trades and (b) the total turnover (i.e., traded value).

We first analyze the high-frequency data for the NSE Nifty index, which we sampled at 1-min intervals to generate the time-series $I(t)$. From $I(t)$ we compute the logarithmic return $R_{\Delta t}(t)$, defined in (5.1). These return distributions calculated using different time intervals may have varying width, owing to differences in their volatility, defined as $\sigma_{\Delta t}^2 \equiv \langle R^2 \rangle - \langle R \rangle^2$, where $\langle \ldots \rangle$ denotes the time average over the given time period. Hence, to be able to compare the distributions, we need to normalize the returns $R(t)$ by dividing them by the volatility $\sigma_{\Delta t}$. However, this leads to systematic underestimation of the tail of the normalized return distribution. This is because, even when a single return $R(t)$ is very large, the scaled return is bounded by $\sqrt{N}$, as the same large return also contributes to the variance $\sigma_{\Delta t}$. To avoid this, we remove the contribution of $R(t)$ itself from the volatility, and the new rescaled volatility is defined as

$$\sigma_{\Delta t}(t) = \sqrt{\frac{1}{N-1} \sum_{t' \neq t} \{R(t', \Delta t)\}^2 - \langle R(t', \Delta t) \rangle^2} \,, \tag{5.6}$$

as described in [4]. The resulting *normalized* return is given by,

$$r(t, \Delta t) \equiv \frac{R - \langle R \rangle}{\sigma_{\Delta t}(t)} \,. \tag{5.7}$$

Prior to obtaining numerical estimates of the distribution parameters, we carry out a test for the nature of the return distribution, that is whether it follows a power law or an exponential or neither. For this purpose we use a statistical tool that is independent of the quantitative value of the distribution parameters. Usually, it is observed that the tail of the return distribution decays at a slower rate than the bulk. Therefore, the determination of the nature of the tail depends on the choice of the lower cut-off u of the data used for fitting a theoretical distribution. To observe this dependence on the cut-off u, we calculate the TP and TE statistics [46, 47] as a function of u, comparing the behavior of the tail of the empirical distributions with power law and exponential functional forms, respectively. These statistics converge to zero if the underlying distribution follows a power law (TP) or exponential (TE), regardless of the value of the exponent or the scale parameter. On the other hand, they deviate from zero if the observed return distribution differs from the target theoretical distribution (power law for TP and exponential for TE).

5.1.3.1 TP Statistic

Consider the power-law distribution

$$F(x) = 1 - P_c(x) = 1 - (u/x)^\alpha, \quad \text{for} \quad x \geq u, \tag{5.8}$$

where u is the lower cut-off, and α is the power-law exponent for the distribution. For a finite sample $x_1, \ldots, x_n$, the TP statistic, $TP(x_1, \ldots, x_n)$, is defined such that it converges to zero asymptotically for large n [46, 47]. If the underlying distribution for a sample differs from the power-law form given in (5.8), TP is seen to deviate from zero. This statistic is based on the first two normalized statistical log-moments of the power-law distribution

$$E_1 = E\left[\log \frac{X}{u}\right] = \int_u^\infty \log \frac{x}{u} dF(x) = \frac{1}{\alpha}, \tag{5.9}$$

and

$$E_2 = E\left[\log^2 \frac{X}{u}\right] = \int_u^\infty \log^2 \frac{x}{u} dF(x) = \frac{2}{\alpha^2}, \tag{5.10}$$

where, $E[z]$ represents the mathematical expectation of z. The TP statistic is then defined as

$$TP = \left[\frac{1}{n}\sum_{k=1}^n \log \frac{x_k}{u}\right]^2 - \frac{1}{2n}\sum_{k=1}^n \log^2 \frac{x_k}{u}, \tag{5.11}$$

which tends to zero as $n \to \infty$. The estimation of the standard deviation for the TP statistic is provided by the standard deviation of the sum

$$\left(\frac{E_2}{2} - 2E_1^2\right) + \frac{1}{n}\sum_{k=1}^n \left[2E_1 \log \frac{x_k}{u} - \frac{1}{2}\log^2 \frac{x_k}{u}\right]. \tag{5.12}$$

5.1.3.2 TE Statistic

Consider the exponential distribution

$$F(x) = 1 - P_c(x) = 1 - \exp(-(x-u)/d), \quad \text{for} \quad x \geq u, \tag{5.13}$$

where u is the lower cut-off, and $d(> 0)$ is the scale parameter of the distribution. For a finite sample $x_1, \ldots, x_n$, the TE statistic, $TE(x_1, \ldots, x_n)$, is defined such that it converges to zero asymptotically for large n [46]. If the underlying distribution for a sample differs from the exponential form given in (5.13), TE is seen to deviate from zero. This statistic is based on the first two normalized statistical (shifted) log-moments of the exponential distribution

$$E_1 = E\left[\log\left(\frac{X}{u} - 1\right)\right] = \int_u^\infty \log\left(\frac{x}{u} - 1\right) dF(x)$$

$$= \log\frac{d}{u} - \gamma, \tag{5.14}$$

where $\gamma = 0.577215$ is the Euler constant, and

$$E_2 = E\left[\log^2\left(\frac{X}{u} - 1\right)\right] = \int_u^\infty \log^2\left(\frac{x}{u} - 1\right) dF(x)$$

$$= \left(\log\frac{d}{u} - \gamma\right)^2 + \frac{\pi^2}{6}. \tag{5.15}$$

As before, $E[\ldots]$ denotes the mathematical expectation. The TE statistic is then defined as

$$TE = \frac{1}{n}\sum_{k=1}^n \log^2\left(\frac{x_k}{u} - 1\right) - \left[\frac{1}{n}\sum_{k=1}^n \log\left(\frac{x_k}{u} - 1\right)\right]^2 - \frac{\pi^2}{6}, \tag{5.16}$$

which tends to zero as $n \to \infty$. The estimation of the standard deviation for the TE statistic is provided by the standard deviation of the sum

$$\frac{1}{n}\sum_{k=1}^n \left[\log\left(\frac{x_k}{u} - 1\right) - E_1\right]^2. \tag{5.17}$$

Figure 5.7 shows visually the deviation of the empirical data from the power law and exponential distributions. The TP and the TE statistics are plotted as functions of the lower cut-off u for 1-min returns of the NSE Nifty index. The TP statistic shows a large deviation till $u \leq 1$, after which it converges to zero indicating power-law behavior for large u. Correspondingly, the TE statistic excludes an exponential model for $u \geq 1$ as well as for very low values of u, although over the intermediate range $2 \times 10^{-1} < u < 6 \times 10^{-1}$ an exponential approximation may be possible.

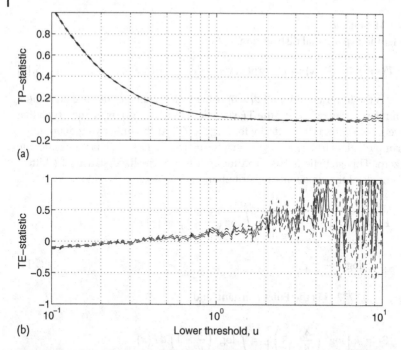

Figure 5.7 (a) TP statistic and (b) TE statistic as a function of the lower cut-off u for positive returns of the NSE with time interval $\Delta t = 1$ min. The broken lines indicate plus or minus one standard deviation of the statistics.

Figure 5.8 shows the cumulative distribution of the normalized returns for $\Delta t = 1$ min. For both positive and negative tails, there is an asymptotic power-law behavior. The power-law regression fit for the region $r \geq 2$ give exponents for the positive and the negative tails estimated as

$$\alpha = \begin{cases} 2.98 \pm 0.09 & \text{(positive tail)} \\ 3.37 \pm 0.10 & \text{(negative tail)} \end{cases}. \tag{5.18}$$

Note that, to avoid artifacts due to the data measurement process in the calculation of return distribution for $\Delta t < 1$ day, we have removed the returns corresponding to overnight changes in the index value.

We also perform an alternative estimation of the tail index of the above distribution by using the Hill estimator [43], which is the maximum likelihood estimator of α. For finite samples, however, the expected value of the Hill estimator is biased and depends crucially on the choice of the number of order statistics used for calculation. We have used the bootstrap procedure [48] to reduce this bias and to choose the optimal number of order statistics for calculating the Hill estimator, described in detail below. We found $\alpha \simeq 3.22$ and 3.47 for the positive and the negative tails, respectively.

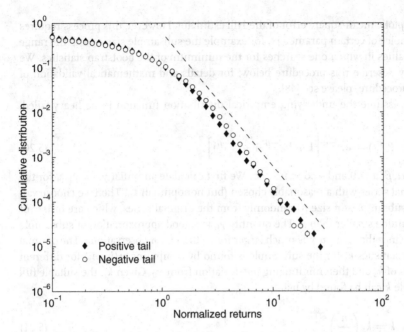

Figure 5.8 The cumulative distribution of the normalized 1-min return for the NSE Nifty index. The broken line indicates a power law with exponent $\alpha = 3$.

5.1.3.3 Hill Estimation of Tail Exponent

The Hill estimator gives a consistent estimate of the tail exponent α from random samples of a distribution with an asymptotic power-law form. First, we arrange the returns in decreasing order such that $r_1 > \cdots > r_n$. Then the Hill estimator (based on the largest $k + 1$ values) is given as

$$\gamma_{k,n} = \frac{1}{k} \sum_{i=1}^{k} \log \frac{r_i}{r_{k+1}}, \qquad (5.19)$$

for $k = 1, \cdots, n - 1$. The estimator $\gamma_{k,n} \to \alpha^{-1}$ when $k \to \infty$ and $k/n \to 0$. However, for a finite time-series, the expectation value of the Hill estimator is biased, that is it will consistently over or underestimate α. Further, γ depends critically on our choice of k, the order statistics used to compute the Hill estimator.

If the form of the distribution function from which the random sample is chosen is known, then the bias and the stochastic error variance of the Hill estimator can be calculated. From this, the optimum k value can be obtained such that the asymptotic mean-square error of the Hill estimator is minimized. Increasing k reduces the variance because more data are used, but increases the bias because the power law is assumed to hold only in the extreme tail. Unfortunately, the distribution for the empirical data is not known and hence this procedure has to be replaced by an asymptotically equivalent data driven process.

One such method is the subsample bootstrap method. This method can be used to estimate an optimal number for the order statistics ($\bar{k}$) that will reduce the asymptotic mean-square error of the Hill estimator. However, this process requires the choice of certain parameters, for example the subsample size n_s and the range of k values in which one searches for the minimum of the bootstrap statistics. We briefly describe this procedure below; for details and mathematical validation of this procedure, please see [48].

We assume the underlying empirical distribution function to be heavy-tailed, viz.,

$$P_c(x) = ax^{-\alpha} \left[1 + bx^{-\beta} + o(x^{-\beta}) \right],\qquad(5.20)$$

with $\alpha, \beta, a > 0$ and $-\infty < b < \infty$. We first calculate an initial $\gamma_0 = \gamma_{k_0, n}$ for the original series with a reasonably chosen (but nonoptimal) k_0. Then we choose various subsamples of size n_s randomly from the original series, which are orders of magnitude smaller than n. The quantity γ_0 is a good approximation of subsample α^{-1}, since the error in γ is much larger for n_s than for n observations. The optimal order statistics $\bar{k}_s$ for the subsample is found by computing $\gamma(k_s, n_s)$ for different values of k_s and then minimizing the deviation from γ_0. Given $\bar{k}_s$, the suitable full sample $\bar{k}$ can be found by using

$$\bar{k} = \bar{k}_s \left(\frac{n}{n_s} \right)^{\frac{2\beta}{2\beta + \alpha}} .\qquad(5.21)$$

Here the initial estimate of α is taken to be $1/\gamma_0$. Further, we have considered $\beta = \alpha$, as done by Hall [49], although the results are not very sensitive to the choice of β. Once $\bar{k}$ is calculated, the final estimate of the tail index is given by $\alpha = 1/\gamma_{\bar{k}, n}$.

For calculating the initial γ_0 we have chosen k_0 to be 0.5% of the sample size n. One thousand subsamples, each of size $n_s = n/40$, are randomly picked from the full dataset. To obtain optimal k_s, we confine ourselves to 4% of the subsample size n_s. To find the stochastic error in our estimation of α, we have computed the 95% confidence interval as given by $\pm 1.96[1/(\alpha^2 m)]^{1/2}$. Although a Jackknife algorithm can also be used to calculate this error bound, the results obtained using this method will be close to that obtained using the bootstrap method over many realizations [48], as we have done here.

5.1.3.4 Temporal Variations in the Return Distribution

To investigate the effect of *intra-day* variations in market activity, we analyze the 1-min return time-series by dividing it into two parts, one corresponding to returns generated in the opening and the closing hours of the market, and the other corresponding to the intermediate time period. In general, it is known that the average intra-day volatility of stock returns follows a U-shaped pattern [50, 51] and one can expect this to be reflected in the nature of the fluctuation distribution for the opening and closing periods, as opposed to the intervening period. We indeed find the index fluctuations for these two datasets to be different (Figure 5.9). In particular, the cumulative distribution tail for the opening and closing hour returns show a power-law scaling with exponent close to 3, whereas for the intermediate period we

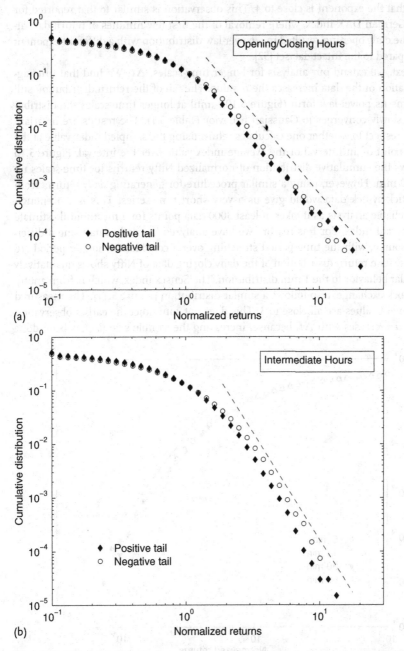

Figure 5.9 Intra-day variation in the cumulative distribution of the normalized 1-min return for the NSE Nifty index: return distribution during (a) the opening and closing hours (the broken line indicates a power law with exponent $\alpha = 3$) and (b) the intermediate time period (the broken line indicates a power law with exponent $\alpha = 4$).

see that the exponent is close to 4. This observation is similar to that reported for the German DAX index, where removal of the first few minutes of return data after the daily opening resulted in a power-law distribution with a different exponent compared to the intact dataset [52].

Next, we extend our analysis for longer time-scales, Δt. We find that time aggregation of the data increases the α value. The tail of the return distribution still retains its power-law form (Figure 5.10), until at longer time-scales the distribution slowly converges to Gaussian behavior (Table 5.1). The results are invariant with respect to whether one calculates return using the sampled index value at the end point of an interval or the average index value over the interval. Figure 5.10 shows the cumulative distribution of normalized Nifty returns for time-scales up to 60 min. However, using a similar procedure for generating *daily* returns from the tick-by-tick data would give us a very short time-series. This is not enough for reliable analysis as it takes at least 3000 data points for a meaningful estimate of the tail index. For this reason, we have analyzed the daily data using a different source, with the time period stretching over a considerably longer period (16 years). The return distribution of the daily closing data of Nifty shows qualitatively similar behavior to the 1 min distribution. The Sensex index, which is from another stock exchange, also follows a similar distribution (Figure 5.11). The measured exponent values are all close to 3. This does not contradict the earlier observation that α increases with Δt, because, increasing the sample size (as has been done

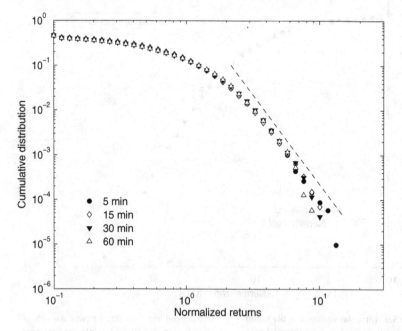

Figure 5.10 The negative tail of the cumulative distribution of the NSE Nifty index returns for different time intervals Δt up to 60 min. The broken line indicates a power law with exponent $\alpha = 4$.

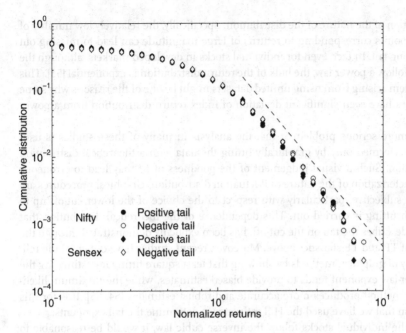

Figure 5.11 The cumulative distribution of the normalized 1-day return for the NSE Nifty and BSE Sensex index. The broken line indicates a power law with exponent $\alpha = 3$.

Table 5.1 Comparison of the power-law exponent α of the cumulative distribution function for various index returns. Power-law regression fits are done in the region $r \geq 2$. The Hill estimator is calculated using the bootstrap algorithm.

Index	Δt	Power-law fit		Hill estimator	
		Positive	Negative	Positive	Negative
Nifty (2003–2004)	1 min	2.98 ± 0.09	3.37 ± 0.10	3.22 ± 0.03	3.47 ± 0.03
	5 min	4.42 ± 0.37	3.44 ± 0.21	4.51 ± 0.03	4.84 ± 0.03
	15 min	5.58 ± 0.88	3.96 ± 0.27	6.25 ± 0.03	4.13 ± 0.04
	30 min	5.13 ± 0.41	3.92 ± 0.45	5.65 ± 0.03	4.30 ± 0.03
	60 min	5.99 ± 1.52	4.42 ± 0.65	7.85 ± 0.03	5.11 ± 0.04
Nifty (1990–2006)	1 day	3.10 ± 0.34	3.18 ± 0.28	3.33 ± 0.14	3.37 ± 0.14
Sensex (1991–2006)	1 day	3.33 ± 0.77	3.45 ± 0.25	2.93 ± 0.15	3.84 ± 0.12

for $\Delta t = 1$ day) improves the estimation of α. This underlines the invariance of the nature of market fluctuations with respect to time aggregation, interval used and different exchanges.

The much shorter dataset of the BSE 500 daily returns shows a significant departure from power-law behavior, essentially following an exponential distribution. This is not surprising, as looking at data over shorter periods can result in misidentification of the nature of the distribution. Specifically, the relatively low number of data points corresponding to returns of large magnitude can lead to missing out the long tail. In fact, even for individual stocks in developed markets, although the tails follow a power law, the bulk of the return distribution is exponential [53]. This problem arising from using limited datasets might be one of the reasons why some studies have seen significant deviation of index return distribution from a power law.

A more serious problem is that the analysis in many of these studies is usually performed only by graphically fitting the data with a theoretical distribution function. Such a visual judgement of the goodness of fit may lead to erroneous characterization of the nature of fluctuation distribution. Graphical procedures are often subjective, particularly with respect to the choice of the lower cut-off up to which fitting is carried out. This dependence of the theoretical distribution that best describes the tail on the cut-off, has been explicitly demonstrated through the use of TP and TE statistics above. Moreover, recent studies have criticized the reliability of graphical methods by showing that least square fitting for estimating the power-law exponent tends to provide biased estimates, while the maximum likelihood method produces more accurate and robust estimates [54, 55]. It is for this reason that we have used the Hill estimator to determine the tail exponents.

If the individual stocks follow the inverse cubic law, it would be reasonable to suppose that the index, which is a weighted average of several stocks, will also behave similarly, provided the different stocks move in a correlated fashion [16]. As the price movements of stocks in an emerging market are even more correlated than in developed markets [56], it is expected that the returns for stock prices and the index should follow the same distribution. Therefore, the demonstration of the inverse cubic law for the index fluctuations in the Indian market is consistent with our preceding demonstration [36] showing that the individual stock prices in this market follow the same behavior.

On the whole, our study points out the remarkable robustness of the nature of the fluctuation distribution for markets. While, in the period under study, the NSE had begun operation and rapidly increased in terms of activity, the BSE had existed for a long time prior to this period and showed a significant decrease in market share. However, both showed very similar fluctuation behavior. This indicates that, at least in the Indian context, the distribution of returns is invariant with respect to markets. The fact that the distribution is quantitatively the same as in developed markets, implies that it is also probably independent of the state of the economy. In addition, our observation that the intra-day return distribution of Indian market index shows properties similar to that reported for developed markets, suggests that even at this level of detail the fluctuation behavior of the two kinds of markets are rather similar. Therefore, our results indicate that although markets may differ from each other in terms of (i) the details of their components, (ii) the nature of interactions and (iii) their susceptibility to news from outside the market, there may be universal mechanisms responsible for generating market fluctuations as indicated by the observation of invariant properties. The rigorous demonstration of such a universal law for market behavior is significant for the physics of strongly interacting complex systems, as it suggests the existence of robust features that are independent of individual details of different systems.

5.2
Distribution of Trading Volume and Number of Trades

Apart from stock price and index fluctuations, one can also measure market activity by looking at the trading volume (the number of shares traded), $V(t)$, and the number of trades, $N(t)$. To obtain the corresponding distributions, we normalize these variables by subtracting the mean and dividing by their standard deviation, such that, $v = \frac{V - \langle V \rangle}{\sqrt{\langle V^2 \rangle - \langle V \rangle^2}}$ and $n = \frac{N - \langle N \rangle}{\sqrt{\langle N^2 \rangle - \langle N \rangle^2}}$. Figure 5.12 shows the distribution of these two quantities for several stocks, based on daily data for the Bombay Stock Exchange (BSE). As is evident, the distribution is very similar for the different stocks, and the nature of the decay is significantly different from a power law. To better characterize the distribution, we have also looked at the intra-day distributions for volume and number of trades, based on high-frequency data from the NSE. Figure 5.13 shows the distributions of the two quantities for trading conducted on a particular stock in 5 minute intervals. Analysis of data for other stocks show qualitatively similar results. As is clear, both of these distributions are nonmonotonic, and are suggestive of a log-normal form. The fact that these distributions are very similar to each other is not surprising in view of the almost linear relationship between the two (Figure 5.13c). This supports previous observation in major US stock markets that statistical properties of the number of shares traded and the number of trades in a given time interval are closely related [58].

For US markets, power-law tails have been reported for the distribution of both the number of trades [57] and the volume [58]. It has also been claimed that these features are observed on the Paris Bourse, and therefore, these features are as universal as the inverse cubic law for price returns distribution [59, 60]. However, anal-

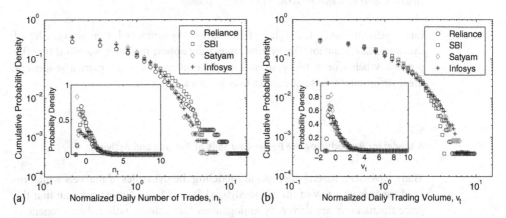

(a) Normalized Daily Number of Trades, n_t (b) Normailzed Daily Trading Volume, v_t

Figure 5.12 Cumulative distribution of the number of trades (top (a)) and the volume of shares traded (top (b)) in a day for four stocks at BSE between July 12, 1995 and January 31, 2006.

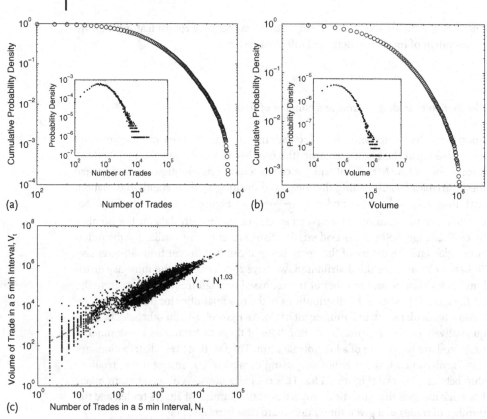

Figure 5.13 Cumulative distribution of the number of trades (a) and the volume of shares traded (b) for a particular stock (Reliance) in 5-minute intervals at NSE between January 1, 2003 to March 31, 2004. (c) shows an almost linear relation between the number of trades in a 5-minute interval and the corresponding trading volume. The broken line indicates the best fit on a doubly logarithmic scale.

ysis of other markets [15, 61] have failed to see any evidence for the universality of the power-law behavior. Our results appear to support the latter assertion that the power-law behavior in this case may not be universal, and the particular form of the distribution of these quantities may be market specific.

5.3
A Model for Reproducing the Power Law Tails of Returns and Activity

There have been several attempts at modeling the dynamics of markets which reproduce at least a few of the above stylized facts. Many of them assume that the price fluctuations are driven by endogenous interactions rather than exogenous factors such as, arrival of news affecting the market and variations in macroeconomic indicators. A widely used approach for such modeling is to consider the market movement to be governed by explicit interactions between agents who are buying and selling assets [18–20, 62, 63]. While this is appealing from the point of view of statistical physics, resembling as it does interactions between spins arranged over a specified network, it is possible that in the market the mediation between agents is done through means of a globally accessible signal, namely the asset price. This is analogous to a mean-field-like simplification of the agent-based model of the market, where each agent is taking decisions based on a common indicator variable. Here, we present a model of market dynamics where the agents do not interact directly with each other, but respond to a global variable defined as price. The price in turn is determined by the relative demand and supply of the underlying asset that is being traded, which is governed by the aggregate behavior of the agents (each of whom can buy, sell or hold at any given time). In the model described here, the trading occurs in a two-step process, with each agent first deciding whether to trade or not at that given instant based on the deviation of the current price from an agent's notion of the "true" price (given by a long-time moving average). This is followed by the agents who have decided to trade choosing to either buy or sell based on the prevalent demand-supply ratio measured by the logarithmic return.

A simplified view of a financial market is that it consists of a large number of agents (say, N) trading in a single asset. During each instant the market is open, a trader may decide to either buy, sell or hold (i.e., remain inactive) based on its information about the market. Thus, considering the time to evolve in discrete units, we can represent the state of each trader i by the variable $S_i(t)$ ($i = 1, \ldots, N$) at a given time instant t. It can take values +1, −1 or 0 depending on whether the agent buys or sells *a unit quantity* of asset or decides not to trade at time t, respectively. We assume that the evolution of price in a free market is governed only by the relative supply and demand for the asset. Thus, the price of the asset at any time t, $P(t)$, will rise if the number of agents wishing to buy it (i.e., the demand) exceeds the number wishing to sell it (i.e., supply). Conversely, it will fall when supply outstrips demand. Therefore, the relation between prices at two successive time instants can be expressed as

$$P_{t+1} = \frac{1 + M_t}{1 - M_t} P_t , \tag{5.22}$$

where $M_t = \sum_i S_i(t)/N$ is the net demand for the asset, as the state of agents who do not trade is represented by 0 and do not contribute to the sum. This functional form of the time-dependence of price has the following desirable feature: when everyone wants to sell the asset ($M_t = -1$), its price goes to zero, whereas if everyone wants to buy it ($M_t = 1$), the price diverges. When the demand equals supply, the price remains unchanged from its preceding value, indicating an equilibrium situation. The multiplicative form of the function not only ensures that price can never be negative, but also captures the empirical feature of the magnitude of stock price fluctuations in actual markets being proportional to the price. Note that, if the ratio of demand to supply is an uncorrelated stochastic process, price will follow a geometric random walk, as originally suggested by Bachelier [6]. The exact form of the price function (5.22) does not critically affect our results, as we shall discuss later.

Having determined how the price of the asset is determined based on the activity of traders, we now look at how individual agents make their decisions to buy, sell or hold. As mentioned earlier, we do not assume direct interactions between agents, nor do we consider information external to the market to be affecting agent behavior. Thus, the only factor governing the decisions made by the agents at a given time is the asset price (the current value as well as its history up to that time). First, we consider the condition that prompts an agent to trade at a particular time (i.e., $S_i = \pm 1$), rather than hold ($S_i = 0$). The fundamental assumption that we shall use here is that this decision is based on the deviation of the current price at which the asset is being traded from an individual agents notion of the "true" value of the asset. Observation of order book dynamics in markets has shown that the life-time of a limit order is longer, the farther it is from the current bid-ask [64]. In analogy to this we can say that the probability of an agent to trade at a particular price will decrease with the distance of that price from the "true" value of the asset. This notion of the "true" asset price is itself based on information about the price history (as the agents do not have access to any external knowledge related to the value of an asset) and thus can vary with time. The simplest proxy for estimating the "true" value is a long-time moving average of the price time-series, $\langle P_t \rangle_\tau$, with the averaging window size, τ, being a parameter of the model. Our use of the moving average is supported by previous studies that have found the long-time moving average of prices to define an effective potential that is seen to be the determining factor for empirical market dynamics [65].

In light of the above discussion, a simple formulation for the probability of an agent i to trade at time t is

$$P(S_i(t) \neq 0) = \exp\left(-\mu \left|\log \frac{P_t}{\langle P_t \rangle_\tau}\right|\right), \tag{5.23}$$

where μ is a parameter that controls the sensitivity of the agent to the magnitude (i.e., absolute value) of the deviation from the "true" value. This deviation is expressed in terms of a ratio so that, there is no dependence on the scale of measurement. For the limiting case of $\mu = 0$, we get a binary state model, where each agent trades at every instant.

Once an agent decides to trade based on the above dynamics, it has to choose between buying and selling a unit quantity of the asset. We assume that this process is fully dictated by the principle of supply and demand, with agents selling (buying) if there is an excess of demand (supply) resulting in an increase (decrease) of the price in the previous instant. Using the logarithmic return as the measure for price movement, we can use the following simple form for calculating the probability that an agent will sell at a given time t

$$P(S_i(t) = -1) = \frac{1}{1 + \exp\left(-\beta \log \frac{P_t}{P_{t-1}}\right)}. \tag{5.24}$$

The form of the probability function is adopted from that of the Fermi function used in statistical physics, for example for describing the transition probability of spin states in a system at thermal equilibrium. The parameter β, corresponding to "inverse temperature" in the context of Fermi function, is a measure of how strongly the information about price variation influences the decision of a trader. It controls the slope of the function at the transition region where it increases from 0 to 1, with the transition getting sharper as β increases. In the limit $\beta \rightarrow \infty$, the function is step-like, such that every trading agent sells (buys) if the price has risen (fallen) in the previous instant. In the other limiting case of $\beta = 0$, the trader buys or sells with equal probability, indicating an insensitivity to the current trend in price movement. The results of the model are robust with respect to variation in β and we shall consider below only the limiting case of $\beta = 0$.

We now report the results of numerical simulations of the model discussed above, reproducing the different stylized facts mentioned earlier. For all runs, the price is assumed to be 1 at the initial time ($t = 0$). The state of every agent is updated at a single time-step or iteration. To obtain the "true" value of the asset at $t = 0$, the simulation is initially run for τ iterations during which the averaging window corresponds to the entire price history. At the end of this step, the actual simulation is started, with the averaging being done over a moving window of fixed size τ.

The variation of the asset price as a result of the model dynamics is shown in Figure 5.14a, which looks qualitatively similar to price (or index) time-series for real markets. The moving average of the price, that is considered to be the notional "true" price for agents in the model, is seen to track a smoothed pattern of price variations, coarse-grained at the time-scale of the averaging window, τ. The price fluctuations, as measured by the normalized logarithmic returns (Figure 5.14b), show large deviations that are significantly greater than that expected from a Gaussian distribution.

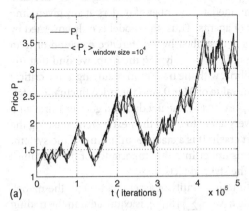
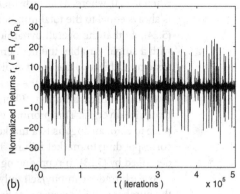

Figure 5.14 (a) Price time-series, along with the moving average of price calculated over a window of size τ, and (b) the corresponding logarithmic returns, normalized by subtracting the mean and dividing by the standard devi- ation, for a system of $N = 20\,000$ agents. The model is simulated with parameter values $\mu = 100$, $\beta = 0$ and averaging window size, $\tau = 10^4$ iterations.

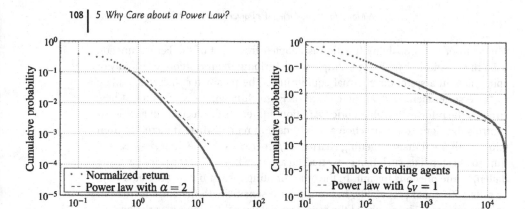

Figure 5.15 Cumulative distributions of (a) normalized returns and (b) trading volume measured in terms of the number of traders at a given time, for a system of $N = 20\,000$ agents. The model is simulat-ed for $T = 200\,000$ iterations, with parameter values $\mu = 100$, $\beta = 0$ and averaging window size, $\tau = 10^4$ iterations. Each distribution is obtained by averaging over 10 realizations.

We now examine the nature of the distribution of price fluctuations by focusing on the cumulative distribution of returns, that is $P(r_t > x)$, shown in Figure 5.15a. We observe that it follows a power law having exponent $\alpha \simeq 2$ over an intermediate range with an exponential cut-off. The quantitative value of the exponent is seen to be unchanged over a large range of variation in the parameter μ and does not depend at all on β. At lower values of μ (viz., $\mu < 10$) the return distribution becomes exponential.

The dynamics leading to an agent choosing whether to trade or not is the crucial component of the model that is necessary for generating the non-Gaussian fluctuation distribution. This can be explicitly shown by considering the special case when $\mu = 0$, where, as already mentioned, the number of traders at any given time is always equal to the total number of agents. Thus, the model is only governed by (5.24), so that the overall dynamics is described by a difference equation or map with a single variable, the net demand (M_t). Analyzing the map, we find that the system exhibits two classes of equilibria, with the transition occurring at the critical value of $\beta = 1$. For $\beta < 1$, the mean value of M is 0, and the price fluctuations follow a Gaussian distribution. When β exceeds 1, the net demand goes to 1 implying that price diverges. This prompts every agent to sell at the next instant, pushing the price to zero, analogous to a market crash. It is a stable equilibrium of the system, corresponding to market failure. This underlines the importance of the dynamics described by (5.23) in reproducing the stylized facts.

As each trader can buy/sell only a unit quantity of asset at a time in the model, the number of trading agents at time t, $V_t = \sum_i |S_i(t)|$, is equivalent to the trading volume at that instant. The cumulative distribution of this variable, shown in Figure 5.15b, has a power-law decay which is terminated by a exponential cut-off due to the finite number of agents in the system. The exponent of the power law, ζ_V, is

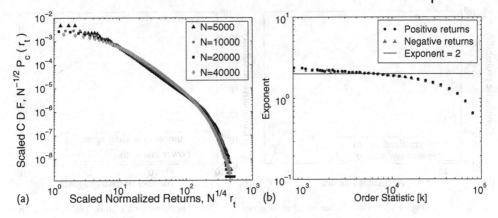

Figure 5.16 (a) Finite size scaling of the distribution of normalized returns for systems varying between $N = 5000$ and $40\,000$ agents, and (b) estimation of the corresponding power-law exponent by the Hill estimator method for a system of $N = 20\,000$ agents. The model is simulated with parameter values $\mu = 100$, $\beta = 0$ and averaging window size, $\tau = 10^4$ iterations.

close to 1, indicating a Zipf's law[3][1] distribution for the number of agents who are trading at a given instant. As in the case of the return distribution exponent, the quantitative value of the exponent ζ_V is seen to be unchanged over a large range of variation in the parameter μ. The power-law nature of this distribution is more robust, as at lower values of μ (viz., $\mu < 10$), when the return distribution shows exponential behavior, the volume distribution still exhibits a prominent power-law tail.

It is well known that for many systems, their finite size can affect the nature of distributions of the observed variables. In particular, we note that the two distributions considered above have exponential cut-offs that are indicative of the finite number N of agents in the system. In order to explore the role of system size in our results, we perform finite size scaling of the return distribution to verify the robustness of the power-law behavior. This is done by carrying out the simulation at different values of N and trying to see whether the resulting distributions collapse onto a single curve when they are scaled properly. Figure 5.16a shows that for systems between $N = 5000$ and $40\,000$ agents, the returns fall on the same curve, when the abscissa and ordinate are scaled by the system size, raised to an appropriate power. This implies that the power law is not an artifact of finite size systems, but should persist as larger and larger number of agents are considered.

To get a quantitatively accurate estimate of the return exponent, we use the method described by Hill [43]. This involves obtaining the Hill estimator, $\gamma_{k,n}$, from a set of finite samples of a time-series showing power-law distributed fluctuations, with n indicating the number of samples. Using the original time-series $\{x_i\}$, we create a new series by arranging the entries in decreasing order of their magnitude and labeling each entry with the order statistic k, such that the magnitude of x_k is

3) Zipf's law is a special case of the power law where the cumulative probability distribution function $P_c(x) \sim 1/x$.

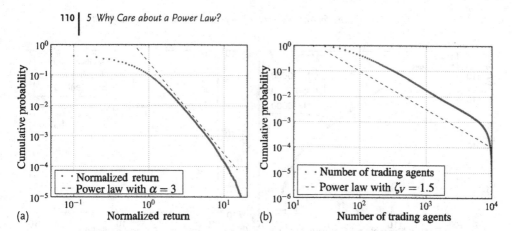

Figure 5.17 Distribution of (a) normalized returns and (b) number of trading agents, for a system of $N = 10\,000$ agents when the parameter μ is randomly selected for each agent from a uniform distribution between [10, 200]. The exponents for the power law seen in both curves agree with the corresponding values seen in actual markets. The model is simulated with parameter values $\beta = 0$ and averaging window size, $\tau = 10^4$ iterations.

larger than that of x_{k+1}. The Hill estimator is calculated as

$$\gamma_{k,n} = \frac{1}{k} \sum_i^k \log \frac{x_i}{x_{k+1}}, \tag{5.25}$$

where $k = 1, \ldots, n-1$. It approaches the inverse of the true value of the power-law exponent as $k \to \infty$ and $\frac{k}{n} \to 0$. Figure 5.16b shows the estimated value of the return distribution exponent, α (i.e., $\gamma_{k,n}^{-1}$) calculated for returns obtained for a system of size $N = 20\,000$ agents. This confirms our previous estimate of $\alpha \simeq 2$ based on least square fitting of the data.

5.3.1
Reproducing the Inverse Cubic Law

So far we have worked in the situation when all the parameter values are constant and uniform for all agents. However, in the real world, agents tend to differ from one another in terms of their response to the same market signal, for example in their decision to trade in a high risk situation. We capture this heterogeneity in agent behavior by using a random distribution of the parameter μ that controls the probability that an agent will trade when the price differs from the "true" value of the asset. A low value of the parameter represents an agent who is relatively indifferent to this deviation. On the other hand, an agent who is extremely sensitive to this difference and refuses to trade when the price goes outside a certain range around the "true" value, is a relatively conservative market player with a higher value of μ.

Figure 5.17 shows the distributions for the return and number of traders when μ for the agents is distributed uniformly over the interval [10, 200] (we have verified that small variations in the bounds of this interval do not change the results). While the power-law nature is similar to that for the constant parameter case seen earlier, we note that the exponent values are now different and quantitatively match those seen in real markets. In particular, the return distribution reproduces the inverse cubic law, that has been found to be almost universally valid. Surprisingly, we find that the same set of parameters which yield this return exponent, also result in a cumulative distribution for the trading volume (i.e., number of traders) with a power-law exponent $\zeta_V \simeq 1.5$ that is identical to that reported for several markets [58]. Thus, our model suggests that heterogeneity in agent behavior is the key factor behind the observed distributions. It predicts that when the behavior of market players become more homogeneous, as for example, during a market crash event, the return exponent will tend to decrease. Indeed, earlier work [66] has found that during crashes, the exponent for the power-law tail of the distribution of relative prices has a significantly different value from that seen at other times. From the results of our simulations, we predict that for real markets, the return distribution exponent α during a crash will be close to 2, the value obtained in our model when every agent behaves identically.

How sensitive are the above results to the specific forms of the dynamics we have used ? One can test the robustness of these results with respect to the way asset price is defined in the model. We have considered several variations of (5.22), including a quadratic function, viz.,

$$P_{t+1} = \left(\frac{1 + M_t}{1 - M_t} \right)^2 P_t , \qquad (5.26)$$

and find the resulting nature of the distributions and the volatility clustering property to be unchanged. The space of parameter values can also be explored for checking the general validity of the results. As already mentioned, the parameter β does not seem to affect the nature, or even the quantitative value of the exponents, of the distributions. The robustness of the results has also been verified with respect to the averaging window size, τ. We find the numerical values of the exponents to be unchanged over the range $\tau = 10^4 - 10^6$.

It may be pertinent here to discuss the relevance of our observation of an exponential return distribution in the model at lower values of the parameter μ. Although the inverse cubic law is seen to be valid for most markets, it turns out that there are a few cases, such as the Korean market index KOSPI, for which the return distribution is reported to have an exponential form [29]. We suggest that these deviations from the universal behavior can be due to the existence of a high proportion of traders in these markets who are relatively indifferent to large deviations of the price from its "true" value. In other words, the presence of a large number of risk takers in the market can cause the return distribution to have exponentially decaying tails. The fact that for the same set of parameter values, the cumulative distribution of number of traders still shows a power-law decay with exponent 1, prompts us to further predict that, despite deviating from the universal form of the return distribution, the trading volume distribution of these markets will follow a power-law form with ζ_V close to 1.

References

1 Newman, M.E.J. (2005) Power laws, Pareto distributions and Zipf's law. *Contemporary Physics*, **46**, 323–351.

2 Clauset, A., Shalizi, C.R., and Newman, M.E.J. (2009) *SIAM Review* **51**(4), 661–703.

3 Mantegna, R.N. and Stanley, H.E. (1999) *Introduction to Econophysics*, Cambridge University Press, Cambridge.

4 Bouchaud, J.P. and Potters, M. (2003) *Theory of Financial Risk and Derivative Pricing*, Second edition, Cambridge University Press, Cambridge.

5 Chatterjee, A. and Chakrabarti, B.K. (eds) (2006) *Econophysics of Stock and other Markets*, Springer, Milan.

6 Bachelier, L. (1900) *Annales Scientifiques de l'École Normale Supérieure Sér*, **3**(17), 21–86.

7 Osborne, M. (1964) The Random Character of Stock Market Prices, chapter *Brownian motion in the stock market*, The MIT Press Cambridge, MA, pp. 100–128.

8 Fama, E.F. (1965) *The Journal of Business*, **38**, 34.

9 Mandelbrot, B. (1963) *The Journal of Business*, **36**(4), 394–419.

10 Mantegna, R.N. and Stanley, H.E. (1995) *Nature*, **376**, 46–49.

11 Gopikrishnan, P., Meyer, M., Nunes Amaral, L.A., and Stanley, H.E. (1998) *Eur. Phys. J. B*, **3**, 139–140.

12 Jansen, D.W. and deVries, C.G. (1991) *The Review of Economics and Statistics*, **73**(1), 18–24.

13 Lux, T. (1996) *Applied Financial Economics*, **6**(6), 463–475.

14 Plerou, V., Gopikrishnan, P., Nunes Amaral, L. A., Meyer, M., and Stanley, H. E. (1999) *Phys. Rev. E*, **60**(6), 6519–6529.

15 Farmer, J.D. and Lillo, F. (2004) *Quantitative Finance*, **4**(1), C7–C11.

16 Gopikrishnan, P., Plerou, V., Nunes Amaral, L.A., Meyer, M., and Stanley, H.E. (1999) *Phys. Rev. E*, **60**(5), 5305–5316.

17 Oh, G., Um, C.-J., and Kim, S. (2006) *Physics*, 0601126.

18 Lux, T. and Marchesi, M. (1999) *Nature*, **397**, 498–500.

19 Bornholdt, S. (2001) *Int. J. Mod. Phys. C*, **12**(5), 667–674.

20 Iori, G. (2002) *Journal of Economic Behavior & Organization*, **49**(2), 269–285.

21 Chowdhury, D. and Stauffer, D. (2004) *Eur. Phys. J. B*, **8**(3), 477–482.

22 Sinha, S., Pan, R.K. (2006) How a hit is born: The emergence of popularity from the dynamics of collective choice, (eds A. Chatterjee, A. Chakraborti, B.K. Chakrabarti), *Econophysics and Sociophysics: Trends and Perspectives*, Wiley-VCH, Berlin, p. 417–448.

23 Mandelbrot, B.B. (1997) *Fractals and Scaling in Finance*, Springer, New York.

24 Bacry, E., Delour, J., and Muzy, J.F. (2001) *Phys. Rev. E*, **64**(2), 026103.

25 Bouchaud, J.P. (2005) *Chaos*, **15**, 026104.

26 Storer, R. and Gunner, S.M. (2002) *Int. J. Mod. Phys. C*, **13**(7), 893–897.

27 Wang, B. and Hui, P. (2001) *Eur. Phys. J. B*, **20**(4), 573–579.

28 Kaizoji, T. and Kaizoji, M. (2003) *Advances in Complex Systems*, **6**(3), 303–312.

29 Yang, J.-S., Chae, S., Jung, W.-S., and Moon, H.-T. (2006) *Physica A*, **363**(2), 377–382.

30 Miranda, L.C. and Riera, R. (2001) *Physica A*, **297**(3–4), 509–520.

31 Gleria, I., Matsushita, R., and Silva, S.D. (2002) *Economics Bulletin*, **7**(3), 1–12.

32 Coronel-Brizio, H.F. and Hernandez-Montoya, A.R. (2005) *Revista Mexicana de Fisica*, **51**(1), 27–31.

33 Rak, R., Drozdz, S., and Kwapien, J. (2007) *Physica A*, **374**, 315–324.

34 Jondeau, E. and Rockinger, M. (2003) *Journal of Empirical Finance*, **10**, 559–581.

35 Matia, K., Pal, M., Salunkay, H., and Stanley, H.E. (2004) *Europhys. Lett.*, **66**(6), 909–914.

36 Pan, R.K. and Sinha, S. (2007) *Europhys. Lett.*, **77**, 58004.

37 http://www.nseindia.com/.

38 *Annual Report and Statistics 2005*, World Federation of Exchanges, 2006, p. 77.

39 http://www.nseindia.com/.

40 http://finance.yahoo.com/.

41 Sinha, S., Pan, R.K. (2006) The Power (Law) of Indian Markets: Analysing NSE and BSE trading statistics, (eds A. Chatterjee, B.K. Chakrabarti), *Econophysics of Stock and Other Markets*, Springer, Milan, p. 24–34.

42 Pan, R.K., Sinha, S. (2008) Inverse-cubic law of index fluctuation distribution in Indian markets, *Physica A*, **387**, 2055–2065.

43 Hill, B.M. (1975) *The Annals of Statistics*, **3**(5), 1163–1174.

44 Drees, H., deHaan, L., and Resnick, S. (2000) *The Annals of Statistics*, **28**(1), 254–274.

45 Annual report and statistics 2005 Technical report World Fedration of Exchanges (2006).

46 Pisarenko, V. and Sornette, D. (2006) *Physica A*, **366**, 387–400.

47 Pisarenko, V., Sornette, D., and Rodkin, M. (2004) *Computational Seismology*, **35**, 138–159.

48 Pictet, O.V., Dacorogna, M.M., and Muller, U.A. (1998) *A practical guide to heavy tails chapter Hill, bootstrap and jackknife estimators for heavy tails*, Birkhauser, pp. 283–310.

49 Hall, P. (1990) *Journal of Multivariate Analysis*, **32**, 177–203.

50 Wood, R.A., McInish, T.H., and Ord, J.K. (1985) *The Journal of Finance*, **40**, 723–739.

51 Harris, L. (1986) *Journal of Financial Economics*, **16**, 99–117.

52 Lux, T. (2001) *Applied Financial Economics*, **11**(3), 299–315.

53 Silva, A.C., Prange, R.E., and Yakovenko, V.M. (2004) *Physica A*, **344**, 227–235.

54 Goldstein, M.L., Morris, S.A., and Yen, G.G. (2004) *Eur. Phys. J. B*, **41**(2), 255–258.

55 Coronel-Brizio, H. and Hernandez-Montoya, A. (2005) *Physica A*, **354**, 437–449.

56 Pan, R.K. and Sinha, S. (2007) *Phys. Rev. E*, **76**(4), 046116.

57 Plerou, V., Gopikrishnan, P., Amaral, L.A.N., Gabaix, X., Stanley, H.E. (2000) Economic fluctuations and anomalous diffusion. *Phys. Rev. E*, **62**, 3023–3026.

58 Gopikrishnan, P., Plerou, V., Gabaix, X., and Stanley, H.E. (2000) Statistical properties of share volume traded in financial markets, *Phys. Rev. E*, **62**, R4493–R4496.

59 Gabaix, X., Gopikrishnan, P., Plerou, V., and Stanley, H.E. (2003) A theory of power-law distributions in financial market fluctuations, *Nature*, **423**, 267–270.

60 Plerou, V. and Stanley, H.E. (2007) *Phys. Rev. E*, **76**(4), 046109.

61 Racz, E., Eisler, Z. and Kertesz, J.(2009) *Phys. Rev. E*, **79**(6), 068101.

62 Chowdhury, D. and Stauffer, D. (1999) A generalised spin model of financial markets. *Eur. Phys. J. B*, **8**, 477–482.

63 Cont, R. and Bouchaud, J.-P. (2000) Herd behavior and aggregate fluctuations in financial markets. *Macroecon. Dyn.*, **4**, 170–196.

64 Potters, M. and Bouchaud, J.-P. (2003) More statistical properties of order books and price impact. *Physica A*, **324**, 133–140.

65 Alfi, V., Coccetti, F., Marotta, M., Pietronero, L., Takayasu, M. (2006) Hidden forces and fluctuations from moving averages: A test study. *Physica A*, **370**, 30–37.

66 Kaizoji T. (2006) A precursor of market crashes: Empirical laws of Japan's internet bubble. *Eur. Phys. J. B*, **50**, 123–127.

6
The Log-Normal and Extreme-Value Distributions

6.1
The Log-Normal Distribution

The log-normal distribution is one of the most common functional forms seen in the economic context [1]. As an idea, it is very simple to comprehend. It is the distribution for a set of values, whose logarithms have a normal or Gaussian distribution. As the logarithmic transformation converts a product into a sum, it is easy to see that this implies that a multiplicative random walk (i.e., where a variable evolves as a successive product of random numbers) will generate a log-normal distribution, just as the Gaussian distribution is the outcome of an additive random walk. The log-normal distribution has been suggested to describe well many important economic distributions, most notably, the distribution of companies by their size and the bulk of the personal income distribution. In fact, because of the heavy tailed nature of the distribution, it is possible that many of the apparent power laws claimed in the econophysics literature are probably better described by the log-normal form.

6.2
The Law of Proportionate Effect

Robert Gibrat's doctoral thesis on economic inequality (1931) pioneered the use of the log-normal form in studying distributions arising in the economic domain, such as that of income. According to Gibrat, the income of an individual (or a firm) can be thought of as arising from a combination of a large number of independent processes, each of which may have evolved over a long time. We will consider processes which evolve in discrete time, that is $t = 1, 2, 3, \ldots$. At a given time t, the change in a variable of interest, x (say), is a random fraction of a function of its value at the previous instant:

$$x_t - x_{t-1} = \epsilon_t f(x_{t-1}) . \qquad (6.1)$$

The random variables ϵ are the output of an uncorrelated stochastic process and do not depend on x. The *law of proportionate effect* of Gibrat corresponds to the case

Econophysics. Sitabhra Sinha, Arnab Chatterjee, Anirban Chakraborti, and Bikas K. Chakrabarti
Copyright © 2011 WILEY-VCH Verlag GmbH & Co. KGaA, Weinheim
ISBN: 978-3-527-40815-3

when $f(x) = x$, so that the change in the variable is a random fraction of its value at the previous instant. Taking the logarithm of the variable, we can express its value at any instant n as

$$\ln x_n = \ln x_0 + \sum_{t=1}^{n} \ln(1 + \epsilon_t) \simeq \ln x_0 + \sum_{t=1}^{n} \epsilon_t , \qquad (6.2)$$

where x_0 is the value of the variable at some initial time $t = 0$. The approximation shown above holds when $|\epsilon| \ll 1$. Then, using the central limit theorem, we can see that $\ln x$ will follow a normal distribution, thereby making x log-normally distributed.

The probability distribution function of the lognormal distribution is

$$p(x) = \frac{1}{x\sqrt{2\pi\sigma^2}} \exp[-(1/2\sigma^2)(\ln x - \mu)^2] , \qquad (6.3)$$

where $x > 0$, and the mean and variance of the distribution are given by $E(x) = \exp(\mu + \frac{1}{2}\sigma^2)$ and $\mathrm{var}(x) = (\exp(\sigma^2) - 1)\exp(2\mu + \sigma^2)$ (see Figures 6.1 and 6.2). Note that, the log-normal distribution also arises from the sequential broken stick process where an unit interval is broken into multiple random parts in successive steps. If we stop the process after the interval has been broken into n parts, we find that the distribution of the lengths of these parts approaches a log-normal form as n becomes large [2].

An alternative method of representing log-normally distributed variables is to perform a logarithmic binning. Unlike the usual method of estimating the probability distribution from the frequency distribution by making a histogram of the data with bins of equal width, here we consider bins that have non-uniform width

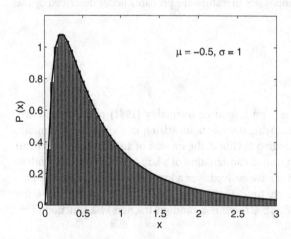

Figure 6.1 The probability distribution of a random variable generated using a log-normal distribution with $\mu = -0.5$ and $\sigma = 1$. The mean of the distribution is 1 and the variance is $e - 1$ (where $e = 2.7182818\ldots$ is Euler's number). The curve indicates the theoretical distribution function obtained from (6.3).

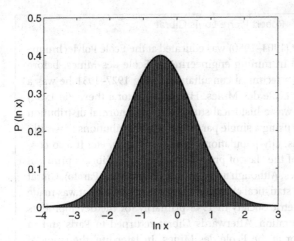

Figure 6.2 The probability distribution of the logarithm of the variable whose distribution is shown in Figure 6.1. As is clearly seen, the normal distribution with mean $\mu = -0.5$ and standard deviation $\sigma = 1$ fits the distribution accurately.

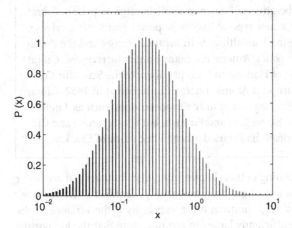

Figure 6.3 The data shown in Figure 6.1 is represented using logarithmic binning, i.e., where the histogram of the random variables is constructed by considering bins of non-uniform width. Here the bin width is considered to be an exponential function of the variable.

which increase with the value of the variable. For log-normally distributed data if the bin widths increase exponentially as a function of the mean value of the bin, the resulting histogram shows a bell-shaped curve characteristic of the Gaussian distribution when plotted on a semi-logarithmic graph with a logarithmic axis on the abscissa and a normal axis along the ordinate (see Figure 6.3).

Robert Pierre Louis Gibrat

Robert Pierre Louis Gibrat (1904–1980) was educated at the Ecole Polytechnique in Paris and specialized in mining engineering at Ecole des Mines, before beginning his career as a technical consultant. During 1927–1931 he was a professor at St. Etienne Ecole des Mines. His 1931 doctoral thesis (at Lyon) on economic inequality was a historical study of the log-normal distribution and its use in modeling, using a single parameter, the distributions of wealth, industrial concentrations, city populations, family statistics, etc. It also contained his formulation of the "law of proportionate effect", Gibrat's principal contribution to economics. Although using the term "law" (like Pareto), Gibrat explained the essentially statistical character of the principle. Gibrat was made a Fellow of the Econometric Society in 1948 primarily for the developments arising from this contribution. Afterwards Gibrat returned to Paris and in 1936 became a Professor at the Ecole des Mines. In later life, the study of nuclear energy and the harnessing of hydrolelectric energy of tides and rivers occupied Gibrat. During the Nazi occupation of France, Gibrat occupied high government positions in the puppet French government, for which he spent a year in prison after the liberation of France as a collaborator. In prison, Gibrat developed the theory for a new type of hydraulic power plant. After release, he resumed his engineering consulting, both on tidal energy and the development of French atomic policy. Among the many honors he received, Gibrat was a Knight of the Legion of Honour and was president of the Scientific Committee of EURATOM (European Atomic Energy Community) in 1962. Gibrat had a great passion for languages, not only European ones such as English, German, Italian, Spanish, Norwegian and Finnish, but also Japanese and Chinese, and just before his death had started a systematic study of Sanskrit.

As mentioned in the beginning of this chapter, often distributions that are better described by the log-normal form have been mistakenly identified as having power law tails. In fact this is a very common error, especially if the variance of the log-normal distribution is sufficiently large. To see this, note that the log-normal distribution,

$$P(x) = \frac{1}{x\sigma\sqrt{2\pi}} e^{-(\ln x - \mu)^2/2\sigma^2} , \tag{6.4}$$

can be written as (on taking logarithm on both sides),

$$\ln P(x) = -\frac{(\ln x)^2}{2\sigma^2} + \left(\frac{\mu}{\sigma^2} - 1\right) - \ln\sqrt{2\pi}\sigma - \frac{\mu^2}{2\sigma^2} , \tag{6.5}$$

which is a quadratic curve in a doubly logarithmic plot. However, a sufficiently small part of the curve will appear as a straight line, with the slope depending on which segment of the curve one is focussing attention [3, 4]. This is the origin of most of the power law tails with exponent $\alpha \neq 1$ that have been reported in the econophysics literature.

6.3
Extreme Value Distributions

On October 19, 1987, the Dow Jones Industrial Average dropped by 29.2%, the worst day of trading until that time in at least a century. Based on a theory of normally distributed fluctuations, the probability of such a cataclysmic event happening was less than one in 10^{50}. Such extreme events are rare, but occur much more often than conventional theories of markets would lead us to believe. These events are difficult to predict, but must be taken into account into any proposal for assessing the risk of market investments. Is it possible to arrive at a theory that better approximates their observed frequency? The fact that the distribution of fluctuations (e.g., of individual stock price or of the index for the entire market) actually seem to follow a power law at short time-scales should make us expect that large events are indeed more common than expected from a Gaussian distribution. But other schools of thought hold that large crashes are special, and they need to be explained using other distributions. This is where the theory of extreme value distribution comes in. Their application in the financial context not only include major crashes in the stock market, but also collapses of large banks and firms, often following each other in a cascading effect.

The probability of extreme events depends to a large extent on how the probability distribution function $p(x)$ decays to zero as $|x| \to \infty$. The rate of decay necessarily has to be estimated from the empirical data. As (almost by definition) extreme events are rare, this estimation is not a simple task, as even very large quantities of data will give us only limited information about the true probability of a very high return (positive or negative) in the market (say). In this context we can use the methods of extreme value statistics to give us a more reliable estimate of the risk involved.

Let us consider the maximum value X for a set of independent random variables $x_1, x_2, \ldots, x_n$, which are obtained from identical distributions. From the perspective of financial markets, one can think of X as $|r_{min}|$, the highest negative return. Only three distributions can be the asymptotic limit of the distribution for X, which are collectively referred to as *extreme value distributions*. The cumulative distribution functions for the three are as follows:

Gumbel distribution:

$$P(X \leq x) = \exp(-e^{(x-\mu)/\sigma}),$$

Frechet distribution:

$$P(X \leq x) = \exp(-[(x - \mu)/\sigma]^{-\xi}) \quad \text{if} \quad x \geq \mu, = 0, \text{ otherwise},$$

Weibull distribution:

$$P(X \leq x) = \exp(-[(\mu - x)/\sigma]^{-\xi}) \quad \text{if} \quad x \leq \mu, = 0, \text{ otherwise}.$$

Thus, the Frechet and Weibull distributions are concentrated on the positive and negative real numbers, while Gumbel distributed variables can have any real number value. The Frechet and Weibull distributions are related to the Gumbel form

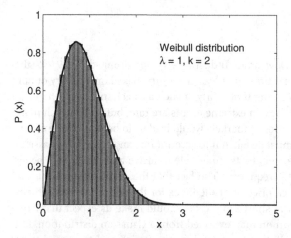

Figure 6.4 The probability distribution of a random variable generated using a Weibull distribution function with scale parameter $\lambda = 1$ and shape parameter $k = 2$.

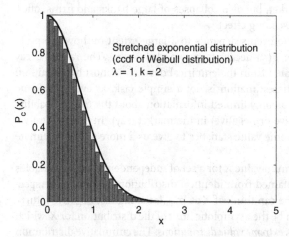

Figure 6.5 The complementary cumulative distribution function for the data shown in Figure 6.4, which is the stretched exponential distribution function with scale parameter $\lambda = 1$ and shape parameter $k = 2$.

by a log-transformation of the variable, that is $x' \rightarrow \log(\mu - x)$ or $\log(x - \mu)$. Also, note that the complementary cumulative distribution function for Weibull is the well-known stretched exponential function $e^{-(x/\lambda)^k}$, where λ and k are often referred to as scale and shape parameters of the distribution, respectively (Figures 6.4 and 6.5).

6.3.1
Value at Risk

Measuring risk is a vital criterion for most present-day financial instruments. For example, the Black–Scholes formula for option pricing uses a single input for empirical market data that is a measure of the risk, measured as standard deviation of price fluctuations. Indeed, in calculating the risk for a given market portfolio, often the weighted combination of the standard deviations of all stock prices comprising it is used. Thus, it is assumed that the standard deviations contain all relevant information about risk. However, it is not an intuitive measure of risk. Rather than beeing thought of in terms of amount of money lost it is expressed in terms of units of deviations from a mean. More importantly, the fluctuations occurring below the expected return may not have the same likelihood as those above, whereas using the standard deviation as a measure of risk assumes symmetric deviations. Therefore, another measure of risk was deemed necessary, one that would estimate the loss associated with a given small probability of deviation occurrence. In other words, it is the possibility of losing a certain amount of money over a given holding period of an investment. Thus, higher risk will imply a higher loss at the given probability. This is the purpose that the quantity Value at Risk (VaR) is designed to serve. When the returns are symmetric about the mean, the information conveyed by VaR is exactly the same as standard deviation except for a scaling factor.

E.J. Gumbel and W. Weibull

Emil J. Gumbel (1891–1966) was a German mathematician and political writer. One of the most important figures in the German pacifist movement, Gumbel was the leading chronicler of the numerous political murders that were common in the Weimar Republic. He uncovered the secret rearmament of Germany after the First World War and identified groups that carried out acts of terror, including the National Socialists. As Professor of Statistics at the University of Heidelberg, he became the focal point for a prolonged controversy with the largely anti-Republican professors and the pro-Nazi students. Forced into exile, he became an important figure in the German political and intellectual exiled community in France and later, the US.

Gumbel's doctoral dissertation was an exploration of methods to determine the population of a nation or region between censuses. This set the tone for much of his research in the 1920s and early 1930s, which mainly concerned population statistics and life expectancy. During the early 1920s he also published a few scholarly articles on statistical problems in physics. Gumbel studied with several leading statisticians, including Georg von Mayr in Munich, and carried out correspondence with Karl Pearson, in England. His years in exile in France were marked by modest professional success. At the University of Lyon, where he was between 1934–1940, he progressed from lecturer to *Maitre des Recherches* and was on track to become a director of research had not WWII intervened. In the late 1930s Gumbel shifted his research focus from social questions to the statistical analysis and prediction of natural and physical phenomena. He developed the theory of extreme values to answer questions like: "How frequently can a flood (or earthquake or rainstorm or wind) of a given magnitude be expected to occur?" From the late 1930s to the early 1950s, he worked exclusively on calculating flood flows (for determining the required strength of dams) and the breaking strength of metals. During the German occupation of France, Gumbel had to flee to Portugal and sailed from there to the US. There, Gumbel initially spent four years at the New School. During the war, he also worked for several government agencies, including the Office of Strategic Services (the precursor to the CIA) and the US Department of Interior's Geological Survey. From 1945 to 1952, except for a brief period as a professor at Brooklyn College, Gumbel did not have any full-time employment. In 1953, he was appointed adjunct professor in engineering in Columbia University, where he continued to work for the rest of his life. The crowning work of Gumbel's career was "Statistics of Extremes". The product of a decade of labor, it was first published in 1958 and immediately won high regard in the international scientific community.

E.H. Waloddi Weibull (1887–1979) was a Swedish physicist and engineer, who applied the ideas of extreme values to the study of strength of materials and in the process came up with one of the extreme value distributions that now carries his name. He joined the Swedish Coast Guard in 1904 as a midshipman. Weibull moved up the ranks eventually becoming a Major in 1940. While in the coast guard he took courses at the Royal Institute of Technology, Stockholm. He graduated in 1924 and became a full professor. Weibull obtained his doctorate from the University of Uppsala in 1932. He worked as a consulting engineer for various Swedish and German industries. His first scientific paper was on the propagation of shock waves generated by an explosion (1914). As part of this interest, he took part in voyages to the Mediterranean Sea, the Caribbean and the Pacific Ocean aboard a research ship to develop a technique for determining the type and thickness of ocean bed sediments by using data obtained from explosive charges. In 1939, he published his first paper on the connection between statistics of extreme values and the phenomenon of rupture in solids. E.J. Gumbel later showed that the distribution proposed by Weibull was identical to one of the three extreme value distributions. Further, he proved that if the failure of a system depends on the probability of the failure of any one of its multiple parts (the concept that a chain is only as strong as its weakest link), then the time to first failure is best modeled by this distribution. In 1941, Weibull became a research professor in Technical Physics at the Royal Institute of Technology in Stockholm. In later life, he continued to publish on many aspects of strength of materials and fatigue. He died on October 12, 1979 in Annecy, France.

References

1 Crow, E.L. and Shimizu, K. (1988) *Log-normal distributions: Theory and Applications*, CRC Press, New York.

2 Aitchison, J. and Brown, J.A.C. (1957) *The Lognormal Distribution*, Cambridge University Press, Cambridge, Mass.

3 Newman, M.E.J. (2005) *Contemporary Physics*, **46**, 323–351.

4 Mitzenmacher, M. (2003) *Internet Mathematics*, **1**, 226–251.

7
When a Single Distribution is not Enough?

"In a condition of society and under an industrial organization which places labor completely at the mercy of capital, the accumulations of capital will necessarily be rapid, and an unequal distribution of wealth is at once to be observed."

– Leland Stanford

7.1
Empirical Data on Income and Wealth Distribution

The distribution of wealth among individuals in an economy has been an important area of research in economics, for more than a hundred years [1, 2]. It was Wilfredo Pareto [3] who first quantified the high-end of the income distribution in a society and found it to follow a power law, quoted simply as

$$P(m) \sim m^{-(1+\nu)} , \tag{7.1}$$

where P gives the normalized number of people with income m. The power law in income and wealth distribution is named after Pareto and the exponent ν is called the Pareto exponent. A historical account of Pareto's data and that from recent sources can be found in [5]. However, there are different forms of this law quoted in the literature. Mandelbrot [6] distinguished between different versions:

1. If $Q(m)$ is the percentage of individuals with an income greater than m, the *strong Pareto law* states that

$$Q(m) = \begin{cases} (m/m_0)^{-\nu} & \text{for } m > m_0 \\ 1 & \text{for } m < m_0 , \end{cases} \tag{7.2}$$

where m_0 is a scale factor. In the *strongest form* of the Pareto law, $\nu = 3/2$, which is the average value of ν in Pareto's original data.

2. However, the value of ν has a wider distribution, and referred to as the *weak Pareto law* which states that the law holds in the asymptotic limit, that is

$$Q(m) \to (m/m_0)^{-\nu} \quad \text{as} \quad m \to \infty . \tag{7.3}$$

Econophysics. Sitabhra Sinha, Arnab Chatterjee, Anirban Chakraborti, and Bikas K. Chakrabarti
Copyright © 2011 WILEY-VCH Verlag GmbH & Co. KGaA, Weinheim
ISBN: 978-3-527-40815-3

In all cases above, $Q(m) = \int_m^\infty P(m)\,dm$.

Although Pareto [3] and Gini [4] had respectively identified the power-law tail and the log-normal bulk of the income distribution, the demonstration of both features in the same distribution was possibly first demonstrated by Montroll and Shlesinger [7] through an analysis of fine-scale income data obtained from the US Internal Revenue Service (IRS) for the years 1935–1936. It was observed that while the top 2–3% of the population (in terms of income) followed a power law with Pareto exponent $\nu \simeq 1.63$, the rest followed a log-normal distribution. Later work on Japanese personal income data, based on detailed records obtained from the Japanese National Tax Administration, indicated that the tail of the distribution followed a power law with an exponent value that fluctuated from year to year around the mean value of 2 [8]. Further work [9] showed that the power law region described the top 10% or less of the population (in terms of income), while the remaining income distribution was well-described by the log-normal form. While the value of the Pareto index fluctuated significantly from year to year, it was observed that the parameter describing the log-normal bulk, the Gibrat index, remained relatively unchanged. The change of income from year to year, that is the growth rate as measured by the log ratio of the income tax paid in successive years, was observed by Fujiwara *et al.* [10] to be also a heavy tailed distribution, although skewed, and centered about zero. Later work on the US income distribution based on data from the IRS for the years 1997–1998, while still indicating a power-law tail (with $\nu \simeq 1.7$), have suggested that the lower 95% of the population have income whose distribution may be better described by an exponential form [11, 12]. The same observation has been made for income distribution in the UK for the years 1994–1999, where the value was found to vary between 2.0 and 2.3, but the bulk seemed to be well-described by an exponential decay.

It is interesting to note that, when one shifts attention from the income of individuals to the income of companies, one still observes the power law tail. A study of the income distribution of Japanese firms [13] concluded that it follows a power law with $\nu \simeq 1$, which is also often referred to as Zipf's law. Similar observation has been reported for the income distribution of US companies [14].

Compared to the empirical work done on income distribution, relatively few studies have looked at the distribution of wealth, which consist of the net value of assets (financial holdings and/or tangible items) owned at a given point in time. The lack of an easily available data source for measuring wealth, analogous to income tax returns for measuring income, means that one has to resort to indirect methods. Levy and Solomon [18] used a published list of the wealthiest people to generate a rank-order distribution, from which they inferred the Pareto exponent for wealth distribution in the US. References [12] and [19] used an alternative technique based on adjusted data reported for the purpose of inheritance tax to obtain the Pareto exponent for the UK. Another study used a tangible asset (namely house area) as a measure of wealth to obtain the wealth distribution exponent in ancient Egyptian society during the reign of Akhenaten (fourteenth century BC) [20]. More recently, the wealth distribution in India was also observed to follow a power law

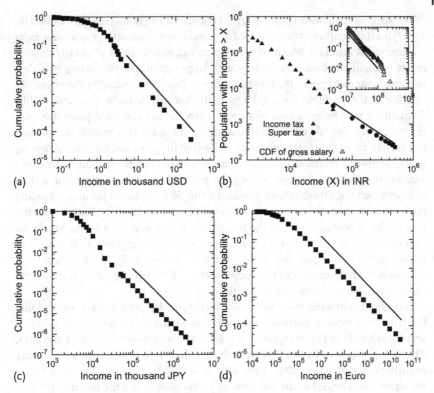

Figure 7.1 (a) Cumulative probability of US personal annual income for IRS data from 2001 (taken from [15]), Pareto exponent $\nu \approx 1.5$ (given by the slope of the solid line). (b) Cumulative income distribution in India during 1929–1930, collected from Income Tax and Super Tax data [16]. The inset shows the cumulative distribution of the employment income for the top 422 salaried Indians (Business Standard survey, 2006) showing a power-law tail with $\nu = 1.75 \pm 0.01$. (c) Cumulative probability distribution of Japanese personal income in the year 2000. The power law region approximately fits to $\nu = 1.96$ (data from [17]). (d) Cumulative probability distribution of firm size (total-assets) in France in the year 2001 for 669 620 firms. The power law region can be approximately fitted with $\nu = 0.84$ (data from [17]).

tail with the exponent varying around 0.9 [21]. The general feature observed in the limited empirical study of wealth distribution is that of a power law behavior for the wealthiest 5–10% of the population, and exponential or log-normal distribution for the rest of the population. The Pareto exponent as measured from the wealth distribution is found to be always lower than the exponent for the income distribution, which is consistent with the general observation that, in market economies, wealth is much more unequally distributed than income [17].

A relatively recent study [23] on the list of the richest people in the world from Forbes data revealed a Pareto law with an average exponent $\nu = 1.49$. Another study [24] reported a power-law tail for the wealth distribution of aristocratic families in medieval Hungary.

The striking regularities (see Figure 7.1) observed in the income distribution [6, 15, 17] for different countries, have led to several new attempts at explaining them on theoretical grounds. Much of the current impetus is from physicists' modeling of economic behavior in analogy with large systems of interacting particles, as treated, for example in the kinetic theory of gases. According to the physicists working on this problem, the regular patterns observed in the income (and wealth) distribution may be indicative of a natural law for the statistical properties of a many-body dynamical system representing the entire set of economic interactions in a society, analogous to those previously derived for gases and liquids. By viewing the economy as a thermodynamic system, one can identify the income distribution with the distribution of energy among the particles in a gas. In particular, a class of kinetic exchange models have provided a simple mechanism for understanding the unequal accumulation of assets. Many of these models, while simple from the perspective of economics, have the benefit of coming to grips with the key factor in socioeconomic interactions that results in very different societies converging to similar forms of unequal distribution of resources (see [1, 2], which consists of a collection of large number of technical papers in this field; see also [3, 26, 27] for some popular discussions and also criticisms).

Considerable investigations with real data during the last ten years revealed that the tail of the income distribution indeed follows the above mentioned behavior and the value of the Pareto exponent v is generally seen to vary between 1 and 3 [13, 21, 29–31]. It is also known that typically less than 10% of the population in any country possesses about 40% of the total wealth of that country and they follow the above law. The rest of the low income population, in fact the majority (90% or more), follow a different distribution which is debated to be either Gibbs [2, 7, 12, 29–31, 33–35] or log-normal [30, 31]. There have also been reports of fitting the low income population to Tsallis function [38] or κ-generalized statistics [39].

References

1 Chatterjee, A., Yarlagadda, S., and Chakrabarti, B.K. (2005) (eds) *Econophysics of Wealth Distributions*, Springer Verlag, Milan.

2 Chakrabarti, B.K., Chakraborti, A., Chatterjee, A. (eds) (2006) *Econophysics and Sociophysics*, Wiley-VCH, Berlin.

3 Pareto, V. (1897) *Cours d'economie Politique*, F. Rouge, Lausanne.

4 Gini, C. (1921) *The Economic Journal*, **31**, 124.

5 Richmond, P., Hutzler, S., Coelho, R., and Repetowicz, P. in [2].

6 Mandelbrot, B.B. (1960) *International Economic Review*, **1**, 79.

7 Montroll, E.W. and Shlesinger, M.F. (1982) *Proc. Natl. Acad. Sci. USA*, **79**, 3380.

8 Aoyama, H., Souma, W., Nagahara, Y., Okazaki, M.P., Takayasu, H., and Takayasu, M. (2000) *Fractals*, **8**, 293.

9 Souma, W. (2000) *Fractals*, **9**, 463.

10 Fujiwara, Y., Souma, W., Aoyama, H., Kaizoji, T., and Aoki, M. (2003) *Physica A*, **321**, 598.

11 Drăgulescu, A.A. and Yakovenko, V.M. (2001) *Eur. Phys. J. B*, **20**, 585.

12 Drăgulescu, A.A. and Yakovenko, V.M. (2001) *Physica A*, **299**, 213.

13 Okuyama, K., Takayasu, M., and Takayasu, H. (1999) *Physica A*, **269**, 125.

14 Axtell, R.L. (2001) *Science*, **293**, 1818.
15 Silva, A.C. and Yakovenko, V.M. in [1].
16 Sinha, S. in [1], p. 177.
17 Fujiwara, Y. in [1].
18 Levy, M. and Solomon, S. (1997) *Physica A*, **242**, 90.
19 Coelho, R., Néda, Z., Ramasco, J.J., and Santos, M.A. (2005) *Physica A*, **353**, 515.
20 Abul-Magd, A.Y. (2002) *Phys. Rev. E*, **66**, 057104.
21 Sinha, S. (2006) *Physica A*, **359**, 555.
22 Samuelson, P.A. (1980) *Economics*, McGraw Hill Int., Auckland.
23 Klass, O.S., Biham, O., Levy, M., Malcai, O., and Solomon, S. (2006) *Economics Letters*, **90**, 290; (2007) *Eur. Phys. J. B*, **55**, 143.
24 Hegyi, G., Néda, Z., and Santos, M.A. (2007) *Physica A*, **380**, 271.
25 Hayes, B. (2002) *American Scientist*, **90**, 400.
26 Hogan, J. (2005) *New Scientist*, 6.
27 Ball, P. (2006) *Nature*, **441**, 686; Editorial, (2006) *Nature*, **441**, 667.
28 Moss de Oliveira, S., de Oliveira, P.M.C., and Stauffer, D. (1999) *Evolution, Money, War and Computers*, B.G. Teubner, Stuttgart, Leipzig.

29 Aoyama, H., Souma, W., and Fujiwara, Y. (2003) *Physica A*, **324**, 352.
30 Di Matteo, T., Aste, T., and Hyde, S.T. (2004) in *The Physics of Complex Systems (New Advances and Perspectives)*, (eds F. Mallamace, H.E. Stanley), Amsterdam, p. 435.
31 Clementi, F. and Gallegati, M. (2005) *Physica A*, **350**, 427.
32 Chakrabarti, B.K. and Marjit, S. (1995) *Ind. J. Phys. B*, **69**, 681.
33 Atkinson, A.B. and Bourguignon, F. (eds) (2000) *Handbook of Income Dtribution*, Elsevier, Amsterdam.
34 Champernowne, D.G. and Cowell, F.A. (1998) *Economic Inequality and Income Distribution*, Cambridge University Press, Cambridge.
35 Kakwani, N. (1980) *Income Inequality and Poverty*, Oxford University Press, Oxford.
36 Ispolatov, S., Krapivsky, P.L., and Redner, S. (1998) *Eur. Phys. J. B*, **2**, 267.
37 Drăgulescu, A.A. and Yakovenko, V.M. (2000) *Eur. Phys. J. B*, **17**, 723.
38 Ferrero, J.C. in [1], p. 159.
39 Clementi, F., Gallegati, M., and Kaniadakis, M. (2007) *Eur. Phys. J B*, **57**, 187.

8
Explaining Complex Distributions with Simple Models

... all models are approximations. Essentially, all models are wrong, but some are useful. However, the approximate nature of the model must always be borne in mind ...

— George Edward Pelham Box

8.1
Kinetic Theory of Gases

The *ideal gas* is a very important subject of study in statistical physics. It constitutes a system where the interaction between the particles or molecules is very weak or negligible. Alternatively, the gas is particularly rarified so that the particles are typically large distances apart, such that the interaction is very small. A typical molecule has its center of mass at $\mathbf{r}$, and let $\mathbf{p}$ denote the momentum of its center of mass. In the absence of external force fields, the energy E of the molecule is given by

$$E = \frac{\mathbf{p}^2}{2m} + E_{\text{int}} , \tag{8.1}$$

where the first term on the right is the kinetic energy of the center of mass, while the second term arises out of the asymmetry (polyatomic molecule) of the molecule, and is simply the internal energy of rotation or vibration with respect to the molecular center of mass. The dilute gas assumption allows us to neglect any potential energy of interaction with other molecules, and thus, E is independent of $\mathbf{r}$. The dilute gas approximation also allows us to treat the translational degrees of freedom classically, while the internal degrees of freedom are usually dealt with using quantum mechanics.

8.1.1
Derivation of Maxwell–Boltzmann Distribution

As for the statistical description, we restrict ourselves to a hard-sphere gas with no intermolecular interactions. The easiest way to realize things is to estimate the

Econophysics. Sitabhra Sinha, Arnab Chatterjee, Anirban Chakraborti, and Bikas K. Chakrabarti
Copyright © 2011 WILEY-VCH Verlag GmbH & Co. KGaA, Weinheim
ISBN: 978-3-527-40815-3

number of collisions that a sample of gas exerts on a planar boundary surface, which is arbitrarily oriented perpendicular to the x-axis. Thus, the collisions per unit area, or pressure is

$$\frac{N}{A} = \rho_N \Delta t \sum_{v_x > 0} v_x f(v_x) ,$$ (8.2)

ρ_N being the number density of particles, Δt is the collision time and $f(v_x)$ is the distribution of the velocity component along x axis. Let m be the mass of each molecule. Now, the change in momentum suffered by the molecules by hitting the wall is given by

$$\Delta p = N2mv_x = 2m\rho_N A\Delta t \sum_{v_x > 0} v_x^2 f(v_x) ,$$ (8.3)

and, hence, the pressure is given by

$$P = \frac{\Delta p}{A\Delta t} = 2m\rho_N \sum_{v_x > 0} v_x^2 f(v_x) .$$ (8.4)

For a stationary gas, we recognize that the sum of all velocities above zero is equal to half the sum of all the velocities from negative to positive infinity:

$$P = 2m\rho_N \sum_{v_x > 0} v_x^2 f(v_x) = 2m\rho_N \frac{1}{2} \sum_{v_x} v_x^2 f(v_x) .$$ (8.5)

The above can be rewritten in terms of macroscopic quantities using the ideal gas law

$$\frac{nRT}{V} = m\rho_N \sum_{v_x} v_x^2 f(v_x) = m\rho_N \langle v_x^2 \rangle ,$$ (8.6)

which generalizes into

$$\langle v_x^2 \rangle = \frac{RT}{M} ,$$ (8.7)

where R is the ideal gas constant and M is the molar weight of the gas. For an ideal gas in three dimensions, $\langle v^2 \rangle = 3\langle v_x^2 \rangle = 3RT/M = 3k_B T/m$, k_B being the Boltzmann constant. Hence the expression for mean kinetic energy of the ideal gas would look like

$$\langle E \rangle = \frac{1}{2}m\langle v^2 \rangle = \frac{3k_B T}{2} .$$ (8.8)

The distribution function $g(v)$ of velocities would be given by

$$g(v)dv = f(v_x) f(v_y) f(v_z)dv_x dv_y dv_z .$$ (8.9)

and the partial derivative of g with respect to one of the axes looks like

$$\frac{\partial g}{\partial v_x} = \frac{dg}{dv}\frac{\partial v}{\partial v_x} = \frac{dg}{dv}\frac{\partial}{\partial v_x}\sqrt{v_x^2 + v_y^2 + v_z^2} = \frac{dg}{dv}\frac{v_x}{v}. \tag{8.10}$$

Rearranging

$$\frac{1}{v}\frac{dg}{dv} = \frac{1}{v_x}\frac{\partial g}{\partial v_x} = \frac{f(v_y)f(v_z)}{v_x}\frac{\partial f(v_x)}{\partial v_x}, \tag{8.11}$$

what follows from the equivalence of the directions is

$$\frac{f(v_y)f(v_z)}{v_x}\frac{\partial f(v_x)}{\partial v_x} = \frac{f(v_z)f(v_x)}{v_y}\frac{\partial f(v_y)}{\partial v_y} = \frac{f(v_x)f(v_y)}{v_z}\frac{\partial f(v_z)}{\partial v_z}. \tag{8.12}$$

Dividing all by g one obtains

$$\frac{1}{f(v_x)v_x}\frac{\partial f(v_x)}{\partial v_x} = \frac{1}{f(v_y)v_y}\frac{\partial f(v_y)}{\partial v_y} = \frac{1}{f(v_z)v_z}\frac{\partial f(v_z)}{\partial v_z} = -k, \tag{8.13}$$

where k is some constant independent of the coordinates. Now we have a first-order differential equation for the velocity coordinates, and integrating one obtains $f(v_x) = B\exp[-kv_x^2/2]$. To determine the value of the integrating constant B, we note that for any probability distribution, the integral of the distribution over the entire region of space it encompasses must be unity. Thus,

$$1 = \int_{-\infty}^{\infty} f(v_x)dv_x = B\int_{-\infty}^{\infty}\exp[-kv_x^2/2]dv_x, \tag{8.14}$$

which yields $B = \sqrt{\frac{k}{2\pi}}$, and thus $f(v_x) = \sqrt{\frac{k}{2\pi}}\exp[-kv_x^2/2]$. Thus, (8.9) can be rewritten as

$$g(v) = \left(\frac{k}{2\pi}\right)^{3/2}\exp[-kv_x^2/2]\exp[-kv_y^2/2]\exp[-kv_z^2/2]. \tag{8.15}$$

The mean kinetic energy is thus

$$\langle E\rangle = \frac{3m}{2}\left(\frac{k}{2\pi}\right)^{3/2}\int_{-\infty}^{\infty} v_x^2\exp[-kv_x^2/2]dv_x$$

$$\times\int_{-\infty}^{\infty}\exp[-kv_y^2/2]dv_y\int_{-\infty}^{\infty}\exp[-kv_z^2/2]dv_z, \tag{8.16}$$

which simplifies to

$$\langle E\rangle = \frac{3m}{2}\left(\frac{k}{2\pi}\right)^{3/2}\frac{2\pi}{k}\int_{-\infty}^{\infty} v_x^2\exp[-kv_x^2/2]dv_x = \frac{3m}{2k}. \tag{8.17}$$

Equating that with the mean value of kinetic energy, we get $k = \frac{m}{k_B T}$. Thus,

$$f(v_x) = \left(\frac{m}{2\pi k_B T}\right)^{1/2} \exp\left[-\frac{mv_x^2}{2k_B T}\right]. \tag{8.18}$$

The distribution of molecular speeds is then given by

$$g(v) = 4\pi \left(\frac{m}{2\pi k_B T}\right)^{3/2} v^2 \exp\left[-\frac{mv^2}{2k_B T}\right], \tag{8.19}$$

where $v^2 = v_x^2 + v_y^2 + v_z^2$. Subsequently, one can also find the distribution function for energy $E = \frac{1}{2}mv^2$ as

$$h(E) = 2\pi \left(\frac{1}{\pi k_B T}\right)^{3/2} E^{1/2} \exp\left(-\frac{E}{k_B T}\right). \tag{8.20}$$

Boltzmann and Gibbs

Ludwig Boltzmann (1844–1906) was an Austrian physicist famous for his contributions to the foundation of statistical mechanics and statistical thermodynamics. He was one of the most important supporter of atomic theory during the time that model was still highly controversial. Boltzmann's most important scientific contributions were in kinetic theory. He is known for the Maxwell–Boltzmann distribution for molecular speeds in a gas. He developed an equation to describe the dynamics of an ideal gas, now known as the Boltzmann equation. Another big contribution came as a logarithmic relation between entropy and probability. In the Maxwell–Boltzmann statistics and the Boltzmann distribution over energies lie the foundations of classical statistical mechanics, and provide a remarkable insight into the meaning of temperature.

Josiah Willard Gibbs (1839–1903) was an American theoretical physicist, chemist, and mathematician. He is responsible for the theoretical foundation of chemical thermodynamics and physical chemistry. In mathematics, he invented vector analysis, and his most important contributions are in conceptualizing chemical potential and free energy. The Gibbsian ensemble sets a foundation in statistical mechanics, while the phase rule is an essential part of thermodynamics. His work and legacy influenced as well as laid the foundation for later important contributions, not only in physics and chemistry, but also in economics. Apart from Irving Fisher, who was strongly influenced by his work, Nobel laureate Paul Samuelson acknowledged the influence of classical thermodynamics developed by Gibbs.

8.1.2
Maxwell–Boltzmann Distribution in D Dimensions

Here we show that for integer or half-integer values of the parameter n the gamma distribution

$$\gamma_n(\xi) = \Gamma(n)^{-1}\xi^{n-1}\exp(-\xi) . \tag{8.21}$$

$\Gamma(n)$ is the Gamma function which represents the distribution of the rescaled kinetic energy ξ ($\xi = E/k_B T$) for a classical mechanical system in $D = 2n$ dimensions. T represents the absolute temperature of the system and k_B is the Boltzmann constant.

The normalized probability distribution in momentum space is simply $f(\mathbf{P}) = \prod_i (2\pi m k_B T)^{-D/2}\exp(-\mathbf{p}_i^2/2mk_B T)$, where $\mathbf{P} = \{\mathbf{p}_1, \ldots, \mathbf{p}_N\}$ represents the momentum vectors of the N particles. Thus, since the kinetic energy distribution factorizes as a sum of single particle contributions, the probability density factorizes as a product of single particle densities, each one of the form

$$f(\mathbf{p}) = \frac{1}{(2\pi m k_B T)^{D/2}}\exp\left(-\frac{\mathbf{p}^2}{2mk_B T}\right), \tag{8.22}$$

where $\mathbf{p} = (p_1, \ldots, p_D)$ is the momentum of a generic particle. It is convenient to introduce the momentum modulus p of a particle in D dimensions

$$p^2 \equiv \mathbf{p}^2 = \sum_{k=1}^{D} p_k^2 , \tag{8.23}$$

where the p_ks are the Cartesian components, since the distribution (8.22) depends only on $p \equiv \sqrt{\mathbf{p}^2}$. One can then integrate the distribution over the $D - 1$ angular variables to obtain the momentum modulus distribution function, with the help of the formula for the surface of a hypersphere of radius p in D dimensions

$$S_D(p) = \frac{2\pi^{D/2}}{\Gamma(D/2)}p^{D-1} . \tag{8.24}$$

One obtains

$$f(p) = S_D(p)f(\mathbf{p}) = \frac{2}{\Gamma(D/2)(2mk_B T)^{D/2}}p^{D-1}\exp\left(-\frac{p^2}{2mk_B T}\right). \tag{8.25}$$

The corresponding distribution for the kinetic energy $E = p^2/2m$ is therefore

$$f(E) = \left[\frac{dp}{dE}f(p)\right]_{p=\sqrt{2mE}} = \frac{1}{\Gamma(D/2)k_B T}\left(\frac{E}{k_B T}\right)^{D/2-1}\exp\left(-\frac{E}{k_B T}\right). \tag{8.26}$$

Comparison with the gamma distribution (8.21) shows that the Maxwell–Boltzmann kinetic energy distribution in D dimensions can be expressed as

$$f(E) = (k_B T)^{-1}\gamma_{D/2}(E/k_B T) . \tag{8.27}$$

The distribution for the rescaled kinetic energy,

$$\xi = E/k_B T ,$$ (8.28)

is just the gamma distribution of order $D/2$

$$f(\xi) = \left[\frac{dE}{d\xi} f(E)\right]_{E=\xi k_B T} = \frac{1}{\Gamma(D/2)} \xi^{D/2-1} \exp(-\xi) \equiv \gamma_{D/2}(\xi) .$$ (8.29)

8.2
The Asset Exchange Model

In principle, any model of the distribution of wealth [1] defined by a process of exchange of assets could be called an "asset exchange model". Generally, the economic activity in these models involve interaction between two individuals resulting in the redistribution of their assets. Of course, "individual" here could mean a single person or a conglomerate such as a company. "Asset" generally refers to anything that contributes to the overall wealth of the economy, and could be cash or any other physical asset.

A model for evolution of wealth distribution in an economically interacting population was introduced by Ispolatov *et al.* [2], where specific amounts of assets are exchanged between two individuals. Two different cases were studied: a "random" exchange where either of the individuals are likely to gain, and a "greedy" exchange, where the richer individual always gains. Again, two types of exchanges were considered: the "additive" exchange where a fixed amount of asset transfers from an individual to the other, and a "multiplicative" exchange, where the amount of asset exchanged is a finite fraction of the wealth of one of the individuals. This work reports a variety of distribution for the asset, depending on the exact trading rules. For the additive type, a random exchange produces a Gaussian distribution of wealth, while greedy exchange produces a Fermi-like scaled wealth distribution. For the multiplicative exchanges, the random exchange produces a steady state, while the greedy exchange produces a continuously evolving power law wealth distribution. Studies [7] have also shown that if the amount of money transfered is a small fixed quantity, or a fixed fraction of the trading pair's average wealth, or a random fraction of the average wealth of the entire population, and even in cases of debt or bankruptcy, the equilibrium distribution is exponential.

Hayes [3] proposed two models of asset exchange in a closed, nonevolving economy based on simple exchange rules: *yard-sale* (YS) and *theft-and-fraud* (TF). In the YS model, the amount of wealth exchanged is a finite fraction of that of the poorer trader, and the resultant distribution corresponds to a monopoly, where all the wealth accumulates with one trader, as reported in an earlier study [4]. In the TF model, the trading pair randomly chooses the loser, and the amount of wealth exchanged is a random fraction of the donor. Thus the rich trader has more to lose while the poor trader has more to gain. The resultant equilibrium distribution is exponential.

Some extensions of these models [5, 6] produce distributions which are of considerable interest. In an asymmetric exchange model [6], the wealth dynamics are defined by

$$
w_i(t+1) = \begin{cases} w_i(t) + \epsilon \left(1 - \tau \left[1 - \frac{w_i(t)}{w_j(t)}\right]\right) w_j(t), & \text{if } w_i(t) \le w_j(t) \\ w_i(t) + \epsilon w_j(t), & \text{otherwise}, \end{cases}
$$

(8.30)

where ϵ is a random number between 0 and 1. Here, $\tau = 0$ corresponds to the random exchange model [7] while $\tau = 1$ corresponds to the minimum exchange model [3, 4]. In general, the relation between agents is asymmetric and the richer agent dictates the terms of the trade. τ is known as the "thrift" parameter, and it measures the degree to which the richer agent is able to use its power. If one considers a uniform distribution of τ among agents between 0 and 1, one observes a power-law distribution for larger wealth, with Pareto exponent 1.5 [6].

M.N. Saha and S.N. Bose

Meghnad Saha (1893–1956) was an Indian astrophysicist. His best-known work concerned the thermal ionization of elements, and it led him to formulate what came to be known as the Saha equation. This equation is one of the basic tools for interpretation of the spectra of stars in astrophysics. By studying the spectra of various stars, one can find their temperature and subsequently, using Saha's equation, determine the ionization state of the various elements making up the star. M. N. Saha and B. N. Srivastava's book "A treatise on heat", India Press, Allahabad (1931), is the first textbook describing a possible application of kinetic theory of gas to income and wealth distribution in a society.

Satyendra Nath Bose (1894–1974) was an Indian physicist, specializing in mathematical physics. He is best known for his work on quantum mechanics in the early 1920s. His derivation of Planck's radiation law without referring to classical physics provided the foundation for the Bose–Einstein statistics and subsequently to the theory of the Bose–Einstein condensate. Particles obeying Bose–Einstein statistics are named boson, after him. The importance of his work is several fold, and has laid the foundation for several Nobel prize winning works.

8.3
Gas-Like Models

In 1960, Mandelbrot wrote, "There is a great temptation to consider the exchanges of money which occur in economic interaction as analogous to the exchanges of energy which occur in physical shocks between molecules. In the loosest possible terms, both kinds of interactions *should* lead to *similar* states of equilibrium. That is, one *should* be able to explain the law of income distribution by a model similar to that used in statistical thermodynamics: many authors have done so explicitly, and all the others of whom we know have done so implicitly." [8]. Unfortunately Mandelbrot did not provide any reference to this material!

The study of pairwise money transfer and the resulting statistical distribution of money has almost no counterpart in modern economics. Econophysicists initiated a new direction here. The search theory of money [9] is somewhat related, but this work was largely influenced by [10] studying the dynamics of money. A probability distribution of money among the agents was only recently obtained numerically within the search theoretical approach [11].

In analogy to two-particle collisions with a resulting change in their individual kinetic energy (or momenta), income exchange models may be based on two-agent interactions. Here two randomly selected agents exchange money by some pre-defined mechanism. Assuming the exchange process does not depend on previous exchanges, the dynamics follows a Markovian process

$$
\begin{pmatrix} m_i(t+1) \\ m_j(t+1) \end{pmatrix} = \mathcal{M} \begin{pmatrix} m_i(t) \\ m_j(t) \end{pmatrix}
\tag{8.31}
$$

where $m_i(t)$ is the income of agent i at time t and the collision matrix $\mathcal{M}$ defines the exchange mechanism.

In this class of models, one considers a closed economic system where total money M and total number of agents N is fixed. This corresponds to a situation where no production or migration occurs and the only economic activity is confined to trading. Each agent i, individual or corporate, possesses money $m_i(t)$ at time t. In any trading, a pair of traders i and j exchange their money [2, 7, 12, 13], such that their total money is (locally) conserved and none end up with negative money ($m_i(t) \geq 0$, i.e., debt not allowed)

$$
m_i(t+1) = m_i(t) + \Delta m ; \quad m_j(t+1) = m_j(t) - \Delta m .
\tag{8.32}
$$

Following local conservation

$$
m_i(t) + m_j(t) = m_i(t+1) + m_j(t+1) ,
\tag{8.33}
$$

time (t) changes by one unit after each trading. The simplest model considers a random fraction of total money to be shared [7]

$$
\Delta m = \epsilon_{ij}[m_i(t) + m_j(t)] - m_i(t) ,
\tag{8.34}
$$

where ϵ_{ij} is a random fraction ($0 \leq \epsilon_{ij} \leq 1$) changing with time or trading. The steady-state ($t \to \infty$) distribution of money is a Gibbs distribution

$$
P(m) = (1/T) \exp(-m/T) ; \quad T = M/N .
\tag{8.35}
$$

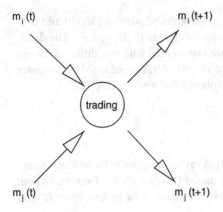

Figure 8.1 Schematic diagram of the trading process. Agents i and j redistribute their money in the market: $m_i(t)$ and $m_j(t)$, their respective money before trading, changes over to $m_i(t+1)$ and $m_j(t+1)$ after trading.

Hence, no matter how uniform or justified the initial distribution is, the eventual steady state corresponds to the Gibbs distribution where most of the people have very little money. This follows from the conservation of money and additivity of entropy

$$P(m_1)P(m_2) = P(m_1 + m_2) . \tag{8.36}$$

This steady state result is quite robust and realistic as well. In fact, several variations of the trading, and of the "lattice" (on which the agents can be put and each agent trades with its "lattice neighbors" only), whether compact, fractal or small-world like [14], leaves the distribution unchanged. Some other variations like random sharing of an amount $2m_2$ only (not of $m_1 + m_2$) when $m_1 > m_2$ (trading at the level of lower economic class in the trade), lead to an even more drastic situation: all the money in the market drifts to one agent and the rest become truly paupers [3, 4].

If one allows for debt in such simple models, things look very different. From the point of view of individuals, debt can be viewed as negative money. As an agent borrows money from a bank, its cash balance M increases, but at the expense of cash obligation or debt D which is a negative money. Thus the total money of the agent $M_b = M - D$ remains the same. Thus if the boundary condition $m_i \geq 0$ is relaxed, $P(m)$ never stabilizes and keeps spreading in a Gaussian manner towards $m = +\infty$ and $m = -\infty$. Total money is conserved, and some agents become richer at the expense of others going into debt, so that $M = M_b + D$.

Xi et al. [15] imposed a constraint on the total debt of all agents in the system. Banks set aside a fraction R of the money deposited into bank accounts, whereas the remaining $1 - R$ can be loaned further. If the initial amount of money in the system is M_b, then, with repeated loans and borrowing, the total amount of positive money available to the agents increases to $M = M_b/R$, where $1/R$ is called the money-multiplier, and this extra money comes from the increase of the total debt in the system. The maximal total debt is $D = M_b/R - M_b$ and is limited by the factor R. For maximal debt, the total amounts of positive (M_b/R) and negative $(M_b(1-R)/R)$ money circulate among the agents in the system. The distributions of positive and negative money are exponential with two different money temperatures $T_+ = M_b/RN$ and $T_- = M_b(1-R)/RN$, as confirmed by computer simulations [15]. Similar results were also observed elsewhere [16].

8.3.1
Model with Uniform Savings

In any trading, savings come naturally [17]. A saving propensity factor λ was therefore introduced in the random exchange model [13] (see [7] for a model without savings), where each trader at time t saves a fraction λ of its money $m_i(t)$ and trades randomly with the rest

$$m_i(t+1) = \lambda m_i(t) + \epsilon_{ij}\left[(1-\lambda)(m_i(t)+m_j(t))\right], \tag{8.37}$$

$$m_j(t+1) = \lambda m_j(t) + (1-\epsilon_{ij})\left[(1-\lambda)(m_i(t)+m_j(t))\right], \tag{8.38}$$

where

$$\Delta m = (1-\lambda)\left[\epsilon_{ij}\{m_i(t)+m_j(t)\} - m_i(t)\right], \tag{8.39}$$

and ϵ_{ij} is a random fraction, coming from the stochastic nature of the trading.

The market (noninteracting at $\lambda = 0$ and 1) becomes "interacting" for any nonvanishing $\lambda(< 1)$: For fixed λ (same for all agents), the steady state distribution $P(m)$ of money is exponentially decaying on both sides with the most-probable money per agent shifting away from $m = 0$ (for $\lambda = 0$) to M/N as $\lambda \to 1$ (Figure 8.2). This self-organizing feature of the market, induced by sheer self-interest of saving by each agent without any global perspective, is quite significant as the fraction of paupers decrease with saving fraction λ and most people end up with some finite fraction of the average money in the market (for $\lambda \to 1$, the socialists' dream is achieved with just people's self-interest of saving!). Interestingly, self-organization also occurs in such market models when there is restriction in the commodity market [25]. Although this fixed saving propensity does not give the Pareto-like power-law distribution, the Markovian nature of the scattering or trading processes (8.36) is effectively lost. Indirectly through λ, the agents get to know (start interacting with) each other and the system cooperatively self-organizes towards a most-probable distribution ($m_p \neq 0$) (see Figure 8.2).

There have been a few attempts to analytically formulate this problem [18] but no analytic expression has yet been found. It has also been claimed through heuristic arguments (based on numerical results) that the distribution is a close approximate form of the gamma distribution [19]

$$P(m) = \frac{n^n}{\Gamma(n)}m^{n-1}\exp(-nm), \tag{8.40}$$

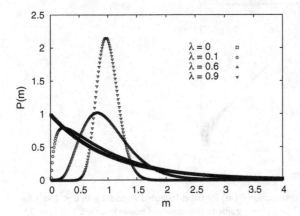

Figure 8.2 Steady state money distribution $P(m)$ for the model with uniform savings. The data shown are for different values of λ: 0, 0.1, 0.6, 0.9 for a system size $N = 100$. All datasets shown are for average money per agent $M/N = 1$.

where $\Gamma(n)$ is the gamma function whose argument n is related to the savings factor λ as

$$n = 1 + \frac{3\lambda}{1 - \lambda} . \tag{8.41}$$

This result has also been supported by numerical results in [20]. However, a later study [21] analyzed the moments, and found that moments up to the third order agree with those obtained from the form of (8.41), and discrepancies start from the fourth order onwards. Hence, the actual form of the distribution for this model still remains to be found.

It seems that a very similar model was proposed by Angle [22, 23] several years back in sociology journals. Angle's one parameter inequality process model (OPIP) is described by the equations:

$$m_i(t + 1) = m_i(t) + D_t w m_j(t) - (1 - D_t) w m_i(t)$$
$$m_j(t + 1) = m_i(t) + (1 - D_t) w m_i(t) - D_t w m_j(t) , \tag{8.42}$$

where w is a fixed fraction and D_t takes value 0 or 1 randomly. The numerical simulation results of OPIP fit well with gamma distributions.

In the gas-like models with uniform savings, the distribution of wealth shows a self-organizing feature. A peaked distribution with a most-probable value indicates an economic scale. Empirical observations in homogeneous groups of individuals as in waged income of factory laborers in the UK and US [24], and data from a population survey in the US among students of different school and colleges produce similar distributions [23]. This is a simple case where a homogeneous population (say, characterized by a unique value of λ) has been identified.

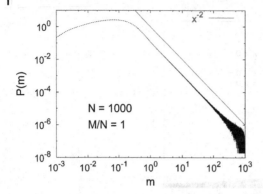

Figure 8.3 Steady-state money distribution $P(m)$ for the distributed λ model with $0 \leq \lambda < 1$ for a system of $N = 1000$ agents. The x^{-2} is a guide to the observed power law, with $1 + \nu = 2$. Here, the average money per agent $M/N = 1$.

8.3.2
Model with Distributed Savings

In a real society or economy, the interest of saving varies from person to person, which implies that λ is a very inhomogeneous parameter. To move a step closer to this real situation, one considers the saving factor λ to be widely distributed within the population [26–28]. The evolution of money in such a trading can be written as

$$m_i(t+1) = \lambda_i m_i(t) + \epsilon_{ij}\left[(1-\lambda_i)m_i(t) + (1-\lambda_j)m_j(t)\right], \tag{8.43}$$

$$m_j(t+1) = \lambda_j m_j(t) + (1-\epsilon_{ij})\left[(1-\lambda_i)m_i(t) + (1-\lambda_j)m_j(t)\right]. \tag{8.44}$$

The trading rules are similar to the previous rules, except that

$$\Delta m = \epsilon_{ij}(1-\lambda_j)m_j(t) - (1-\lambda_i)(1-\epsilon_{ij})m_i(t), \tag{8.45}$$

where λ_i and λ_j are the saving propensities of agents i and j. The agents have fixed (over time) saving propensities, distributed independently, randomly and uniformly (white) within an interval 0 to 1: agent i saves a random fraction λ_i ($0 \leq \lambda_i < 1$) and this λ_i value is quenched for each agent, that is λ_i are independent of trading or t.

The distribution $P(m)$ of money M is found to follow a strict power-law decay, which fits to Pareto law with $\nu = 1.01 \pm 0.02$ (Figure 8.3). This power law is extremely robust. For a distribution

$$\rho(\lambda) \sim |\lambda_0 - \lambda|^\alpha, \quad \lambda_0 \neq 1, \quad 0 < \lambda < 1, \tag{8.46}$$

of quenched λ values among the agents, the Pareto law with $\nu = 1$ is universal for all α. The data in Figure 8.3 corresponds to $\lambda_0 = 0$, $\alpha = 0$.

The role of the agents with high saving propensity ($\lambda \to 1$) is crucial: the power law behavior is truly valid up to the asymptotic limit if 1 is included. Indeed, had we

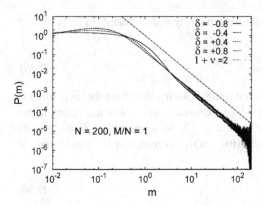

Figure 8.4 Steady-state money distribution $P(m)$ in the model for $N = 200$ agents with λ distributed as $\rho(\lambda) \propto \lambda^{\alpha}$ with different values of α. For all cases, the average money per agent $M/N = 1$.

assumed $\lambda_0 = 1$ in (8.46), the Pareto exponent ν immediately switches over to $\nu = 1 + \alpha$. Of course, $\lambda_0 \neq 1$ in (8.46) leads to the universality of the Pareto distribution with $\nu = 1$ (irrespective of λ_0 and α). Obviously, $P(m) \sim \int_0^1 P_\lambda(m)\rho(\lambda)d\lambda \sim m^{-2}$ for $\rho(\lambda)$ given by (8.46) and $P(m) \sim m^{-(2+\alpha)}$ if $\lambda_0 = 1$ in (8.46) (for large m values).

Patriarca et al. [32] studied the correlation between the saving factor λ and the average money held by an agent whose savings factor is λ. This numerical study revealed that the product of this average money and the unsaved fraction remains constant, or in other words, the quantity $\langle m(\lambda)\rangle(1 - \lambda)$ is a constant. This result turns out to be the key to the formulation of a mean field analysis to the model [33, 34].

In a recent mean field approach [33, 34], one can calculate the distribution for the ensemble average of money for the model with distributed savings. It is assumed that the distribution of money of a single agent over time is stationary, which means that the time-averaged value of money of any agent remains unchanged independent of the initial value of money. Taking the ensemble average of all terms on both sides of (8.43), one can write

$$\langle m_i \rangle = \lambda_i \langle m_i \rangle + \langle \epsilon \rangle \left[(1 - \lambda_i)\langle m_i \rangle + \left\langle \frac{1}{N}\sum_{j=1}^{N}(1 - \lambda_j)m_j \right\rangle \right]. \tag{8.47}$$

The last term on the right is replaced by the average over the agents, where it is assumed that any agent (ith agent here) on the average, interacts with all others in the system, which is the *mean field* approach.

Writing

$$\overline{\langle (1 - \lambda)m \rangle} \equiv \left\langle \frac{1}{N}\sum_{j=1}^{N}(1 - \lambda_j)m_j \right\rangle \tag{8.48}$$

and since ϵ is assumed to be distributed randomly and uniformly in $[0, 1]$, so that $\langle \epsilon \rangle = 1/2$, (8.47) reduces to

$$(1 - \lambda_i)\langle m_i \rangle = \overline{\langle (1 - \lambda)m \rangle} \ . \tag{8.49}$$

Since the right side is free of any agent index, it seems that this relation is true for any arbitrary agent, that is $\langle m_i \rangle (1 - \lambda) = $ constant, where λ is the saving factor of the ith agent. What follows is, $d\lambda = $ const. $\frac{dm}{m^2}$. An agent with a (characteristic) saving propensity factor (λ) ends up with wealth (m) such that one can in general relate the distributions of the two

$$P(m)dm = \rho(\lambda)d\lambda \ . \tag{8.50}$$

Therefore, the distribution in m is bound to be of the form

$$P(m) \propto \frac{1}{m^2} \ , \tag{8.51}$$

for uniform distribution of savings factor λ, that is $\nu = 1$. This analysis can also explain the nonuniversal behavior of the Pareto exponent ν, that is $\nu = 1 + \alpha$ for $\rho(\lambda) = (1 - \lambda)^\alpha$. Thus, this mean field study explains the origin of the universal $(\nu = 1)$ and the nonuniversal $(\nu \neq 1)$ Pareto exponents in the distributed savings model.

These model income distributions $P(m)$ compare very well with the wealth distributions of various countries: data suggests Gibbs-like distribution in the low-income range (more than 90% of the population) and Pareto-like in the high-income range [29–31] (less than 10% of the population) of various countries. In fact, we compared one model simulation of the market with saving propensity of the agents distributed following (8.46), with $\lambda_0 = 0$ and $\alpha = -0.7$ [26]. The qualitative resemblance of the model income distribution with the real data for Japan and the US in recent years is quite intriguing. In fact, for negative α values in (8.46), the density of traders with low saving propensity is higher and since a $\lambda = 0$ ensemble yields a Gibbs-like income distribution (8.35), we see an initial Gibbs-like distribution which crosses over to Pareto distribution with $\nu = 1.0$ for large m values. The position of the crossover point depends on the value of α. It is important to note that any distribution of λ near $\lambda = 1$, of finite width, eventually gives the Pareto law for large m limit. The same kind of crossover behavior (from Gibbs to Pareto) can also be reproduced in a model market of mixed agents where $\lambda = 0$ for a finite fraction p of the population and λ is distributed uniformly over a finite range near $\lambda = 1$ for the rest $1 - p$ fraction of the population.

In recent years, several papers discuss these models at length and provide rigorous analysis of the models and related ones [35–39] using a variety of approaches like Fokker–Planck equations and generalized Boltzmann transport equations. Several issues regarding the structure and dynamics of such models are now known.

References

1 Chatterjee, A., Yarlagadda, S., and Chakrabarti, B.K. (eds) (2005) *Econophysics of Wealth Distributions*, Springer, Milan.

2 Ispolatov, S., Krapivsky, P.L., and Redner, S. (1998) *Eur. Phys. J. B*, **2**, 267.

3 Hayes, B. (2002) *American Scientist*, **90**, 400.

4 Chakraborti, A. (2002) *Int. J. Mod. Phys. C*, **13**, 1315.

5 Sinha, S. (2003) *Phys. Scripta T*, **106**, 59.

6 Sinha, S. in [1], p. 177.

7 Drăgulescu, A.A. and Yakovenko, V.M. (2000) *Eur. Phys. J. B*, **17**, 723.

8 Mandelbrot, B.B. (1960) *International Economic Review*, **1**, 79.

9 Kiyotaki, N. and Wright, R. (1993) *Am. Econ. Rev.*, **83**, 63.

10 Bak, P., Nørrelykke, S.F., and Shubik, M. (1999) *Phys. Rev. E*, **60**, 2528.

11 Molico, M. (2006) *Int. Econ. Rev.*, **47**, 701.

12 Chakrabarti, B.K. and Marjit, S. (1995) *Ind. J. Phys. B*, **69**, 681.

13 Chakraborti, A. and Chakrabarti, B.K. (2000) *Eur. Phys. J. B*, **17**, 167.

14 Moss de Oliveira, S., de Oliveira, P.M.C., and Stauffer, D. (1999) *Evolution, Money, War and Computers*, B.G. Tuebner, Stuttgart, Leipzig.

15 Xi, N., Ding, N., and Wang, Y. (2005) *Physica A*, **357**, 543.

16 Fischer, R. and Braun, D. (2003) *Physica A*, **321**, 605.

17 Samuelson, P.A. (1980) *Economics*, McGraw Hill Int., Auckland.

18 Das, A. and Yarlagadda, S. (2003) *Phys. Scripta T*, **106**, 39.

19 Patriarca, M., Chakraborti, A., and Kaski, K. (2004) *Phys. Rev. E*, **70**, 016104.

20 Bhattacharya, K., Mukherjee, G., and Manna, S.S. in [1].

21 Repetowicz, P., Hutzler, S., and Richmond, P. (2005) *Physica A*, **356**, 641.

22 Angle, J. (1986) *Social Forces*, **65**, 293.

23 Angle, J. (2006) *Physica A*, **367**, 388.

24 Willis, G. and Mimkes, J. cond-mat/0406694.

25 Chakraborti, A., Pradhan, S., and Chakrabarti, B.K. (2001) *Physica A*, **297**, 253.

26 Chatterjee, A., Chakrabarti, B.K., and Manna, S.S. (2004) *Physica A*, **335**, 155.

27 Chatterjee, A., Chakrabarti, B.K., and Manna, S.S. (2003) *Phys. Scripta T*, **106**, 36.

28 Chakrabarti, B.K. and Chatterjee, A. (2004) in *Application of Econophysics*, (ed. H. Takayasu), Springer, Tokyo, pp. 280.

29 Drăgulescu, A.A. and Yakovenko, V.M. (2001) *Physica A*, **299**, 213.

30 Levy, M. and Solomon, S. (1997) *Physica A*, **242**, 90.

31 Aoyama, H., Souma, W., and Fujiwara, Y. (2003) *Physica A*, **324**, 352.

32 Patriarca, M., Chakraborti, A., Kaski, K., and Germano, G. in [1], pp. 93–110.

33 Mohanty, P.K. (2006) *Phys. Rev. E*, **74**, 011117.

34 KarGupta, A. (2006) in *Econophysics and Sociophysics*, (eds B.K. Chakrabarti, A. Chakraborti, A. Chatterjee), Wiley-VCH, Berlin, p. 161.

35 Düring, B. and Toscani, G. (2007) *Physica A*, **384**, 493.

36 Düring, B., Matthes, D., and Toscani, G. (2008) *Physica A*, **78**, 056103.

37 Matthes, D. and Toscani, G. (2008) *J. Stat. Phys.*, **130**, 1087.

38 Matthes, D. and Toscani, G. (2008) *Kinetic and related Models*, **1**, 1.

39 Comincioli, V., Della Croce, L., and Toscani, G. (2009) *Kinetic and related Models*, **2**, 135.

9

But Individuals are not Gas Molecules …

"Science can only ascertain what is, but not what should be, and outside of its domain value judgments of all kinds remain necessary."

– Albert Einstein

9.1

Agent-Based Models: Going beyond the Simple Statistical Mechanics of Colliding Particles

One might actually wonder how the mathematical theories and laws which try to explain the physical world of electrons, protons, atoms, and molecules can be applied to understand the complex social structure and economic behavior of human beings. This is especially so because human beings are complex and widely varying in their nature, properties, or characteristics, whereas, for example, electrons are identical (that is, we do not have electrons of multiple types) and indistinguishable. Each electron has a charge of 1.60218×10^{-19} C, a mass of 9.10938×10^{-31} kg, and is a spin-1/2 particle. Moreover, such properties of electrons are universal (identical for all electrons), and this amount of information is sufficient to explain many physical phenomena concerning electrons, once you know the interactions. But is such little information sufficient to infer the complex behavior of human beings? Is it possible to quantify the nature of the interactions between human beings? The answers to these questions are on the negative. Nevertheless, during the past decade physicists have made attempts at studying problems in economics, the "social science that analyses and describes the consequences of choices made concerning scarce resources".

We have already seen in the previous chapters models of wealth exchange between individuals or economic entities, referred to as kinetic wealth exchange models (KWEMs), since they provide a description of wealth flow in terms of stochastic wealth exchange between agents, resembling the energy transfer between the molecules of a fluid. We do realize that even though KWEMs have been the subject of intensive investigations, their economic interpretation is still an open problem. It is important to keep in mind that in the framework of a KWEM the agents should

Econophysics. Sitabhra Sinha, Arnab Chatterjee, Anirban Chakraborti, and Bikas K. Chakrabarti
Copyright © 2011 WILEY-VCH Verlag GmbH & Co. KGaA, Weinheim
ISBN: 978-3-527-40815-3

not be related to the "rational" agents of neoclassical economics: an interaction between two agents does not represent the effect of decisions taken by two economic
agents who have full information about the market and behave rationally in order
to maximize their utility. The description of wealth flow provided by KWEMs takes
into account the stochastic element, which does not respond by definition to any
rational criterion. Also, some terms employed in the study of KWEMs, such as saving propensity and risk aversion, should be taken with caution since they might
seem to imply a decisional aspect behind the behavior of agents. However, it is
interesting to note that very recently Chakrabarti and Chakrabarti [1] have put forward a microeconomic formulation of the above models using the utility function
as a guide to the behavior of agents in the economy, which might actually bring an
even greater cross-exchange of ideas.

Of course, the behavior of most of the complex systems found in natural and
social environments can be characterized by the competition among interacting
agents for scarce resources and their adaptation to the environment [2, 3]. The different agents could be diverse in form and in capability, for example, cells in an
immune system to great firms in a business center according to the system considered. In these dynamically evolving complex systems the nature of agents and their
manners also differ. In order to have a deeper understanding of the interactions
of the large number of agents, one should study the capabilities of the individual
agents. For simplicity, an agent's behavior may be thought of as a collection of rules
governing "responses" to "stimuli". For example, a typical response of an animal
(prey) when it sees a predator, is that it should run, or if the stock indices fall, then
a financial agent should take immediate action (sell/buy), and so on. Therefore,
in order to model any complex dynamically adaptive system, a major concern is
the selection and representation of the stimuli and responses, since the behavior
and strategies of the component agents are thus determined. In a behavioral model, the rules of action are a straightforward way to describe agents' strategies. One
studies the behavior of the agents by looking at the rules acting sequentially. Then
one considers "adaptation", which is described in biology as a process by which an
organism tries to fit itself into its environment on the basis of Charles Darwin's
theory of evolution: the survival of the fittest and natural selection. The timescales
over which the agents adapt will, of course, vary from one system to another. Thus,
for example, adaptive changes in the immune system take hours to days, adaptive
changes in a financial firm usually take months to years, and the adaptive changes
in the ecosystem require years to several millennia. It is noteworthy that in complex
adaptive systems, a major part of the environment of a particular agent includes
other adaptive agents. Thus, a considerable amount of an agent's effort goes into
the adaptation to other agents. This feature is the main source of the interesting
temporal patterns that these complex adaptive systems produce. For example, in
financial markets, human beings react with strategy and foresight by considering
outcomes that might result as a consequence of their behavior. This brings in a
new dimension to the system, namely "rational" actions, which are not innate to
agents in natural environments. To handle this new dimension, the use of game
theory has become quite natural. It helps in making decisions when a number of

rational agents are involved under conditions of conflict and competition. However, game theory and other conventional theories in economics study patterns in behavioral equilibrium that induce no further interaction. These consistent patterns are quite different from the temporal patterns that complex adaptive systems produce. Hence, models of physical systems that exhibit self-organization and cooperative phenomena can also be used to model such complex systems. In order to model these features of complex systems, one goes beyond the use of simple statistical mechanics, and in many cases, devise multi-agent models to mimic the behavior of complex systems. Interestingly, in traditional economic models there is usually the "representative agent" who has boundless foresight and capabilities (termed as "perfect rationality"), and uses the "utility maximization" principle to act economically. Though the economic activities of the agents are driven by various considerations like utility maximization, in the KWEMs mentioned earlier, the eventual exchanges of money in any trade can be simply viewed as money/wealth conserving two-body scatterings, as in the entropy maximization-based kinetic theory of gases. Recently, an equivalence between the two maximization principles have been quantitatively established [1]. This can be regarded as a very important step in putting the two fields closer. The "rational agent" paradigm in economics is accompanied by another reductionism which goes like this: Since there is a single way to act perfectly rationally, all the agents should display exactly the same behavior, and so a "representative agent" would be sufficient to model. To summarize, the typical format of current economic models is that of a single agent or firm maximizing its utility or profit over a finite or infinite period. Now, concepts like "bounded rationality", and also the multi-agent models that have originated from simple statistical physics considerations, have allowed one to go beyond the prototype theories with a representative agent in traditional economics. The recent failure of economists to anticipate the collapse of markets worldwide since 2007 over a short period of time has led to some voices from within the field of economics itself suggesting that new foundations for the discipline are required [4]. There are several review articles on multi-agent models [5, 6]. In this chapter, we study a few models which have some of these features and are attempting to address such issues.

9.2
Explaining the Hidden Hand of Economy: Self-Organization in a Collection of Interacting "Selfish" Agents

9.2.1
Hidden Hand of Economy

An economic market is perhaps the most commonly encountered self-organizing system or network, whose dynamics profoundly affects us all. Its dynamics is no doubt very intriguing. In fact Adam Smith first made the formal notice of a self-organizing aspect, which he called the 'invisible hand' effect, of the market consisting of selfish agents [7, 8]. Although it still remains illusive to demonstrate that a pure competitive market, consisting of agents pursuing pure self-interest, can self-organize or reach (dynamic) equilibrium, the mainstream economists seem to consider it to be true in principle (a matter of faith?) [7].

In 1759, Adam Smith published his first work, "The Theory of Moral Sentiments", which he continued to revise until his death in 1790. Published in 1776, "An Inquiry into the Nature and Causes of the Wealth of Nations" was Adam Smith's *magnum opus*, which was in fact an account of political economy written at the dawn of the industrial revolution. It is widely considered to be the first modern work in the field of economics. There are three main concepts that Adam Smith expanded upon in his work that form the foundation of free market economics: division of labor, pursuit of self-interest, and freedom of trade. Although "The Wealth of Nations" is widely regarded as Smith's most influential work, Smith himself considered the "The Theory of Moral Sentiments" as a much superior work. It was actually in this book (The Theory of Moral Sentiments) that Smith first referred to the "invisible hand". Smith wrote:

> ". . . In spite of their natural selfishness and capacity, though they mean only their own convenience, though the sole end which they propose . . . be the gratification of their own vain and insatiable desires, they divide with the poor the produce of all their improvements. They are led by an invisible hand to make nearly the same distribution of the necessaries of life, which would have been made, had the earth been divided into equal portions among all its inhabitants, and thus without intending it, without knowing it, advance the interest of the society."

Adam Smith's primary aim was to explain the source of mankind's ability to form moral judgements, in spite of natural inclinations of a person towards "self-interest".

In various statistical physics models of interacting systems or networks, such self-organization has indeed been demonstrated to emerge in the global aspects of the system which consists of a large number of simple dynamical elements having local (in time and space) interactions and dynamics [9]. It was also noted by the economists a long time back that these dynamics, which takes the system to equilibrium, is greatly facilitated by "paper money" (rather than the direct commodity exchanges as in barter economy) which does not have any value of its own, but rather can be considered as a good 'lubricant' in the econodynamics [8]. Consistent with this analogy, it was also seen that when the (paper) money supply is changed, it does not just scale the commodity prices up (for increased money supply) or down (for decreased money supply), the (self-organizing) dynamics toward equilibrium is seriously affected [10].

9.2.2
A Minimal Model

We present here a self-organizing feature of a model market's dynamics, where the agents "trade" for a single commodity with their money [11]. We demonstrate that the model, apart from having self-organizing behavior, has a crucial role for the money supply in the market, and that its self-organizing behavior is significantly affected when the money supply becomes less than optimal. We also observed that this optimal money supply level of the market depends on the amount of "frustration" or scarcity in the commodity market. In our model, each agent having a commodity less than the "subsistence" level trades with any other having more than the "subsistence" level, in exchange for money. Specifically, we consider a closed economic system consisting of N economic agents, each economic agent i has, at any time, money m_i and commodity q_i, such that ($\sum_{i=1}^{N} m_i = M$ and $\sum_{i=1}^{N} q_i = Q$), where N, M and Q are fixed. The "subsistence" commodity level for each agent is q_0. Hence, at any time, agents having $q_i < q_0$ will attempt to trade, using their money m_i at that time, with agents having commodity more than q_0, and will purchase to make their commodity level q_0 (and no further), if the money permits. The agents with $q_i > q_0$ will sell off the excess amount to such "hungry" agents having $q_i < q_0$, and will attempt to maximize their wealth (money). This dynamic is local in time ("daily") and it stops eventually when no further trade is possible satisfying the above criteria. We introduce an "annual", or long-time, dynamic when some random fluctuations occur in the money and commodity of all the agents. Annually, each agent gets a minor reshuffle of their money and commodity (perhaps due to the noise in the stock market, and in the harvest due to the changes in the weather respectively). This (short- and long-time) combined dynamic is similar to that of the "sand-pile" models studied extensively in recent times [9]. The price of the commodity does not change in our model with the money supply M in the market; it remains fixed here (at unity). We look for the steady-state features of this market; in particular, the distributions $P(m)$ and $P(q)$ of the money and commodity respectively among the agents. We have investigated how many agents $P(q_0)$ can satisfy their basic needs through this dynamic, that is, can reach the subsistence level q_0, as a function of the money supply M for both unfrustrated ($g < 1$) (where $g = q_0/\langle q \rangle$ and $\langle q \rangle = Q/N$ is the average commodity in the market) and frustrated ($g > 1$) cases of the commodity market. We observed that an optimum supply M_0 of money is required for evolving the market towards the maximum possible value of $P(q_0)$. This optimum value of money M_0 is observed to decrease with an increase in g in the market.

9.2.2.1 Unlimited Money Supply and Limited Supply of Commodity

Here, we consider the money supply M in the market to be infinitely large and it therefore drops out from any consideration. The dynamic is then entirely governed by the commodity distribution among agents: for agents with $q_i < q_0$, the attempt will be to find another trade partner having $q_i > q_0$; and having found such partners, through a random search in the market, trades occur for mutual benefit (for the selling agent we still consider the extra money from trade to be important). The system thus evolves towards its steady-state, as the fixed-point feature of the short-time or daily dynamics gets affected by the random noise reshuffling in the commodity of each agent. This reshuffling essentially induces a Gibbs-like distribution [12, 13]. The trade dynamics are clearly motivated or "directed". We look for the combined effect on the steady-state distribution of commodity $P(q)$, which is independent of the initial commodity distribution among the agents.

For the unfrustrated case ($g < 1$), where *all* the agents can be satisfied, the typical distributions $P(q)$ are shown in Figure 9.1 for different values of g. We see that the $P(q)$ is Gibbs-like ($P(q) = A\exp(-q/\langle q\rangle)$ and $A = 1 - g$), for $q > q_0$, while $P(q_0) = g$ (as shown in the inset). One can easily explain these observations using the fact that the cumulative effect of the long-time randomization gives Gibbs distribution ($\exp(-q/\langle q\rangle)$) for all q. We then estimate the final steady-state distribution $P(q)$ from the additional effect of the short-time dynamics on this (long-time dynamics induced) Gibbs distribution. All the agents with $q < q_0$ manage here to acquire a q_0-level of commodity (as $g < 1$ and everybody has enough money to purchase the required amount). Their number is then given by $N_- = \int_0^{q_0} \exp(-q)dq$. They require the total amount of commodity $q_0 N_-$. The amount of commodity already available with them is given by $Q_- = \int_0^{q_0} q\exp(-q)dq$. The excess amount required $Q_{demand} = q_0 N_- - Q_-$ has to come from the agents having $q > q_0$. The average of the excess amount of commodity of the agents who are above the q_0-line is given by $\langle q_{excess}\rangle = (1 - Q_- - (1 - N_-)q_0)/(1 - N_-)$. The number of agents who supply the Q_{demand} amount is given by $N_+ = Q_{demand}/\langle q_{excess}\rangle$. This gives $P(q_0) = N_- + N_+ = g$. We can easily determine the prefactor of the final steady-state distribution $P(q)$ for $q > q_0$, $A = 1 - g$ from the conservation of total number of agents and total commodity.

For the frustrated case ($g > 1$), the results are shown in Figure 9.1. A similar calculation for $P(q_0)$ is done as follows: $N_+ = \int_{q_0}^{\infty} \exp(-q)dq$ is the num-

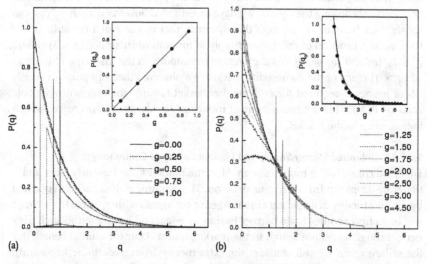

(a) q   (b) q

Figure 9.1 (a) The distributions of commodity $P(q)$ for different values of g for $N = 1000$, $Q = 1$, $M = 100$ ($M > M_0(g)$), for the unfrustrated case ($g < 1$). The steady-state distribution of commodity $P(q)$ is Gibbs-like: $P(q) = A\exp(-q/\langle q\rangle)$ with $A = 1 - g$, for $q > q_0$. The inset shows the linear variation of $P(q_0)$ with $g(P(q_0) = g)$. (b) The distributions of commodity $P(q)$ for different values of g for $N = 1000$, $Q = 1$, $M = 100$ ($M > M_0(g)$), for the frustrated case ($g > 1$). The variation of $P(q_0)$ with g is shown in the inset where the theoretical estimate ($P(q) = g\exp(-g)/(g - 1 + \exp(-g))$) is also indicated by the solid line.

ber of people above the q_0-line who will sell off their excess amount of commodity to come to q_0-level, $Q_+ = \int_{q_0}^{\infty} q \exp(-q) dq$ is the commodity of the agents above the q_0-level. Then the supplied amount of commodity to the agents below the q_0-line is $Q_{supply} = Q_+ - q_0 N_+$. The average of the deficit commodity, $\langle q_{deficit} \rangle = ((1-N_+)q_0 - 1 + Q_+)/(1-N_+)$. Hence, the number of agents who will reach q_0-level from below is $N_- = Q_{supply}/\langle q_{deficit} \rangle$, so that $P(q_0) = N_+ + N_- = g \exp(-g)/(g - 1 + \exp(-g))$. A comparison of this estimate for $P(q_0)$ with g is also shown in the inset of Figure 9.1. It may be mentioned that in absence of the strict Gibbs distribution for $P(q)$ ($q < q_0$), the above expression for $P(q_0)$ is somewhat approximate.

9.2.2.2 Limited Money Supply and Limited Supply Of Commodity

When the money supply is limited, the self-organizing behavior is significantly affected and the fraction of agents who can secure q_0 amount of commodity for themselves, $P(q_0)$, does not always reach its maximum possible value (as suggested by the amount of commodity available in the market). For all values of g, as we increase the money supply in the market M, $P(q_0)$ increases and then after a certain amount M_0, it saturates. In Figure 9.2, we have shown how the quantity $P(q_0)$ varies with M for different values of g (for $g > 1$ only, as we are more interested in the frustrated case). We define M_0 to be the optimum amount of money sup-

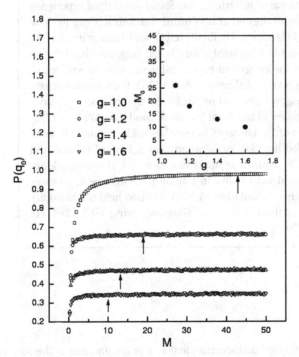

Figure 9.2 The variation of quantity $P(q_0)$ with M for different values of g in the frustrated cases ($g > 1$). The inset shows the variation of M_0 with g.

ply needed for the smooth functioning of the market. We also observed that this optimal money supply level of the market depends on the amount of frustration g in the commodity market. In the inset, the variation of M_0 with g is shown for the frustrated case ($g > 1$) only.

Adam Smith
Adam Smith (1723–1790) was a Scottish economist and philosopher. He is known primarily as the author of two treatises: *The Theory of Moral Sentiments* in 1759, and *An Inquiry into the Nature and Causes of the Wealth of Nations* (commonly known as *The Wealth of Nations*) in 1776. Smith is also known for his explanation of how rational self-interest and competition, following moral obligations in a society, can ultimately lead to economic well-being and prosperity. His work provided a solid base for modern economics and provided one of the best-known rationales for "free trade". Hence, he is also widely acknowledged as the "father of economics". Smith attended the University of Glasgow at the age of fourteen and studied moral philosophy under the supervision of Francis Hutcheson, where he developed his concepts and passion for liberty, reason, and free speech. In 1740, Smith left the University of Glasgow to attend Balliol College, Oxford. Smith considered the teaching at Glasgow to be far superior when compared to Oxford, and found his Oxford experience intellectually stifling. Toward the end of his Oxford stay, Smith began to show symptoms of a nervous breakdown. He finally left Oxford University in 1746, before the formal end of his scholarship. Smith had originally intended to study theology and enter the clergy, but he was influenced by the writings of David Hume, and finally opted a different route. In 1762, the academic senate of the University of Glasgow conferred on Smith the title of Doctor of Laws (LL.D.). He was elected fellow of the Royal Society of London in 1773. In 1775, he was elected a member of the Literary Club established by Samuel Johnson, whose members included the likes of Edmund Burke, Edward Gibbon and Joshua Reynolds. In 1778, Smith was appointed to a post as commissioner of customs in Scotland and went to live in Edinburgh, where he and others founded the Royal Society of Edinburgh in 1783. He also held the honorary position of Lord Rector of the University of Glasgow during 1787–1789. He died in Edinburgh in 1790.

9.3
Game Theory Models

Game theory, a branch of applied mathematics that was originally used in the social sciences (most notably economics, political science, international relations), is now applied to far-related subjects like biology, computer science, and philosophy.

Although some developments have taken place, it can considered that the field of game theory became firmly established with the 1944 book "Theory of Games and Economic Behavior" by John von Neumann and Oskar Morgenstern. This theory was then developed extensively in the 1950s by many scholars, most notably John Nash.

Game theory tries to capture behavior in "strategic" situations, in which an individual's success in making choices depends on the choices of others. It was initially developed to analyze competitions in which one individual does better at the expense of another ("zero sum" games). Now it has expanded to treat a wide class of interactions and situations. Traditionally, game theory attempts to find equilibria (e.g., the "Nash equilibrium" proposed by John Nash) in the games, where each player of the game adopts a strategy that they are unlikely to change. These equilibrium concepts originated differently depending on the field of application, though the "solution" concepts are usually based on what is required by norms of "rationality". There have been several criticisms of the methodology, and debates continue on the particular equilibrium concepts, and the applications of the mathematical models in general. However, economists have long used and applied successfully game theory to analyze a wide array of economic phenomena like auctions, bargaining, fair division, and oligopolies.

What is a game? The term "game" represents a competition or conflict between two or more members, where each opponent acts in a rational manner, trying to resolve the conflict in their own favor. The competitors in the game are usually referred to as *players*.

What is game theory? Game theory deals with decision making of a number of rational opponents under conditions of conflict and competition. It is applicable to situations that satisfy the following conditions:

1. The number of players is finite.
2. There is a conflict of interests between the players.
3. The players act rationally and intelligently.
4. The rules governing the choices are specified and known to all the players.
5. Each player has a finite set of possible courses of action.
6. The players make individual decisions without communicating directly and simultaneously select their respective courses of action.
7. The payoff (outcome) is fixed and determined in advance.

Concepts in game theory A *strategy* for a player is defined as a set of rules or alternative courses of action available to him in advance, by which a player decides the course of action he should adopt. Strategy may be either *pure* or *mixed*. If the players select the same strategy each time, then it is referred to as *pure strategy*. In this case, each player knows exactly what the other is going to do. In other words, it is a deterministic situation, where the objective of the players is to maximize gains or minimize losses, that is based on "utility maximization". When the players use a combination of strategies and each player is always kept guessing as to which course of action is to be selected by the other player at a particular occasion, then this is known as *mixed strategy*. Thus, there is a probabilistic situation and the objective of the player is to maximize expected gains or to minimize losses. In simple words, a mixed strategy can be considered a selection among pure strategies with fixed probabilities. A course of action or play which puts the player in the most "preferred" position, no matter what the strategies of his competitors are, is called an *optimum strategy*, and any deviation from this optimum strategy must result in a decreased payoff for the player.

When there are two competitors playing a game, it is called a *two person game*, and when the number of competitors are N (where $N > 2$), it is called an N-person game. When the sum of the amounts won by all winning agents is equal to the sum of the amounts lost by all losing agents, that is, the sum of gains and losses in the game vanish, it is called a *zero sum game*. On the contrary, if the sum of gains or losses is non-zero, we call it a *non-zero sum game*. When there is communication between the participants, they may reach an agreement and increase their payoffs through some form of *collaborative game*, otherwise it is a *non-collaborative game*.

The normal form of the game is generally represented by a matrix which shows the strategies and payoffs, along with the players. A two-person game is conveniently represented by a matrix as shown in Figure 9.3. In the example shown in Figure 9.3, there are two players: One chooses the row while the other chooses the column. Each player has exactly two strategies, specified by the number of rows and the number of columns. The payoffs are given inside the table. The first number is the payoff received by the Player 1 (row player in our example); the second is the payoff for the Player 2 (column player in our example). If Player 1 plays Up and Player 2 plays Left, then Player 1 gets a payoff of 4, and Player 2 gets 3. The other entries in the table can be interpreted similarly. When a game is represented in the normal form, it is assumed that each player acts simultaneously or, at least, without knowing the actions of the other.

	Player 2 chooses Left	Player 2 chooses Right
Player 1 chooses Up	4, 3	−1, −1
Player 1 chooses Down	0, 0	3, 4
	Normal form or payoff matrix of a 2-player, 2-strategy game	

Figure 9.3 Example of a two-person game expressed in normal form.

More generally, we can represent by any function that associates a payoff for each player with every possible combination of actions. The matrix, which shows the outcome of the game as the players A and B select their particular strategies, is known as the *payoff matrix*. It is important to assume that each player knows not only their own list of possible courses of action, but also of the opponent. Let player A have m courses of action $(A_1, A_2, \ldots, A_m)$ and player B have n courses of action $(B_1, B_2, \ldots, B_n)$. Therefore, the total number of possible outcomes is mn (the numbers m and n need not be equal).

The payoff matrices for players A and B are written as follows (with $b_{ij} = -a_{ij}$ for a zero-sum game):

Player A's payoff matrix

$$
\begin{array}{c}
\text{Player } B \\
\begin{array}{cccc}
B_1 & B_2 & \cdots & B_n
\end{array} \\
\text{Player } A \quad
\begin{array}{c}
A_1 \\ A_2 \\ \vdots \\ A_m
\end{array}
\begin{bmatrix}
a_{11} & a_{12} & \cdots & a_{1n} \\
a_{21} & a_{22} & \cdots & a_{2n} \\
\vdots & \vdots & \ddots & \vdots \\
a_{m1} & a_{m2} & \cdots & a_{mn}
\end{bmatrix}
\end{array}
$$

Player B's payoff matrix

$$
\begin{array}{c}
\text{Player } B \\
\begin{array}{cccc}
B_1 & B_2 & \cdots & B_n
\end{array} \\
\text{Player } A \quad
\begin{array}{c}
A_1 \\ A_2 \\ \vdots \\ A_m
\end{array}
\begin{bmatrix}
b_{11} & b_{12} & \cdots & b_{1n} \\
b_{21} & b_{22} & \cdots & b_{2n} \\
\vdots & \vdots & \ddots & \vdots \\
b_{m1} & b_{m2} & \cdots & b_{mn}
\end{bmatrix}
\end{array}
$$

The concept of Nash equilibrium   We say that a pair of strategies is in *Nash equilibrium* if A's choice is optimum given B's choice, and B's choice is optimum given A's choice. However, neither of them know what the other person will do when they have to make their own choice of strategy. However, each has an expectation of what the other's choice would be. A Nash equilibrium may be interpreted as a pair of expectations about each person's choice such that when the other person's choice is revealed, neither individual wants to change their behavior. Note that a game can have more than one Nash equilibrium. On the other hand, there are also games which have no Nash equilibrium of the above type.

Finally, note that there may be *repeated* games. If the game is to be played repeatedly by the same players, new strategic possibilities are open to each player, so that if one player chooses to defect in one round, the other player can choose to defect in the next round. Thus, the opponent can be "punished" for "bad" behavior. Thus, in a repeated game, each player has the opportunity to establish a reputation for "cooperation", and thereby encourage the other players to do the same. Whether this kind of strategy will be advantageous, however, depends on whether the game is going to be played a *fixed* number of times or an *indefinite* number of times. If the game is going to be repeated an indefinite number of times, then one has a way of influencing the opponent's behavior; for example, if A refuses to cooperate one time, B can refuse to cooperate next time also.

The prisoner's dilemma   The prisoner's dilemma is one of the traditionally studied games, originally formulated by M. Flood and M. Dresher in 1950. It was A.W. Tucker who formalized the game with prison sentence payoffs and gave it the name "prisoner's dilemma". In its traditional form, the prisoner's dilemma may be thought to be as follows:

> Two suspects, Bunty and Babli (names borrowed from two characters in a popular Bollywood Hindi movie "Bunty Aur Babli"), are arrested by the police. The police have insufficient evidence for a conviction. Therefore, having separated both the prisoners, they visit each one of them to offer the same deal: "You may choose to confess or remain silent. If you confess and your accomplice remains silent we will drop all charges against you and use your testimony to ensure that your accomplice does serious time. Likewise, if your accomplice confesses while you remain silent, your accomplice will go free while you do the time. If you both confess, we get two convictions, but we will see to it that you both get early parole. If you both remain silent, we will have to settle for token sentences on petty theft charges. If you wish to confess, you must leave a note with the jailer before our return tomorrow morning." What should Bunty and Babli do?

If it is assumed that each player only cares about "maximizing utility" by *minimizing* his or her own time in jail, then the prisoner's dilemma forms a classic non-zero sum game in which two players may cooperate with each other, or not cooperate by betraying the other player. It is interesting to note that the unique equilibrium for this game is the Pareto-suboptimal[1] solution. Thus, the rational choice makes the two players to both defect (betray), even though each player's individual reward would have been greater if they both cooperated. Note that in the classic form of the game, cooperating is strictly dominated by defecting, such that the only possible equilibrium for the game is the case where both players defect. Irrespective of what one player does, the other player will always gain a greater payoff by defecting. Since in all situations defecting is more beneficial than cooperating, all "rational" players will defect, all other things being equal.

1) Pareto optimality or Pareto efficiency, is an important concept in economics with broad applications in game theory and the social sciences. The term is named after Vilfredo Pareto, the Italian economist who used the concept in his studies of economic efficiency, but was more famous for his studies of income distribution. A situation is said to be Pareto efficient when any change to make any person better off is not possible without making someone else worse off. Given a set of alternative allocations of, say, goods or income for a set of individuals, a change from one allocation to another that can make at least one individual better off without making any other individual worse off is called a Pareto improvement. An allocation is defined as Pareto efficient or Pareto-optimal when further Pareto improvements are not possible.

9.3.1
Minority Game and its Variants (Evolutionary, Adaptive and so on)

9.3.1.1 El Farol Bar Problem

In 1994, Brian Arthur introduced the "El Farol Bar" problem as a paradigm of complex economic systems. Based on a bar in Santa Fe, New Mexico, the problem was defined originally as follows:

> "*N* people decide independently each week whether to go to a bar that offers enter-tainment on a certain night. For concreteness, let us set *N* at 100. Space is limited, and the evening is enjoyable if things are not too crowded – specifically, if fewer than 60 % of the possible 100 are present. There is no way to tell the numbers coming for sure in advance, therefore a person or agent: goes – deems it worth going – if he expects fewer than 60 to show up, or stays home if he expects more than 60 to go."
>
> – W. B. Arthur (see [14])

Thus, the problem is simply stated as follows: there is a finite group of people, all of whom want to go to the El Farol Bar on every Thursday night. However, the El Farol is quite small, and it is not pleasing to go there if it is over-crowded. So, the preferences of the population can be described as follows:

- If less than 60 % of the population go to the bar, they all have a better time than if they stayed at home.
- If 60 % or more of the population go to the bar, they all have a worse time than if they stayed at home.

As a matter of fact, it is necessary for everyone to decide simultaneously whether they will go to the bar or not. Before making their own decisions, they cannot wait and see how many others go on a particular Thursday.

An important aspect of the problem is that, if there was an obvious method that all individuals could use to decide, then it would be possible to find a "deductive" solution to the problem. However, irrespective of what method each individual uses to decide whether they will go to the bar or not, if everyone uses the same method, then the method is bound to fail. Hence, from the point of view of the individual, the problem is ill-defined and no deductive rational solution exists.

Situations like the above represented by this problem highlight two specific reasons why perfect deductive reasoning might fail to provide clear solutions to such theoretical problems:

- It is simply a question of the limitations of the mind. Beyond a certain level of complexity, logical capacity fails to cope.
- It is that, in complex strategic situations, individuals cannot always rely on persons they are interacting with to behave under assumptions of perfect rationality.

Here, the individuals are forced to choose their strategies by guessing the likely behavior of their opponents. These types of problems become ill-defined and cannot be solved rationally. The question that arises is: "How does one best model bounded rationality in economics when perfect rationality fails?" Thus, the problem itself actually provides a useful framework to explore models of bounded rationality in general. By "bounds on rationality" we mean, for example, the agents:

- do not have perfect information about their environment; in general they will only acquire information through interaction with the dynamically changing environment
- do not have a perfect model of their environment
- have limited computational power, so they can't work out all the logical consequences of their knowledge
- have other resource limitations (e.g., memory).

Arthur investigated this model of the El Farol Bar problem through the use of computational experiments. He designed artificial agents and simulated their dynamic interaction over time. In Arthur's computer simulations, the size of the population, N, was set to 100 and the enjoyable capacity of the El Farol bar, C, was set to 60. Arthur then created a set of diverse forecasting models, or predictors, which map attendance histories to a predicted bar attendance for the coming week. These models or predictors were distributed uniformly and randomly, such that each agent was assigned with a non-transferable set of K forecasting models. Each simulation experiment was then run for 100 periods with the combined runs totalling 10 000 periods. An example of the attendance rates from a typical run of 100 periods can be seen in Figure 9.4. From the results of these computer experiments one notes that, given the initial conditions and the fixed set of forecasting models or predictors available to each simulated agent, the dynamics are completely deterministic. Two observations become immediately apparent:

- First, mean attendance always converges to the capacity of the bar.
- Second, on average 40 % of the active predictors forecasted attendance to be higher than the capacity level and 60 % below.

Arthur expanded on these observations by noting that, "the predictors self-organize into an equilibrium pattern".

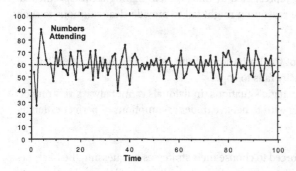

Figure 9.4 Bar attendance in the first 100 weeks as simulated by Arthur [14].

9.3.1.2 Basic Minority Game

As Arthur wrote in the Foreword of the book written by Challet, Zhang and Marsili [15]:

"The Minority Game grew out of my El Farol Bar problem... To me, El Farol was not a problem of how to arrive at a coordinated solution (although the Minority Game very much is). I saw it as a conundrum for economics: How do you proceed analytically when there is no deductive, rational solution? ...The physics community took it up, and in the hands of Challet, Marsili and Zhang, it inspired something different than I expected the Minority Game. El Farol emphasized (for me) the difficulties of formulating economic behavior in ill-defined problems. The Minority Game emphasizes something different: the efficiency of the solution. This is as it should be. The investigation reveals explicitly how strategies co-adapt and how efficiency is related to information. This opens an important door to understanding financial markets."

The basic minority game [16] consists of N (odd natural number) agents that can perform only two actions, for example, "buy" or "sell" commodities/assets, denoted by 0 or 1, at a given time t. An agent wins the game if it is one of the members of the minority group. All the agents have access to a finite amount of public information, which is a common bit string "memory" of the M most recent outcomes. Thus, the agents are said to exhibit "bounded rationality" [3]. Consider, for example, memory $M = 2$; then there are $P = 2^M = 4$ possible "history" bit strings: 00, 01, 10 and 11. A "strategy" consists of a response, that is, 0 or 1, to each possible history of bit strings; therefore, there are $G = 2^P = 2^{2^M} = 16$ possible strategies which constitute the "strategy space", as shown in Figure 9.5. At the beginning of the game, each agent randomly picks k strategies, and after the game assigns one "virtual" point to a strategy which would have predicted the correct outcome. The actual performance r of the player is measured by the number of times the player wins, and the strategy with the highest number of virtual points is the best strategy. A record of the number of agents who have chosen a particular action, say, "selling" denoted by 1, $A_1(t)$ as a function of time is kept (see Figure 9.6). The fluctuations in the behavior of $A_1(t)$ actually indicate the system's total utility. For example, we can have a situation where only one player is in the minority and all the other players lose. The other extreme case is when $(N - 1)/2$ players are in the minority and $(N + 1)/2$ players lose. The total utility of the system is obviously greater for the latter case and from this perspective, the latter situation is more desirable. Therefore, the system is more efficient when there are smaller fluctuations around the mean than when the fluctuations are larger.

9.3.1.3 Evolutionary Minority Games

Challet and Zhang [16] also extended the basic minority game described above to include the Darwinist selection: the worst player is replaced by a new one after some time-steps, the new player is a clone of the best player; that is, it inherits all the strategies but with corresponding virtual capitals reset to zero (analogous to a new born baby, though having all the predispositions from the parents, does not inherit their knowledge). To keep a certain diversity, they introduced a mutation

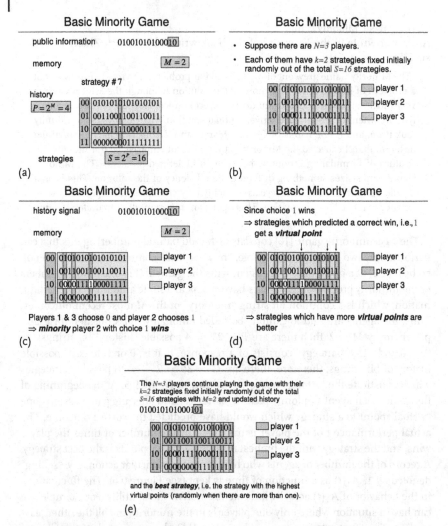

Figure 9.5 Basic minority game. Panel (a) Suppose public information of a bit string is given to all the minority game players. Each player has the memory of 2 bits, that is memory $M = 2$; then there are $P = 2^M = 4$ possible "history" bit strings: 00, 01, 10 and 11. A "strategy" consists of a response, that is, 0 or 1, to each possible history bit string; therefore, there are $G = 2^P = 2^{2^M} = 16$ possible strategies which constitute the "strategy space". Suppose a player chooses strategy number "7", then his response would be "1". Panel (b) At the beginning of the game, each agent randomly picks $k = 2$ strategies. Panel (c) Each player opts for a strategy randomly out of its pool of $k = 2$ strategies and plays. The player in the minority wins. Panel (d) After the game, each player assigns one "virtual" point to a strategy which would have predicted the correct outcome. The actual performance r of the player is measured by the number of times the player wins. Panel (e) The game continues after the history is updated. The strategy with the highest number of virtual points is the best strategy, which the player starts using from then on.

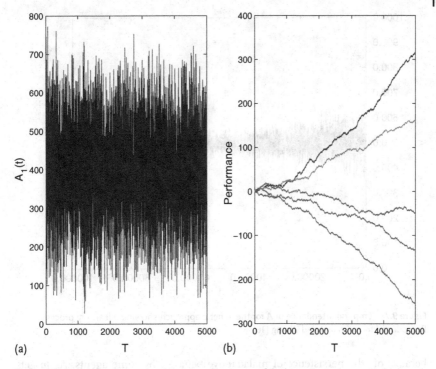

Figure 9.6 Attendance fluctuation (a) and performances of the best player, the worst player, and three randomly chosen players (b) in the basic minority game.

possibility in cloning. They allowed one of the strategies of the best player to be replaced by a new one. Since strategies are not just recycled among the players any more, the whole strategy phase space is available for selection. They expected this population to be capable of "learning" since bad players are weeded out with time, and fighting is among the "best" players. Indeed, in Figure 9.7, they observed that learning emerged in time. Fluctuations are reduced and saturated, which implies that average gain for everybody improved but never reached the ideal limit.

Li, Riolo and Savit [17] studied the minority game in the presence of evolution. In particular, they examined the behavior in games in which the dimension of the strategy space, m, is the same for all agents and fixed for all time. They found that for all values of m not too large, evolution results in a substantial improvement in overall system performance. They also showed that after evolution, results obeyed a scaling relation among games played with different values of m and different numbers of agents, analogous to that found in the non-evolutionary, adaptive games. Best system performance still occurs for a given number of agents at m_c, the same value of the dimension of the strategy space as in the non-evolutionary case, but system performance is now nearly an order of magnitude better than the non-evolutionary result. For $m < m_c$, the system evolves to states in which average agent wealth is better than in the random choice game, despite (and in some sense

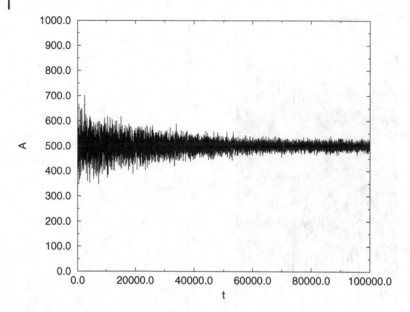

Figure 9.7 Temporal attendance of A for the genetic approach showing a learning process. Reproduced from Challet and Zhang [16].

because of) the persistence of maladaptive behavior by some agents. As m gets large, overall system performance approaches that of the random choice game.

The researchers [18] continued the study of evolution in minority games by examining games in which agents with poorly performing strategies can trade in their strategies for new ones from a different strategy space. In the context of the games, that meant allowing for strategies which use information from different numbers of time lags, m. They found that, in all the games, after evolution, wealth per agent is high for agents with strategies drawn from small strategy spaces (small m), and low for agents with strategies drawn from large strategy spaces (large m). In the game played with N agents, wealth per agent as a function of m was very nearly a step function. The transition is at $m = m_t$, where $m_t \simeq m_c - 1$, and m_c is the critical value of m at which N agents playing the game with a fixed strategy space (fixed m) have the best emergent coordination and the best utilization of resources. They also found that overall system-wide utilization of resources is independent of N. Furthermore, although overall system-wide utilization of resources after evolution varied somewhat depending on some other aspects of the evolutionary dynamics, in the best cases, utilization of resources was on the order of the best results achieved in evolutionary games with fixed strategy spaces.

9.3.1.4 Adaptive Minority Games

Sysi-Aho and others [19–22] presented a simple modification of the basic minority game where the players modify their strategies periodically after every time interval τ, depending on their performances: if a player finds that he is among the fraction n (where $0 < n < 1$) who are the worst performing players, he adapts himself and modifies his strategies. The authors proposed that the agents use different one-point genetic crossover mechanisms (as shown in Figure 9.8), inspired by genetic evolution in biology, to modify the strategies and replace the bad strategies. They study the performances of the agents under different conditions and investigate how they adapt themselves in order to survive or be the best by finding new strategies using the highly effective mechanism. It should be noted that the mechanism of evolution of strategies is considerably different from earlier attempts mentioned in the last sub-section. This is because, in this mechanism, the strategies are changed by the agents themselves and even though the strategy space evolves continuously, its size and dimensionality remain the same.

They also introduced the measure of total utility of the system, which is nothing but the total number of players that win the game, and can be expressed as:

$$U(x_t) = (1 - \theta(x_t - x_M))x_t + \theta(x_t - x_M)(N - x_t), \tag{9.1}$$

where $x_M = (N-1)/2$, x_t is either equal to $A_1(t)$ or $A_0(t)$ and so $x_t \in \{0, 1, 2, \ldots, N\}$ and

$$\theta(x) = \begin{cases} 0 & \text{when} \quad x \le 0 \\ 1 & \text{when} \quad x > 0 \end{cases}. \tag{9.2}$$

When $x_t \in \{x_M, x_M + 1\}$, the total utility of the system is maximum $U_{\max}$ as the highest number of players win. The system is more efficient when the deviations from the maximum total utility $U_{\max}$ are smaller, or in other words, the fluctuations in $A_1(t)$ around the mean become smaller.

Interestingly, the fluctuations disappear totally and the system stabilizes to a state where the total utility of the system is at maximum, since at each time-step the

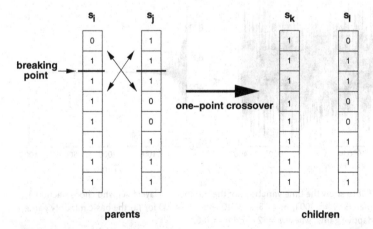

Figure 9.8 Schematic diagram to illustrate the mechanism of one-point genetic crossover for producing new strategies. The strategies s_i and s_j are the parents. We choose the break-ing point randomly and through this one-point genetic crossover, the children s_k and s_l are produced and substitute the parents.

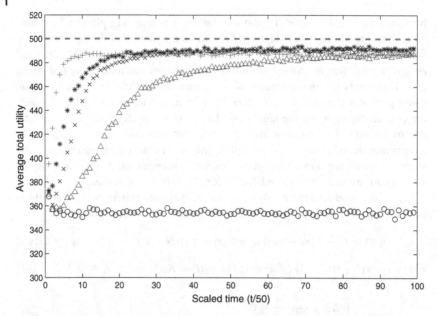

Figure 9.9 Plot to show the variation of the total utility of the system in time for the basic minority game for $N = 1001$, $m = 5$, $s = 10$, $t = 5000$, and the adaptive game for the same parameters, but different values of τ and n. Each point represents a time average of the total utility for separate bins of size 50 time-steps of the game. The maximum total utility $(= (N - 1)/2)$ is shown as a dashed line. The data for the basic minority game is shown in circles. The plus signs are for $\tau = 10$ and $n = 0.6$; the asterisk marks are for $\tau = 50$ and $n = 0.6$; the cross marks for $\tau = 10$ and $n = 0.2$; and triangles for $\tau = 50$ and $n = 0.2$. The ensemble average over 70 different samples was taken in each case.

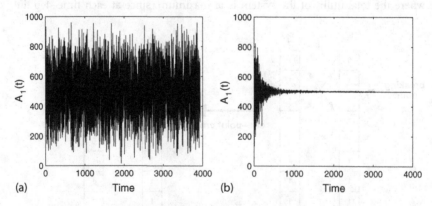

Figure 9.10 Plot to show the time variations of the number of players A_1 who choose action 1, with the parameters $N = 1001$, $m = 5$, $s = 10$ and $t = 4000$ for (a) the basic minority game and (b) the adaptive game, where $\tau = 25$ and $n = 0.6$.

highest number of players win the game (see Figure 9.10). As expected, the behavior depends on the parameter values for the system.

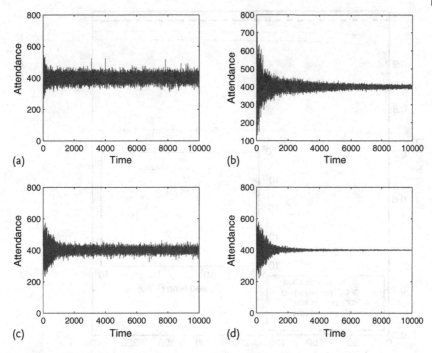

Figure 9.11 Plots of the attendances by choosing parents randomly (a) and (b), and using the best parents in a player's pool (c) and (d). In cases (a) and (c) parents are replaced by children, and in cases (b) and (d) case children replace the two worst strategies. Simulations have been done with $N = 801$, $M = 6, k = 16, t = 40, n = 0.4$ and $T = 10\,000$.

Most importantly, the time of adaptation has to be optimal, as very frequent adaptation and very slow adaptation both lead to bad performances. They also used the utility function to study the efficiency and dynamics of the game as shown in Figure 9.9.

If the parents are chosen randomly from the pool of strategies then the mechanism represents a "one-point genetic crossover", and if the parents are the best strategies then the mechanism represents a "hybridized genetic crossover". The children may replace parents or the two worst strategies and accordingly four different interesting cases arise: (a) one-point genetic crossover with parents "killed", that is, parents are replaced by the children, (b) one-point genetic crossover with parents "saved", that is, the two worst strategies are replaced by the children, but the parents are retained, (c) hybridized genetic crossover with parents "killed" and (d) hybridized genetic crossover with parents "saved". In order to determine which mechanism is the most efficient, we have made a comparative study of the four cases, mentioned above. We plot the attendance as a function of time for the different mechanisms in Figure 9.11.

In Figure 9.12 we show the total utility of the system in each of the cases (a)–(d), where we have plotted results of the average over 100 runs and each point in the

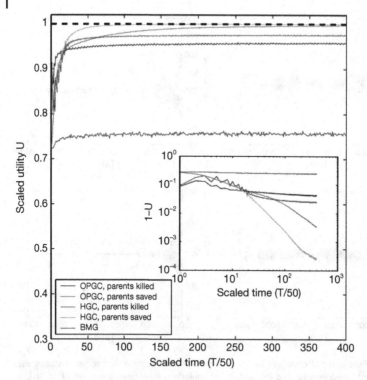

Figure 9.12 Plots of the scaled utilities of the four different mechanisms in comparison with that of the basic minority game. Each curve represents an ensemble average over 100 runs and each point in a curve is a time average over a bin of length 50 time-steps. In the inset, the quantity $(1 - U)$ is plotted against scaled time in the double logarithmic scale. Simulations are done with $N = 801$, $M = 6$, $k = 16$, $t = 40$, $n = 0.4$, and $T = 20\,000$.

utility curve represents a time average taken over a bin of length 50 time-steps. The simulation time is doubled from those in Figure 9.11, in order to expose the asymptotic behavior better. On the basis of Figures 9.11 and 9.12, we find that the case (d) is the most efficient.

9.4
The Kolkata Paise Restaurant Problem

Let there be N restaurants in a city and assume that each can accommodate a fixed number of customers, normalized to unity. Let there be exactly N customers or agents in the city. Though each of the restaurants costs the same, we assume that there is a common preference ranking of these restaurants to the agents. If, in a day, there is more than one agent (prospective customer) arriving in any particular restaurant, one of them is chosen randomly and the rest do not get service. *In Kolkata, there happened to exist very cheap and fixed rate "Paise Restaurants" (also called "Paise Hotels"), popular among the daily laborers in the city. During lunch hours,* the laborers used to walk down (to save the transport costs) to any of these restaurants and would miss the lunch if they arrived at a restaurant where their number is more than the capacity of the restaurant for such cheap lunch. Walking down to the next restaurant would mean failing to report back to the job in time! Paise means the smallest Indian coin and there were indeed some well-known rankings of these restaurants as some of them would offer tastier items compared to the others. These problems, abbreviated the KPR problem, can serve as simple toy models to study learning, formulating strategies for efficient resource utilization.

Herd behavior and herd externality (see [23, 24]) cannot explain the omnipresence of social inefficiency for the KPR problem since we assume that there is no ambiguity among agents over the ranking of the restaurants. The KPR problem is similar to the minority game since it punishes herd behavior and promotes diversity (see [15]). The KPR problem differs from the minority game in the following aspects: (a) while the minority game is a simultaneous move *two choice* problem, the KPR problem is a simultaneous move *many choice* problem, and (b) for the KPR problem there is a common preference ranking of these restaurants and this sort of ranking is absent for the minority game.

It is assumed that while deciding on the choice of the restaurant, there is no mutual interaction among the agents, and they decide simultaneously on the basis of their own past "experiences" and complete history. There are N^N possible outcomes of which most are socially inefficient in the sense that there exists at least one agent not getting lunch and equivalently, there is at least one restaurant without a customer. There are exactly $N!$ (that is, a fraction $\exp[-N]$ as $N \rightarrow \infty$) outcomes that correspond to socially efficient utilization of the restaurants where each agent ends up in a different restaurant. Given that all agents pay the same price (though rankings are different), in general a socially efficient outcome is not individually optimal in this repeated set up. Therefore, even when there is a realization of the efficient outcome in one period, it gets destabilized in the next period and the whole system breaks down to any of the infinitely many ($\exp[N]$; $N \rightarrow \infty$) socially inefficient states. Hence, there can be no fixed point result for the problem and the game continues dynamically forever. The system then moves to an inefficient state characterized by overcrowding at higher ranked restaurants (absence of any "absorbing" state). It is precisely this generic deviation from the efficient outcome that motivated the investigation of some of the optimally utilized states of the system and the average occupation (utilization) fraction of the restaurants on any day. Thus, different versions and modifications of the KPR problem using various deterministic and stochastic strategies, are studied.

9.4.1
One-Shot KPR Game

How does one capture the features of the Kolkata Paise Restaurant problem of *any given day*? That is, the situation where on any given day

a. each agent goes for lunch to any one of the N restaurants

b. depending on the number of agents in each restaurant, only one agent is selected randomly and allowed to have lunch
c. the agent derives a utility from having lunch and this utility level depends on the rank of the restaurant, and if the agent does not get lunch then his utility is zero on that particular day.

One can use the tools available in the game theory literature to model the Kolkata Paise Restaurant problem *of any given day* as a one-shot game. Let us introduce [25] a general one-shot *common preference restaurant game* and identify the conditions under which it represents the *one-shot KPR game*. Let $\mathcal{N} = \{1, \ldots, N\}$ be the set of agents ($N < \infty$) and let the vector $u = (u_1, u_2, \ldots, u_N) \in \text{Re}^N$ represent the utility (in terms of money) associated with each restaurant which is common to all agents. Assume, without loss of generality, that $0 < u_N \leq \ldots \leq u_2 \leq u_1$. These three things (that is, $\mathcal{N}$, $u = (u_1, \ldots, u_N)$ and the ranking $0 < u_N \leq \ldots \leq u_2 \leq u_1$) are used to define the one-shot common preference restaurant game. Formally, $G(u) = (\mathcal{N}, S, \Pi)$ represent a one-shot common preference restaurant game where $\mathcal{N} = \{1, \ldots, N\}$ is the set of agents (players), $S = \{1, \ldots, N\}$ is the (common) strategy space of all agents where a typical strategy $s_i = k$ denotes the strategy that the i-th agent's strategy is to go to the k-th restaurant. In other words, each agent on any given day selects any one of the N restaurants numbered 1 to N, and if any agent i decides to go to restaurant $k \in \{1, \ldots, N\}$ then we represent it as $s_i = k$. Let $\Pi = (\Pi_1, \ldots, \Pi_N)$ be the expected payoff vector. In particular, $\Pi(s) = (\Pi_1(s), \ldots, \Pi_N(s))$ is the expected payoff vector associated with any strategy combination $s = (s_1, \ldots, s_N) \in S^N$ where agent i's payoff is $\Pi_i(s) = u_{s_i}/n_i(s)$ and $n_i(s) = 1 + |\{j \in \mathcal{N} \setminus \{i\} \mid s_i = s_j\}|$. Therefore, the strategy combination $s = (s_1, \ldots, s_N)$ represents the selection of restaurants made by the agents and the expected payoff that an agent $i \in \mathcal{N}$ gets depends not only on which restaurant he has selected, but also on how many other agents have selected the same restaurant. The number of agents selecting the same restaurant as that of agent i under the strategy combination is given by $n_i(s)$. Since it is assumed that the restaurant selects one agent randomly from the set of $n_i(s)$ agents, it follows that the expected payoff of agent i given the strategy combination $s = (s_1, \ldots, s_N)$ is $\Pi_i(s) = u_{s_i}/n_i(s)$. For each selection of a vector $u = (u_1, u_2, \ldots, u_N)$ with the property that $0 < u_N \leq \ldots \leq u_2 \leq u_1$ let us associate a common preference restaurant game $G(u) = (\mathcal{N}, S, \Pi)$. Let us now introduce an added restriction on the common preference restaurant games to represent the one-shot KPR games.

The first feature of the KPR problem is that agents have a common preference ranking over the restaurants. If all restaurants are identical to the agents (that is, if $u_1 = \ldots, u_N$) then the problem is trivial. Hence, to introduce the problem in a non-trivial way it is reasonable to assume that for the one-shot KPR game $u_1 \neq u_N$. The next feature of the KPR problem is that agents prefer getting lunch to not getting lunch. To capture this feature, we assume that the payoff specification of a one-shot KPR game must be such that $u_{k'} > u_k/2$ for all $k, k' \in \{1, \ldots, N\}$ and given $0 < u_N \leq \ldots \leq u_2 \leq u_1$ and $u_1 \neq u_N$ this is equivalent to assuming that $u_N > u_1/2$. Intuitively, if $u_N > u_1/2$ then even the person who has been at the lowest ranked restaurant, would not be willing to alter his strategy. Therefore, *if a one-shot common preference restaurant game* $G(u) = (\mathcal{N}, S, \Pi)$ *is such that* $0 < u_N \leq \ldots \leq u_2 \leq u_1, u_1 \neq u_N$ *and* $u_N > u_1/2$ *then it is a one-shot KPR game.*

It turns out that a strategy combination $s = (s_1, \ldots, s_N)$ leads to a socially efficient (or Pareto efficient) outcome if and only if each agent goes to different restaurants, that is, $s_i \neq s_j$ for all $i, j \in \mathcal{N}$, $i \neq j$. A strategy combination $s = (s_1, \ldots, s_N)$ is a pure strategy Nash equilibrium if no agent has an incentive to deviate from his existing strategy s_i given the strategy of the other players. For the above mentioned version of the one-shot KPR game *the set of pure strategy Nash equilibria coincides with the set of socially efficient allocations.* This is formally established as follows:

A strategy combination $s \in S^N$ for the one-shot common preference restaurant game $G(u) = (\mathcal{N}, S, \Pi)$ leads to a *socially (Pareto) efficient* outcome if and only if $n_i(s) = 1$ for all $i \in N$. That is, given any one-shot common preference restaurant game $G(u)$, a strategy combination leads to a socially efficient outcome if and only if each agent goes to a different restaurant so that the sum of the utilities of the agents is $\sum_{k=1}^{N} u_k$. Let s_{-i} be the strategy combination of all but agent i. A strategy combination $s^* \in S^N$ is a *pure strategy Nash equilibrium* of the one-shot common preference restaurant game $G(u) = (\mathcal{N}, S, \Pi)$ if for all $i \in \mathcal{N}$, $\Pi_i(s^*) \geq \Pi_i(s_i, s^*_{-i})$ for all $s_i \in S$. For a typical one-shot common preference restaurant game $G(u) = (\mathcal{N}, S, \Pi)$, let $SE(G(u)) = \{s^* \in S^N \mid n_i(s^*) = 1 \; \forall \; i \in \mathcal{N}\}$ denote the set of all strategy combinations that lead to socially efficient allocations and let $PN(G(u)) = \{s^* \in S^N \mid \forall \; i \in \mathcal{N}, \Pi_i(s^*) \geq \Pi_i(s_i, s^*_{-i}) \; \forall \; s_i \in S\}$ denote the set of all pure strategy Nash equilibria. Let $\bar{G}(u)$ represent any one-shot KPR game we introduced in Section 9.4.1, that is, any one-shot common preference restaurant game with the property that $0 < u_N \leq \ldots \leq u_1 < 2u_N, u_1 \neq u_N$. We have the following result.

Proposition 9.1

For any one-shot KPR game $\bar{G}(u)$, $SE(\bar{G}(u)) = PN(\bar{G}(u))$.

Proof. Consider any one-shot KPR game and a strategy combination $s^* \in SE(\bar{G}(u))$. Take any agent $i \in \mathcal{N}$ and any deviation $s_i = k \neq s^*_i = k'$. Since the number of agents is equal to the number of restaurants, and since for a strategy combination $s^* \in SE(\bar{G}(u))$, $n_i(s^*) = 1$ for all $i \in \mathcal{N}$, it follows that $n_i(s_i, s^*_{-i}) = 2 > n_i(s^*) = 1$. Therefore, $\Pi_i(s_i, s^*_{-i}) = u_k/2 < u_{k'} = \Pi_i(s^*)$ since for the one-shot KPR game $0 < u_N \leq \ldots \leq u_1 < 2u_N \Leftrightarrow u_k/2 < u_{k'}$ for all $k, k' \in \{1, \ldots, N\}$ and, hence, agent i has no incentive to deviate. Since the selection of i was arbitrary it follows that $s^* \in PN(\bar{G}(u))$ implying $SE(\bar{G}(u)) \subseteq PN(\bar{G}(u))$.

To prove $PN(\bar{G}(u)) \subseteq SE(\bar{G}(u))$ consider any strategy combination s_{-i} for all but agent i. Given s_{-i}, the best response of agent i is to select a best ranked restaurant from the set $U\mathcal{N}_i(s_{-i}) = \{k \in \mathcal{N} \mid s_j \neq k \; \forall \; j \neq i\}$ of unoccupied restaurants. The reasons being: (1) given $u_1/2 < u_N$, $u_k/q < u_{k'}$ for all $k, k' \in \{1, \ldots, N\}$ and all integers $q \geq 2$, implying that, given the strategy of all other agents, agent i strictly prefers to occupy the best unoccupied restaurant as opposed to either occupying a different unoccupied restaurant (which is not the best restaurant) or crowding in an occupied restaurant, and (2) given that the number of agents is equal to the number of restaurants, $\mathcal{UN}_i(s_{-i}) \neq \emptyset$ for any $s_{-i} \in S^{N-1}$ implying that given any strategy of all other agents, agent i can always find an unoccupied restaurant and, hence, agent i can also find the best unoccupied restaurant, since $0 < u_N \leq \ldots \leq u_1 < 2u_N$. Therefore, it is in the interest of each agent to occupy an unoccupied restaurant, and given the fact that no matter what others do, an agent can always find an unoccupied restaurant, it follows that any pure strategy Nash equilibrium for the one-shot KPR game necessarily leads to a socially efficient allocation, that is, $PN(\bar{G}(u)) \subseteq SE(\bar{G}(u))$.

Thus, we have the one-shot common preference restaurant games and identified conditions under which these games also represent the one-shot KPR games. The key features that separate out a one-shot KPR game from other non-KPR one-shot common preference restaurant games are non-triviality ($u_1 \neq u_N$) and the restriction $u_N > u_1/2$, which ensures that it is always in the interest of any agent to choose an unoccupied restaurant.

9.4.2
Simple Stochastic Strategies and Utilization Statistics

It is well known from the repeated game theory literature that if *agents can coordinate their actions* then any pure strategy Nash equilibrium outcome of the one-shot KPR game (which is also a socially efficient outcome) can be sustained as an equilibrium outcome of the KPR problem (the repeated one-shot KPR game) with appropriate punishment schemes (see [26]). However, sustaining such equilibrium in repeated play of the one-shot KPR game would require that the *set of agents is small so that they can coordinate their actions*. Since expecting such coordinated actions from the agents is unrealistic for the KPR problem (where N is large), we assume that with repetition of a one-shot KPR game agents work out their strategies without coordination.

Following Brian Arthur we assume that agents reason inductively (see [27]). Inductive reasoning means that agents learn from past experiences, discard their old strategy for something new if it performs badly and continues with the old strategy if it performs well. However, our objective is not to find out strategies that survive this process of inductive reasoning, but to analyze the statistics of the fraction of restaurants utilized per day associated with each such strategy. It is assumed that all agents take up some (common) stochastic strategy independent of each other. By stochastic strategies we mean that agents choose their strategies in a probabilistic way. At any step this choice of probabilities can depend on the outcome of the previous period. In each case one considers the limit $N \to \infty$ and analyzes the statistics of the fraction f of the restaurants utilized per period. The specific one-shot KPR game $\bar{G}(u)$ that we use is the one where $0 < u_N < \cdots < u_1$ so that agents have a strict ranking and $u_1/2 < u_N$.

9.4.2.1 No Learning (NL) Strategy

On each day t, an agent selects any one restaurant with equal probability. There is no memory or learning and each day the same process is repeated.

If any restaurant is chosen by $m(> 1)$ agents on a particular day t, the restaurant chooses one of them randomly, and the remaining $m - 1$ agents go without any lunch that day. Let there be λN agents ($\lambda = 1$ in the KPR problem) choosing randomly from the set of N restaurants. Then the probability p of choosing any restaurant is $1/N$. The probability $P(m)$ that any restaurant is chosen simultaneously by m agents is therefore given by Poisson distribution

$$P(m) = \left(\begin{array}{c} \lambda N \\ m \end{array} \right) p^m (1 - p)^{\lambda N - m} ; \quad p = \frac{1}{N}$$

$$= \frac{\lambda^m}{m!} \exp(-\lambda) \quad \text{as} \quad N \to \infty . \tag{9.3}$$

That means, the fraction of restaurants not chosen on any day is given by $P(m = 0) = \exp(-\lambda)$, giving the average fraction of restaurants occupied

$$\bar{f} = 1 - \exp(-\lambda) \simeq 0.63 \quad \text{for} \quad \lambda = 1 , \tag{9.4}$$

in the KPR problem. The distribution $D(f)$ of the fraction utilized any day will be a Gaussian around the average given above. See the simulation results in Figure 9.13 for $N = 1000$.

9.4.2.2 Limited Learning (LL) Strategy

Let us consider [25] two types of learning strategies and show that in both cases the utilization fraction $\bar{f}$ goes down compared to the no learning case.

LL(1) *If an agent gets lunch on day t, then the agent goes to the best restaurant on day t + 1. Moreover, if on some day t the agent goes to the best restaurant and fails to get lunch, then on day t + 1, the agent selects a restaurant from the remaining N − 1 restaurants with equal probability.*

When $f_t N$ number of agents get their lunch on any day t and decide to get their lunch the next day $((t + 1)$th day) in the best ranked restaurant, then only one of them will get into the highest ranked restaurant and the others will not get their lunch. The remaining $N - f_t N$ agents will try independently for the remaining $N - 1$ restaurants, following the same strategy as in the no learning case. Then the recursion relation will simply be (following (9.4)):

$$f_{t+1} = 1 - \exp(-\lambda_t) ; \quad \lambda_t = 1 - f_t \tag{9.5}$$

giving the fixed point fraction $\bar{f} \simeq 0.43$. This again compares well with the numerical simulation results (for $N = 1000$) and shows a Gaussian distribution $D(f)$ around the above value of $\bar{f}$ as in Figure 9.13.

LL(2) *If an agent reaches the k-th ranked restaurant on day t then the agent's strategy is the following: (a) if the agent gets lunch on day t then, on day t + 1, the agent*

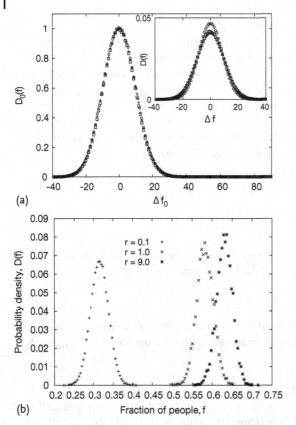

(a)

(b)

Figure 9.13 (a) Numerical simulation results for the normalized (by the respective half widths) distribution $D_0(f)$ of the fraction f of people getting lunch on any day (or fraction of restaurants occupied) against $\Delta f \equiv f - \bar{f}$, where $\bar{f}$ denotes the average occupancies in each case. The inset shows the bare distributions $D(f)$ against Δf. All the numerical simulations have been done for $N = 1000$ agents for 10^6 time-steps. $\bar{f} \simeq 0.63$ in case NL, $\bar{f} \simeq 0.43$ in case LL(1) and $\bar{f} \simeq 0.77$ in case OPR, as estimated analytically (Equations (9.4), (9.5) and (9.8)). The data points are (+), ($\square$), and ($\bigcirc$) for cases NL, LL(1), and OPR respectively. (b) Numerical simulation results for the distribution $D(f)$ of the fraction f of people getting lunch any day (or fraction of restaurants occupied on any day) against f, for $r = 0.1$, 1.0 and 9.0 for case FC. All the numerical simulations have been done for $N = 512$ agents for 10^6 time-steps.

attempts to select a restaurant from the rank set $\{k-1, \ldots, 1\}$ with equal probability and (b) if the agent does not get lunch on day t then, on day $t+1$, the agent attempts to select a restaurant from the set $\{k, \ldots, N\}$ with equal probability.

For this strategy one gets the fixed point fraction $\bar{f} \simeq 0.5$ which shows a Gaussian distribution around $\bar{f}$. Hence, in this case there is also no improvement compared to the NL case. If one can somehow introduce repetition in the strategy of the agent, then an improvement in the utilization fraction can be achieved. In fact, since the reverse choice problem, where for (a) the choice would be $\{k, \ldots, N\}$ and for (b) would be $\{k - 1, \ldots, 1\}$, is identical, we get $\bar{f} = 1 - \bar{f}$, giving $\bar{f} = 0.5$.

9.4.2.3 One Period Repetition (OPR) Strategy

If an agent gets lunch on day t from restaurant k, then the agent goes back to the same restaurant on day $(t + 1)$ and competes for the best restaurant (restaurant 1) on day $(t + 2)$. If the agent fails to get lunch on day t, then the agent tries for a restaurant on day $(t + 1)$ which was vacant on day t, using the same stochastic strategy as in the NL case.

The fraction f_t on any day t of the people occupying any restaurant consists of two parts: fraction x_{t-1} of the people already continuing in their randomly selected restaurant the earlier day and the fraction x_t of people who have chosen today. As such, using (9.4),

$$f_t = x_{t-1} + (1 - x_{t-1})[1 - \exp(-1)] . \tag{9.6}$$

Since the fraction x_t chosen today is given by the fraction $\bar{f}$ in (9.4) where $N(1 - x_{t-1})$ left out agents are choosing randomly out of $N(1 - x_{t-1})$ vacant restaurants. The next day, the fraction f_{t+1} will therefore be given by

$$f_{t+1} = (1 - x_t)(1 - \exp(-1)) + [1 - (1 - x_t)(1 - \exp(-1))][1 - \exp(-1)] . \tag{9.7}$$

If we assume that $f_{t+1} = \bar{f} = f_t$ and $x_t = x$ at the fixed point, then equating (9.6) and (9.7)

$$x = [1 - (1 - x)(1 - \exp(-1))](1 - \exp(-1)) , \tag{9.8}$$

or $x \simeq 0.38$, giving $\bar{f} \simeq 0.77$, as can be seen in the numerical simulation results in Figure 9.13. The fluctuations in the occupation density or utilization fraction $D(f)$ is again Gaussian around the most probable occupation fraction derived here. Therefore, with OPR strategy we have an increase in $\bar{f}$ compared to the NL strategy, suggesting that if the strategies are such that an agent, after getting lunch on a particular day, decides to go to the same restaurant the next day, we have an improvement in utilization fraction.

Thus, the OPR strategy is more consistent when the one-shot common preference restaurant game $G(u)$ is such that $u = (u_1 = \cdots = u_N)$ [25]. It may be noted that if all the restaurants are identical then the dynamics always converge to an absorbing state characterized by social efficiency (with Pavlov's *win-stay, lose-shift (WS-LS)* strategy (see [28])). The reason being the following: even if everybody starts by random choices among all the N restaurants and, initially (say, on day t), several restaurants get more than one prospective customer and choose one of them, on day $t + 1$, the agents not getting lunch on day t will go to a restaurant that was unoccupied on day t and the agents getting lunch on day t will continue to go to the same restaurant. Soon (before, $N + 1$ days), each agent finds one restaurant to get the lunch and the dynamics effectively stops there.

Now let us think of a strategy where the queue length in a restaurant on day t either acts as an inducing device or acts as a rationing device for the strategy to be adopted by an agent on day $t + 1$.

9.4.2.4 Follow the Crowd (FC) Strategy

The probability of selecting a restaurant k on day t + 1 depends on the queue length of agents in restaurant k on day t.

Let us assume that the probability $p_k(t+1)$ of choosing kth restaurant on $(t+1)$th day is given by

$$p_k(t+1) = \frac{q_k(t) + r}{N(1 + r)} \qquad (9.9)$$

where $q_k(t)$ is the queue length (or equivalently, the number of people who arrived at the kth restaurant on day t). r is a parameter that interpolates between the random ($r \to \infty$, independent of history) and condensation ($r \to 0$, strongly dependent only on history) limits. The representative histograms are shown in the Figure 9.13, where the most probable fraction $\bar{f}$ of occupied restaurants converge to a value 0.63 (as obtained earlier) as $r \to \infty$. It is also seen that the probability of a queue length q of any restaurant goes as q^{-1} for $r \to 0$ (exponentially decaying otherwise). This implies that while the utilization fraction has a Gaussian distribution (with $\bar{f} = 0.63$), the queue length distribution on any arbitrary day t has a power law distribution if $r \to 0$ and it is exponentially decaying otherwise.

If the agents follow an *avoid the crowd (AC)* strategy, that is, if $p_k(t) \to 1 - p_k(t)$ (people choosing to avoid the last day's crowded restaurants), the most probable restaurant utilization fraction $\bar{f} \simeq 0.63$ is independent of the value of $r(> 0)$. Here, no power law for the queue length was observed for any value of r. The observations of this subsection are similar to the hospital waiting list size distribution (see [29, 30]), though the Hospital problem is significantly different from the KPR problem (for example, the absence of possibilities of iterative learning and the presence of heterogeneity of choices in the Hospital Waiting List problem).

9.4.3
Limited Queue Length and Modified KPR Problem

Until now we had assumed that on any day t one can have potentially all N agents turning up in one particular restaurant. If the arrival of the agents was sequential, then all agents turning up in one restaurant is tantamount to saying that queues of all sizes are allowed to form in any restaurant and the restaurant then picks up any one agent to serve. All other agents have to go back without having lunch on day t. Here we are assuming that formation of a queue of all sizes is allowed. What if all restaurants have limited capacity to accommodate agents in a queue and an agent, having arrived at a restaurant that has reached its queue limit, gets the option to move to another restaurant (in the same day)? Before concluding, we consider this particular modification of the KPR problem since it has important implications about the robustness of our occupancy statistics $\bar{f}$.

A modified KPR problem: On each day the restaurants are filled up sequentially. We also assume that *each restaurant allows for a queue length q^*, so that the (q^*+1)th person arriving at the kth restaurant has to search for another restaurant with a queue length less than q^*.* When all the customers have arrived, one of the $q_i \leq q^*$ cus-

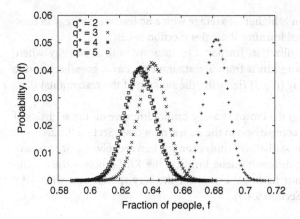

Figure 9.14 Numerical simulation results for the distribution $D(f)$ of the fraction f of people getting lunch any day (or fraction of restaurants occupied on any day) against f, for $q^* = 2, 3, 4, 5$ for the modified KPR problem. All the numerical simulations have been done for $N = 1024$ agents for 10^6 time-steps.

tomers gets lunch at the kth restaurant, while the $q_i - 1$ number of prospective customers do not get any lunch that day. In terms of the dynamics, the most probable restaurant utilization fraction $\bar{f} = 1$ occurs for $q^* = 1$. We find that for $q^* = 2$, the fraction $\bar{f}$ *drastically falls* to 0.68. Also, for $q^* \to N$, $\bar{f} = 0.63$ as discussed and derived earlier. Figure 9.14 shows the result for different values of q^*. Note that $\bar{f}$ approaches the value for the random occupancy case as q^* becomes large. Thus, except for the special case where $q^* = 1$, for all other cases we get conclusions that are closer to the KPR problem derived in the previous section.

For the general KPR problem we have a perpetual dynamic characterized by large fluctuations in the measure of daily misuse, which has a Gaussian distribution of utilization fraction f around the average $\bar{f}$, where $\bar{f}$ differs for different strategies. We have argued this point by showing that the average utilization fraction $\bar{f}$ with no learning and other simple stochastic strategies could be analytically calculated and the distributions are seen to be Gaussian. For the "follow/avoid the crowd" strategies we observe the statistics of the number of people *going without lunch* to have power law fluctuations (see also [15, 27, 29] for similar behavior).

The main findings in terms of utilization fraction comparisons across different strategies are [25]:

a. If we have a one-shot KPR game, then any pure strategy Nash equilibrium leads to a socially efficient outcome, as shown in Section 9.4.1, and, hence, we have full capacity utilization. However, if we have a KPR problem (which is a repeated game where any given one-shot KPR game is repeated) and agents cannot coordinate their actions, then we do not have full capacity utilization for many (common) stochastic strategies. In such a situation, utilization fractions under different stochastic strategies is of prime importance.

b. The utilization fraction is higher on average with a no learning strategy in comparison to some limited learning strategies (Section 9.4.2).

c. The statistics for a utilization fraction can improve with a strategy where each agent, after getting lunch from a restaurant on day t, goes back to the same restaurant on day $(t + 1)$ (ignoring the ranking of the restaurants) (Section 9.4.2.3).

d. Following (or avoiding) the crowd strategy cannot improve on the statistics of utilization fraction in comparison to the no learning case (Section 9.4.2.4).

e. Our findings about the statistics of utilization fraction are robust since we have shown that by adding different queue limits to the KRP problem the conclusions are similar to our findings, except for the very special case where the queue limit is unity (Section 9.4.3).

9.4.4
Some Uniform Learning Strategy Limits

For KPR, most of the strategies will give a low average value of resource utilization (average fraction $\bar{f} \ll 1$), because of the absence of mutual interaction/discussion among the agents. However, a simple, dictated strategy, instructing each agent go to a sequence of the ranked restaurants respectively on the first evening, and then shift by one rank step in the next evening, will automatically lead to the best optimized solution (with $f = \bar{f} = 1$). Also, each one gets in turn to the best ranked restaurant (with periodicity N). The process starts from the first evening itself. It is hard to find a strategy in KPR, where each agent decides independently (democratically) based on past experience and information, to achieve this even after long learning time.

Let the strategy chosen by each agent in the KPR game be such that, at any time t, the probability $p_k(t)$ to arrive at the k-th ranked restaurant is given by

$$p_k(t) = \frac{1}{z}\left[k^\alpha \exp\left(-\frac{n_k(t-1)}{T}\right)\right], \quad z = \sum_{k=1}^{N} k^\alpha \exp\left(-\frac{n_k(t-1)}{T}\right),$$

$$(9.10)$$

where $n_k(t-1)$ gives the number of agents arriving at the k-th ranked restaurant on the previous evening (or time $t-1$), T is a noise scaling factor and α is an exponent. Here for $\alpha > 0$ and $T > 0$, the probability for any agent to choose a particular restaurant increases with its rank k and decreases with the past popularity of the same restaurant (given by the number $n_k(t-1)$ of agents arriving at that restaurant on the previous evening). For $\alpha = 0$ and $T \to \infty$, $p_k(t) = 1/N$ corresponds to the random choice (independent of rank) case. For $\alpha = 0$, $T \to 0$, the agents avoid those restaurants visited last evening and choose again randomly among the rest. For $\alpha = 1$, and $T \to \infty$, the game corresponds to a strictly rank-dependent choice case. We concentrate on these three special limits.

9.4.4.1 Numerical Analysis

Random-choice For the case where $\alpha = 0$ and $T \to \infty$, the probability $p_k(t)$ becomes independent of k and becomes equivalent to $1/N$. For simulation, we take 1000 restaurants and 1000 agents, and on each evening t an agent selects any restaurant with equal probability $p = 1/N$. All averages have been made for 10^6 time-steps. We study the variation of probability $D(f)$ of the agents getting dinner versus their fraction f. The numerical analysis shows that mean and mode of the distribution occurs around $f \simeq 0.63$ and that the distribution $D(f)$ is Gaussian around that (see Figure 9.15).

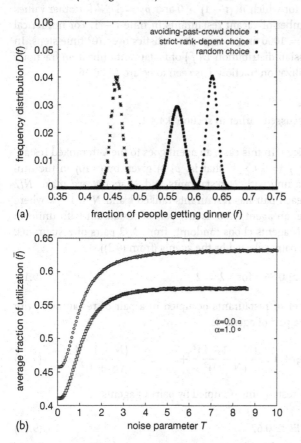

(a)

(b)

Figure 9.15 (a) Numerical simulation results for the distribution $D(f)$ of the fraction f of people getting dinner any evening (or fraction of restaurants occupied on any evening) against f for different limits of α and T. All the simulations have been done for $N = 1000$ (number of restaurants and agents) and the statistics have been obtained after averages over 10^6 time-steps (evenings) after stabilization. (b) Numerical simulation results for the average resource utilization fraction ($\bar{f}$) against the noise parameter T for different values of α (>0).

Strict-rank-dependent choice For $\alpha = 1$, $T \to \infty$, $p_k(t) = k/z$; $z = \sum k$. In this case, each agent chooses a restaurant having rank k with a probability, strictly given by its rank k. Here also we take 1000 agents and 1000 restaurants and average over 10^6 time-steps for obtaining the statistics. Figure 9.15 shows that $D(f)$ is again Gaussian and that its maximum occurs at $f \simeq 0.58 \equiv \bar{f}$.

Avoiding-past-crowd choice In this case, an agent chooses randomly among those restaurants that went vacant the previous evening: with probability $p_k(t) = \exp(-(n_k(t-1))/T)/z$, where $z = \sum_k \exp(-(n_k(t-1))/T)$ and $T \to 0$, one gets $p_k \to 0$ for k values for which $n_k(t-1) > 0$ and $p_k = 1/N'$ for other values of k where N' is the number of vacant restaurants in time $t-1$. For numerical studies we again take $N = 1000$ and average the statistics over 10^6 time-steps. In the Figure 9.15, the Gaussian distribution $D(f)$ of restaurant utilization fraction f is shown. The average utilization fraction $\bar{f}$ is seen to be around 0.46.

9.4.4.2 Analytical Results
Random-choice case Discussed earlier in section 9.4.4.1.

Strict-rank-dependent choice In this case, an agent goes to the k-th ranked restaurant with probability $p_k(t) = k/\sum k$, that is, $p_k(t)$ given by (9.10) in the limit $\alpha = 1$, $T \to \infty$. Starting with N restaurants and N agents, we make $N/2$ pairs of restaurants and each pair has restaurants ranked k and $N + 1 - k$ where $1 \le k \le N/2$. Therefore, an agent chooses any pair of restaurant with uniform probability $p = 2/N$ or N agents choos randomly from $N/2$ pairs of restaurants. Therefore, the fraction of pairs selected by the agents (from (9.3))

$$f_0 = 1 - \exp(-\lambda) \simeq 0.86 \quad \text{for} \quad \lambda = 2 . \tag{9.11}$$

Also, the expected number of restaurants occupied in a pair of restaurants with rank k and $N + 1 - k$ by a pair of agents is

$$E_k = 1 \times \frac{k^2}{(N+1)^2} + 1 \times \frac{(N+1-k)^2}{(N+1)^2} + 2 \times 2 \times \frac{k(N+1-k)}{(N+1)^2} . \tag{9.12}$$

Therefore, the fraction of restaurants occupied by pairs of agents

$$f_1 = \frac{1}{N} \sum_{i=1,\ldots,N/2} E_k \simeq 0.67 . \tag{9.13}$$

Hence, the actual fraction of restaurants occupied by the agents is

$$\bar{f} = f_0 . f_1 \simeq 0.58 . \tag{9.14}$$

Again, this compares well with the numerical observation of the most probable distribution position (see Figure 9.15).

Avoiding-past-crowd choice   We consider here the case where each agent, on any evening (*t*), chooses randomly among the restaurants to which nobody had gone the last evening (*t*−1). This corresponds to the case where $\alpha = 0$ and $T \to 0$ (9.10). Our numerical simulation results for the distribution $D(f)$ of the fraction *f* of utilized restaurants is again Gaussian with the most probable peak at $\bar{f} \simeq 0.46$ (see Figure 9.15). This can be explained in the following way: As the fraction $\bar{f}$ of restaurants visited by the agents during the last evening is avoided by the agents this evening, the number of available restaurants is $N(1 - \bar{f})$ for this evening and is chosen randomly by all the *N* agents. Hence, when fitted to (9.3), $\lambda = 1/(1 - \bar{f})$. Therefore, following (9.3), we can write the equation for $\bar{f}$ as

$$(1 - \bar{f})\left(1 - \exp\left(-\frac{1}{1 - \bar{f}}\right)\right) = \bar{f} \,. \tag{9.15}$$

The solution of this equation gives $\bar{f} \simeq 0.46$. This result agrees well with the numerical results for this limit (see Figure 9.15; $\alpha = 0$, $T \to 0$).

Thus, one can find cases where each agent chooses the *k*-th ranked restaurant with probability $p_k(t)$ given by (9.10). The utilization fraction *f* of those restaurants on every evening can be studied and their distribution is $D(f)$. From numerical studies, their distribution is found to be Gaussian with the most probable utilization fraction $\bar{f} \simeq 0.63$, 0.58 and 0.46 for the cases with $\alpha = 0$, $T \to \infty$, $\alpha = 1$, $T \to \infty$ and $\alpha = 0$, $T \to 0$ respectively. The analytical estimates for $\bar{f}$ in these limits agree very well with the numerical observations.

9.4.5
Statistics of the KPR Problem: A Summary

The KPR problem [25, 31–33] is a repeated game, played between a large number *N* of agents having no interaction amongst themselves. In the KPR problem, prospective customers (agents) choose from *N* restaurants each evening simultaneously (in parallel decision mode); *N* is fixed. Each restaurant has the same price for a meal, but a different rank (agreed upon by all customers) and can serve only one customer any evening. Information regarding the customer distributions for earlier evenings is available to everyone. Each customer's objective is to go to the restaurant with the highest possible rank while avoiding the crowd so as to be able to get dinner there. If more than one customer arrives at any restaurant on any evening, one of them is randomly chosen (each of them are anonymously treated) and is served. The rest do not get dinner that evening.

As we have already mentioned earlier, such problems naturally arise in deciding about the restaurants in any crowded city. A more general example of such a problem would be when society provides hospitals (and beds) in every locality, but the local patients go to hospitals of better rank (commonly perceived) elsewhere, thereby competing with the local patients of those hospitals. Unavailability of treatment in time may be thought of as a lack of service for those people, and consequently, as (social) wastage of service by those unattended hospitals.

A dictator's solution to the KPR problem is the following: the dictator asks every-one to form a queue and then assigns each one a restaurant with rank matching the sequence of the person in the queue on the first evening. Then each person is told to go to the next ranked restaurant in the following evening (for the person in the last ranked restaurant this means going to the first ranked restaurant). This shift proceeds then continuously for successive evenings. This is clearly one of the most efficient solutions (with utilization fraction $\bar{f}$ of the services by the restaurants equal to unity), and the system arrives at this this solution immediately (from the first evening itself). However, in reality this cannot be the true solution of the KPR problem, where each agent decides on his own (in parallel or democratically) every evening, based on complete information about past events. In this game, the customers try to evolve a learning strategy to eventually get dinners at the best pos-sible ranked restaurant, avoiding the crowd. The evolution of these strategies take considerable time to converge, and even then the eventual utilization fraction $\bar{f}$ is far below unity.

The symmetric stochastic strategy chosen by each agent is such that at any time t, the probability $p_k(t)$ to arrive at the k-th ranked restaurant is given by (9.10).

For any natural number α and $T \to \infty$, an agent goes to the k-th ranked restau-rant with probability $p_k(t) = k^\alpha / \sum k^\alpha$; which, in the limit $T \to \infty$ in (9.10) gives $p_k(t) = k^\alpha / \sum k^\alpha$.

If an agent selects any restaurant with equal probability p, then the probability that a single restaurant is chosen by m agents is given by

$$\Delta(m) = \binom{N}{m} p^m (1-p)^{N-m} . \tag{9.16}$$

Therefore, the probability that a restaurant with rank k is not chosen by any of the agents will be given by

$$\Delta_k(m = 0) = \binom{N}{0} (1-p_k)^N ; \quad p_k = \frac{k^\alpha}{\sum k^\alpha}$$

$$\simeq \exp\left(\frac{-k^\alpha N}{\widetilde{N}}\right) \quad \text{as} \quad N \to \infty , \tag{9.17}$$

where $\widetilde{N} = \sum_{k=1}^{N} k^\alpha \simeq \int_0^N k^\alpha dk = N^{\alpha+1}/(\alpha+1)$. Hence,

$$\Delta_k(m = 0) = \exp\left(-\frac{k^\alpha (\alpha+1)}{N^\alpha}\right) . \tag{9.18}$$

Therefore, the average fraction of agents getting dinner in the k-th ranked restau-rant is given by

$$\bar{f}_k = 1 - \Delta_k (m = 0) . \tag{9.19}$$

Naturally, for $\alpha = 0$, the problem corresponding to random choice $\bar{f}_k = 1 - e^{-1}$, giving $\bar{f} = \sum \bar{f}_k / N \simeq 0.63$ and for $\alpha = 1$, $\bar{f}_k = 1 - e^{-2k/N}$ giving $\bar{f} = \sum \bar{f}_k / N \simeq 0.58$.

In the KPR problem, the decision made by each agent each evening t is independent and is based on the information about the rank k of the restaurants and their previous occupancy, given by the numbers $n_k(t-1) \ldots n_k(0)$. For several stochastic strategies, only $n_k(t-1)$ is utilized and each agent chooses the k-th ranked restaurant with probability $p_k(t)$ given by (9.10). The utilization fraction f_k of the k-th ranked restaurants on every evening is studied and their average (over k) distributions $D(f)$ are studied numerically as well as analytically, and one finds [25, 31, 32] their distributions to be Gaussian with the most probable utilization fraction $\bar{f} \simeq 0.63$, 0.58 and 0.46 for the cases with $\alpha = 0$, $T \to \infty$; $\alpha = 1$, $T \to \infty$; and $\alpha = 0$, $T \to 0$ respectively. For the stochastic crowd-avoiding strategy discussed in [33], where $p_k(t+1) = 1/n_k(t)$ for $k = k_0$, the restaurant visited by the agent last evening, and $= 1/(N-1)$ for all other restaurants ($k \neq k_0$), one gets the best utilization fraction $\bar{f} \simeq 0.796$, and the analytical estimates for $\bar{f}$ in these limits agree very well with the numerical observations. Also, the time required to converge to the above value of $\bar{f}$ is independent of N (see Figure 9.16).

The KPR problem has similarity with the Minority Game Problem [15, 27] as in both the games, herding behavior is punished and diversity is encouraged. Also, both involve the learning of the agents from past successes. Of course, KPR has some simple exact solution limits, a few of which are discussed here. The real challenge is, of course, to design algorithms of learning mixed strategies (e.g., from the pool discussed here) by the agents so that the fair social norm emerges eventually (in N^0 or $\ln N$ order time) even when everyone decides on the basis of their own information independently. As we have seen, some naive strategies give better values of $\bar{f}$ compared to most of the "smarter" strategies, like strict crowd-avoiding strategies. This observation, in fact, compares well with earlier observation in minority games (see, for example, [34]).

It may be noted that all the stochastic strategies, being parallel in computational mode, have the advantage that they converge to solutions at smaller time-steps ($\sim N^0$ or $\ln N$), while for deterministic strategies the convergence time is typically of order of N, which renders such strategies useless in the truly macroscopic ($N \to \infty$) limits. However, deterministic strategies are useful when N is small

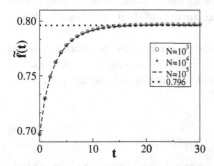

Figure 9.16 Variation of $\bar{f}(t)$ (utilization fraction at time or evening t) for the stochastic crowd-avoiding strategy. Simulations are for $N = 10^3$, 10^4 and 10^5. The convergence to $\bar{f} \simeq 0.796$ in time $\sim O(15)$ is seen to be independent of N.

and rational agents can design appropriate punishment schemes for the deviators (see [26]).

The study of the KPR problem shows that, though not acceptable, a dictated strategy leads to one of the best possible solutions to the problem, with each agent getting dinner at the best ranked restaurant with a period of N evenings, and with best possible value of $\bar{f}(= 1)$, starting from the first evening itself. The parallel decision strategies (employing evolving algorithms by the agents and past information, for example, of $n(t)$), which are necessarily parallel among the agents and stochastic (as in democracy), are less efficient ($\bar{f} \ll 1$; the best one is discussed in [33], giving $\bar{f} \simeq 0.8$ only). Note here that the time required is not dependent on N (see Figure 9.16). We also note that most of the "smarter" strategies lead to much lower efficiency.

9.5
Agent-Based Models for Explaining the Power Law for Price Fluctuations, and so on

9.5.1
Herding Model: Cont–Bouchaud

This model proposed by Cont and Bouchaud (see original paper [35], from which we adapt this part) represents a stock market with N agents, labeled by an integer $1 \leq i \leq N$, trading a single asset, whose price at time t is denoted by $x(t)$. During each time period, an agent may choose either to buy the stock, sell it, or not to trade. The demand for the stock of agent i is represented by a random variable ϕ_i:

1. $\phi_i = +1$ represents a "bull" – an agent willing to buy stock
2. $\phi_i = -1$ represents a "bear" – eager to sell stock
3. $\phi_i = 0$ means that agent i does not trade during a given period.

The random character of individual demands may be due either to heterogeneous preferences or to random resources of the agents, or both. In contrast with many binary choice models in the economics, an agent is allowed to be inactive, that is, not to trade during a given time period t.

Thus, it is assumed that during each time period an agent may either trade one unit of the asset or remain inactive. The demands of each agent i, represented by $\phi_i \in \{-1, 0, +1\}$, gives the aggregate excess demand for the asset at time t and is, therefore,

$$D(t) = \sum_{i=1}^{N} \phi_i(t) \tag{9.20}$$

given the algebraic nature of the ϕ_i. The marginal distribution of the agent's individual demand is assumed to be symmetric:

$$P(\phi_i = +1) = P(\phi_i = -1) = a \quad P(\phi_i = 0) = 1 - 2a \tag{9.21}$$

such that the average aggregate excess demand is zero, that is, the market is considered to fluctuate around equilibrium. A value of $a < 1/2$ allows for a finite fraction of agents not to trade during a given period.

There is a need to relate the aggregate excess demand in a given period to the return or price change during that period. The aggregate excess demand has an impact on the price of the stock, causing it to rise if the excess demand is positive and to fall if it is negative. It is assumed that price change (or return) is proportional to the excess demand:

$$\Delta x = x(t+1) - x(t) = \frac{1}{\lambda} \sum_{i=1}^{N} \phi_i(t) \tag{9.22}$$

where λ is a factor measuring the liquidity or the *market depth* (the excess demand needed to move the price by one unit) and it measures the sensitivity of price to fluctuations in demand. Equation 9.22 may be considered either in absolute terms with $x(t)$ being the price, or as representing *relative* variations of the price, $x(t)$ then considered as the log of the price and its increment as the instantaneous return. The latter has the advantage of guaranteeing the positivity of the price, but for short-run dynamics the two specifications do not differ substantially since the two quantities have the same empirical properties.

In order to evaluate the distribution of stock returns from (9.22), we need to know the *joint* distribution of the individual demands $(\phi_i(t))_{1 \leq i \leq N}$. The simplest case is where the individual demands ϕ_i of different agents are independent identically distributed (*iid*) random variables – referred as the "independent agents" hypothesis. Then the joint distribution of the individual demands is simply the product of individual distributions and the price variation Δx is a sum of N *iid* random variables with finite variance. When the number of terms in (9.22) is large, the *central limit theorem* applied to the sum in (9.22) shows that Δx is well-approximated by a Gaussian distribution, provided the distribution of individual demands has finite variance. Unfortunately, empirically we know that the distributions both of asset returns and of asset price changes have been repeatedly shown to deviate significantly from the Gaussian distribution, exhibiting fat tails and excess kurtosis. Though many distributions verify these conditions, Cont and Bouchaud proposed an exponentially truncated stable distribution so that the tails of the density then have the asymptotic form of an exponentially truncated power law:

$$p(\Delta x) \underset{|\Delta x| \to \infty}{\sim} \frac{C}{|\Delta x|^{1+\mu}} \exp\left(-\frac{\Delta x}{\Delta x_0}\right). \tag{9.23}$$

The exponent μ is found to be close to 1.5 ($\mu \simeq 1.4-1.6$) for a wide variety of stocks and market indices.

Cont and Bouchaud proposed that agents group together in coalitions or *clusters* and that, once a coalition has formed, all its members coordinate their individual demands so that all individuals in a given cluster have the same belief regarding future movements of the asset price. All agents belonging to a given cluster are assumed to have the same demand ϕ_i for the stock. In the context of a stock market, these clusters may correspond, for example, to mutual fund portfolios managed by the same fund manager or to herding among security analysts. The right hand side of (9.22) may, therefore, be rewritten as a sum over clusters:

$$\Delta x = \frac{1}{\lambda} \sum_{a=1}^{k} W_a \phi_a(t) = \frac{1}{\lambda} \sum_{a=1}^{n_c} X_a \tag{9.24}$$

where W_a is the size of cluster a, $\phi_a(t)$ the (common) individual demand of agents belonging to the cluster a, n_c the number of clusters (coalitions) and $X_a = \phi_a W_a$.

One may consider that coalitions are formed through binary links between agents – a link between two agents signifies that they undertake the same action on the market, that is, both of them either buy or sell stocks. For any pair of agents i and j, let p_{ij} be the probability that i and j are linked together. Again, in order to simplify, it is assumed that all links are equally probable: $p_{ij} = p$ (independent of i and j). Then $(N - 1)p$ denotes the average number of agents a given agent is linked to. A choice of $p_{ij} = c/N$ gives in $N \to \infty$ limit, a finite $(N - 1)p$; any other choice satisfying this condition is asymptotically equivalent to this choice. The distribution of coalition sizes in the market is thus completely specified by a single parameter c, which represents the willingness of agents to align their actions (this can also be interpreted as the coordination number, measuring the degree of clustering among agents). Thus, it is actually a *random graph* where we consider agents as vertices of a random graph of size N, and the coalitions as connected components of the graph. The properties of large random graphs in the $N \to \infty$ limit were first studied by Erdös and Renyi. One can show that for $c = 1$ the probability density for the cluster size distribution decreases asymptotically as a power law:

$$P(W) \underset{W \to \infty}{\sim} \frac{A}{W^{5/2}} \tag{9.25}$$

while for values of c close to and smaller than 1 ($0 < 1 - c \ll 1$), the cluster size distribution is cut off by an exponential tail:

$$P(W) \underset{W \to \infty}{\sim} \frac{A}{W^{5/2}} \exp\left(-\frac{(c-1)W}{W_0}\right) \tag{9.26}$$

For $c = 1$, the distribution has an infinite variance while for $c < 1$ the variance becomes finite because of the exponential tail. Note that (9.26) is very similar to (9.23) with $\mu = 3/2$, very close to the empirical results.

Lattice version: Percolation model   The Cont–Bouchaud market model, originally restricted to the mathematically solvable random graph limit, can be simulated on lattices, where each occupied site is a trader and the neighboring occupied sites form clusters, just as in site percolation (see appendix).

In the models (see review by Stauffer [36] and references therein), the sites of a d-dimensional lattice with $2 \leq d \leq 7$ are randomly occupied with probability p and empty with probability $(1 - p)$, and the occupied nearest-neighbors form clusters. Each cluster containing traders decides randomly, to buy (with probability a), sell (also with the same probability a), or to remain inactive (with probability $1 - 2a$). Hence, for any time-step Δt, we first find the existing clusters and the number n_s of clusters, each containing s traders. Then each cluster randomly decides whether to buy, sell, or remain inactive with the above mentioned probabilities. The parameter a is called the "activity" and the increase in activity is equivalent to the increase in the time unit, since a is the fraction of traders which are active per unit time. Thus, small values of a correspond to small time intervals and large values of a (with the maximum of 0.5) correspond to large time intervals.

Then the relative price change for one time-step is considered proportional to the difference of the total demand D and total supply S:

$$R(t) = \ln P(t + \Delta t) - \ln P(t) \propto \sum_s n_s^{\text{buy}} s - \sum_s n_s^{\text{sell}} s , \tag{9.27}$$

where the constant of proportionality is taken to be unity.

The concentration p is either fixed or varies over the interval between zero and unity, or between zero and p_c. The market price is driven by the difference between the total demand and supply; the logarithm of the price changes proportionally to this difference (or to the square-root of the difference). The results are not so good if the price change is no longer proportional to the difference between demand and supply, but, for instance, to the relative difference $(D-S)/(D+S)$ or to a hyperbolic tangent $\tanh[\text{constant} \times (D - S)]$. Moreover, if we take one time-step Δt to be very small so that only one cluster of traders can trade during this time interval, then the probability distribution $P(R)$ is completely symmetric about zero (as in real stock markets) and just follows the distribution n_s of clusters. The distribution, right at $p = p_c$ is $n_s \propto 1/s^\tau$ with $2 < \tau < 2.5$ in two to infinite dimensions. If the time-step Δt is large so that all the traders can trade in each time-step, then the probability distribution $P(R)$ is closer to a Gaussian distribution (see Figure 9.17). When the time-step is in the intermediate range, the price changes are bell-shaped with power law tails. This crossover to Gaussian behavior with the variation of a is also observed in reality.

More variations have been tried and studies have been made to understand the dynamics of financial markets and reproduce stylized facts. Many such attempts are reviewed very well by Stauffer [36].

9.5.2
Strategy Groups Model: Lux–Marchesi

The model of Lux and Marchesi (see the original papers [5, 38, 39], from which we adapt this part) supports the idea that the scaling laws in finance arise from mutual interactions of participants although the "news arrival process" does not have either the power law scaling or any temporal dependence in volatility. However, this is in contradiction to the "efficient market hypothesis" in economics [40], which assumes that all the available information about future earning prospects is instantly

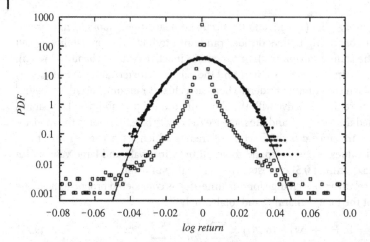

Figure 9.17 Crossover from bell-shaped to Gaussian distribution. Reproduced from Kullmann and Kertesz [37].

processed and it is immediately reflected in the movement of the financial prices. According to the hypothesis, the scaling in the price changes would simply reflect similar scaling in the "news arrival process" that influence them.

In the LM model, the traders are divided into two groups of speculators: (a) "fundamentalists" – who follow the efficient market hypothesis by expecting the price p to follow the so-called fundamental value of the asset p_f, which is the discounted sum of expected future earnings, for example, dividend payments, and (b) "noise traders" or "chartists" – who do not believe in an immediate tendency of the price to revert to its underlying fundamental value. A fundamentalist strategy consists of buying/selling when the actual market price is believed to be below/above the fundamental value. The chartists instead attempt to identify price trends and patterns, and also consider the behavior of other traders as a source of information, which results in a herding tendency. Furthermore, the individuals in the noise trader group itself are distinguished as "optimists", who are more likely to buy if the market is rising than "pessimists". In the model there are movements of individuals from one group to another together with the changes of the fundamental value and the price changes resulting from the market operations of the traders. The dynamics of the LM model is illustrated in Figure 9.18.

There is also some kind of feedback mechanism between group dynamics, and price adjustment occurs in the presence of imbalances between demand and supply.

Formally, the model considers a market with N agents that can be part of two distinct groups of traders: n_f traders are "fundamentalists", who share an exogenous idea p_f of the value of the current price p, and n_c traders are "chartists", who make assumptions on the price evolution based on the observed trend (mobile average). The total number of agents is constant, so that $n_f + n_c = N$ at any time.

At each time-step, the price can be moved up or down with a fixed jump size of ± 0.01 (a tick). The probability to go up or down is directly linked to the excess

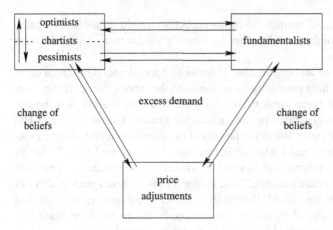

Figure 9.18 Lux–Marchesi model dynamics, following Samanidou *et al.* [5].

demand ED through a coefficient β. The demand of each group of agents is determined as follows:

- Each fundamentalist trades a volume V_f accordingly (through a coefficient γ) to the deviation of the current price p from the perceived fundamental value p_f: $V_f = \gamma(p_f - p)$.
- Each chartist trades a constant volume V_c. Denoting n_+ the number of optimistic (buyer) chartists and n_- the number of pessimistic (seller) chartists, the excess demand by the whole group of chartists is written $(n_+ - n_-)V_c$.

Finally, assuming that there exist some noise traders on the market with random demand μ, the global excess demand is written:

$$ED = (n_+ - n_-)\, V_c + n_f \gamma(p_f - p) + \mu \,. \tag{9.28}$$

The probability that the price goes up (respectively, down) is then defined to be the positive (respectively, negative) part of βED.

As in the Cont–Bouchaud model [35], the authors expect non-trivial features of the price series to result from herding behavior and transitions between groups of traders. A mimicking behavior among chartists is thus proposed. The n_c chartists can change their view on the market (optimistic, pessimistic), their decision being based on a clustering process modeled by an opinion index $x = (n_+ - n_-)/n_c$ representing the weight of the majority. The probabilities π_+ and π_- to switch from one group to another are formally written:

$$\pi_\pm = v\frac{n_c}{N}e^{\pm U} \quad U = \alpha_1 x + \alpha_2 p/v \,, \tag{9.29}$$

where v is a constant, and α_1 and α_2 reflect respectively the weight of the majority's opinion and the weight of the observed price in the chartists' decision. Lastly, transitions between fundamentalists and chartists are also allowed, decided by the comparison of expected returns. We do not explicitly reproduce the other probabilities of transitions of the model, as they are formally similar to $\pi_\pm$ with a different function U.

In their results, the authors are able to derive an approximate differential equation in continuous time governing mean values of the model, thus deriving a stationary solution for these mean values. They are also able to show that the distribution of returns generated by their model have excess kurtosis. Using a Hill estimator, they fit a power law to the fat tails of the distribution and observe exponents grossly ranging from 1.9 to 4.6. The authors also check for hints of volatility clustering: absolute returns and squared returns exhibit a slow decay of autocorrelation, while raw returns do not. Thus, it appears that a model such as this can grossly fit some "stylized facts", though the number of parameters involved, and the quite obscure rules of transition between agents makes the identification of sources of phenomena and calibration to market data very difficult.

9.6
Spin-Based Model of Agent Interaction

Financial markets are subject to long periods of polarized behavior, such as bull-market or bear-market phases, in which the vast majority of market participants seem to almost exclusively choose one action (between buying or selling) over the other. From the point of view of conventional economic theory, such events are thought to reflect the arrival of "external news" that justifies the observed behavior. However, empirical observations of the events leading up to such market phases, as well events occurring during the lifetime of such a phase, have often failed to find significant correlation between news from outside the market and the behavior of the agents comprising the market. In this section, we explore the alternative hypothesis that the occurrence of such market polarizations are due to interactions amongst the agents in the market, and not due to any influence external to it. In particular, Sinha and Raghavendra [41] have presented a model where the market (i.e., the aggregate behavior of all the agents) is observed to become polarized even though individual agents regularly change their actions (buy or sell) on a time-scale much shorter than that of the market polarization phase.

Analysis of the empirical data from different financial markets has led to the discovery of several *stylized facts*, that is, features that are relatively invariant with respect to the particular market under study. One such phenomenon that has been widely reported in financial markets is the existence of *polarized* phases, when the majority of market participants seem to opt exclusively to buy rather than sell (or vice versa) for prolonged periods. Such bull-market (or bear-market) phases, when the market exhibits excess demand (or supply) relative to the market *equilibrium* state, where the demand and supply are assumed to balance each other, are quite common and may be of substantial duration. Such events are less spectacular than episodes of speculative bubbles and crashes [42], which occur over a relatively faster time-scale; however, their impact on the general economic development of nations maybe quite significant, partly because of their prolonged nature. Hence, it is important to understand the reasons for occurrence of such market polarizations.

Conventional economic theory seeks to explain such events as reflections of news external to the market. If it is indeed true that particular episodes of market polarizations can only be understood as responses to specific historical contingencies, then it should be possible to identify the significant historical events that precipitated each polarized phase. However, although *a posteriori* explanation of any particular event is always possible, there does not seem to be any general explanation for such events in terms of extra-market variables, especially one that can be used to predict future market phases.

In contrast to this preceding approach, one can view the market behavior entirely as an emergent outcome of the interactions between the agents comprising the market. While external factors may indeed influence the actions of such agents, and hence the market, they are no longer the main determinants of market dynamics. In this explanatory framework, the occurrence of market polarization can be understood in terms of time evolution of the collective action of agents. It is important to note here that the individual agents are assumed to exercise their free will in choosing their particular course of action (i.e., whether to buy or sell). However, in any real-life situation, an agent's action is also determined by the information it has access to about the possible consequences of the alternative choices available to it. In a free market economy, devoid of any central coordinating authority, the personal information available to each agent may be different. Thus the emergence of market behavior, which is a reflection of the collective action of agents, can be viewed as a self-organized coordination phenomenon in a system of heterogeneous entities.

The simplest model of collective action is one where the action of each agent is completely independent of the others; in other words, agents choose from the available alternatives at random. In the case of binary choice, where only two options are available to each agent, it is easy to see that the emergence of collective action is equivalent to a random walk on a one-dimensional line, with the number of steps equal to the number of agents. Therefore, the result will be a Gaussian distribution, with the most probable outcome being an equal number of agents choosing each alternative. As a result, for most of the time the market will be balanced, with neither excess demand nor supply. While this would indeed be expected in the idealized situation of conventional economic theory, it is contrary to observations in real life indicating strongly polarized collective behavior among agents in a market. In these cases, a significant majority of agents choose one alternative over another, resulting in the market being either in a buying or selling phase. Examples of such strong bimodal behavior has been also observed in contexts other than financial markets, for example, in the distribution of opening gross income for movies released in theaters across the USA [43].

The polarization of collective action suggests that the agents do not choose their course of action completely independently, but are influenced by neighboring agents. In addition, their personal information may change over time as a result of the outcome of their previous choices, for example, whether or not their choice of action agreed with that of the majority.[2] This latter effect is an example of global feedback process that we think is crucial for the polarization of the collective action of agents, and hence, the market.

We now discuss a model suggested by Sinha and Raghavendra [41] for the dynamics of market behavior which takes into account these different effects in the decision process of an agent choosing between two alternatives (e.g., buy or sell) at any given time instant. A phase transition in the market behavior is observed, from an equilibrium state to a far-from-equilibrium state characterized by either excess demand or excess supply under various conditions. However, most strikingly, we observe that the transition to polarized market states occurs when an agent learns to adjust its action according to whether or not its previous choice accorded with that of the majority. One of the striking consequences of this global feedback is that, although individual agents continue to regularly switch between the alternatives available to it, the duration of the polarized phase (during which the collective action is dominated by one of the alternatives) can become extremely long.

We assume that individual agents behave in a rational manner, where rationality is identified with actions that would result in market equilibrium in the absence of interaction between agents. Therefore, for a large ensemble of such non-interacting agents we will observe only small fluctuations about the equilibrium. Here we explore how the situation alters when agents are allowed to interact with each other. The market behavior reflects the collective action of many interacting agents, each deciding to buy or sell based on limited information available to it about the consequences of such action. An example of such limited information available to an agent is news of the overall market sentiment as reflected in market indices such as S&P 500. A schematic diagram of the various influences acting in the market is shown in Figure 9.19.

The model is defined as follows. Consider a population of N agents, whose actions are subject to bounded rationality, that is, they either buy or sell an asset based on information about the action of their neighboring agents and how successful their previous actions were. The fundamental value of the asset is assumed to be unchanged throughout the period. In addition, the agents are assumed to have limited resources, so that they cannot continue to buy or sell indefinitely. However, instead of introducing explicit budget constraints [45], we have implemented gradually diminishing returns for a decision that is taken repeatedly. This is akin to the belief adaptation process in the Weisbuch–Stauffer model of social percolation [46], where making similar choices in successive periods decreases the probability of making the same choice in the subsequent period.

At any given time t, the state of an agent i is fully described by two variables: its choice, S_i^t, and its belief about the outcome of the choice, θ_i^t. The choice can be either *buy* $(= +1)$ or *sell* $(= -1)$, while the belief can vary continuously over a range (initially, it is chosen from a uniform random distribution). At each time-step, every

---

2) This would be the case if, as in Keynes' "beauty contest" analogy for the stock market, agents are more interested in foreseeing how the general public will value certain investments in the immediate future, rather than the long-term probable yields of these investments based on their fundamental value [44].

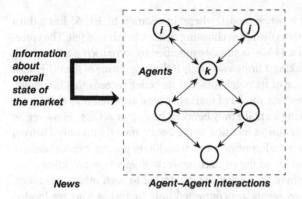

News   **Agent–Agent Interactions**

Figure 9.19 An abstract model of a market. Each agent interacts (interactions indicated by arrows) with a subset of the other agents comprising the market, indicated by the boundary formed from the broken lines. The combined action of all agents results in the overall state of the market. The news of this state is available to all agents, although the information about the individual actions of all agents may not be accessible to any one agent.

agent considers the average choice of its neighbors at the previous instant, and if this exceeds its belief, then it makes the same choice; otherwise, it makes the opposite choice. Then, for the i-th agent, the choice dynamics is described by:

$$S_i^{t+1} = \text{sign}(\sum_{j \in \mathcal{N}} J_{ij} S_j^t - \theta_i^t) , \qquad (9.30)$$

where $\mathcal{N}$ is the set of neighbors of agent $i(i = 1, \dots, N)$, and sign $(z) = +1$, if $z > 0$, and $= -1$, otherwise. The degree of interaction among neighboring agents, J_{ij}, is assumed to be a constant $(= 1)$ for simplicity and normalized by $z(= |\mathcal{N}|)$, the number of neighbors. In a lattice, $\mathcal{N}$ is the set of spatial nearest neighbors and z is the coordination number, while in the mean field approximation, $\mathcal{N}$ is the set of all other agents in the system and $z = N - 1$.

The individual belief, θ evolves over time as:

$$\theta_i^{t+1} = \begin{cases} \theta_i^t + \mu S_i^{t+1} + \lambda S_i^t , & \text{if } S_i^t \neq \text{sign}(M^t) , \\ \theta_i^t + \mu S_i^{t+1} , & \text{otherwise} , \end{cases} \qquad (9.31)$$

where $M^t = (1/N) \sum_j S_j^t$ is the fractional excess demand and describes the overall state of the market at any given time t. The adaptation rate μ governs the time-scale of diminishing returns, over which the agent switches from one choice to another in the absence of any interactions between agents. The learning rate λ controls the process by which an agent's belief is modified when its action does not agree with that of the majority at the previous instant. As mentioned earlier, the desirability of a particular choice is assumed to be related to the fraction of the community choosing it. Hence, at any given time, every agent is trying to coordinate its choice with that of the majority. Note that, for $\mu = 0, \lambda = 0$, the model reduces to the well-known zero-temperature, random field Ising model (RFIM) of statistical physics.

One can also consider a 3-state model, where, in addition to ± 1, S_i^t has a third state, 0, which corresponds to the agent choosing neither to buy nor sell. The corresponding choice dynamics, (9.30), is suitably modified by introducing a threshold, with the choice variable taking a finite value only if the magnitude of the difference between the average choice of its neighbors and its belief exceeds this threshold. This is possibly a more realistic model of markets where an agent may choose not to trade, rather than making a choice only between buying or selling. However, as the results are qualitatively almost identical to the 2-state model introduced before, in the following section we shall confine our discussion to the latter model only.

As the connection topology of the contact network of agents is not known, we consider both the case where the agents are connected to each other at random, as well as, the case where agents are connected only to agents who are located at spatially neighboring locations. Both situations are idealized, and in reality is likely to be somewhere in between. However, it is significant that in both of these very different situations we observe market polarization phases which are of much longer duration compared to the timescale at which the individual agents switch their choice state (S).

9.6.1
Random Network of Agents and the Mean Field Model

We choose the z neighbors of an agent at random from the $N - 1$ other agents in the system. We also assume this randomness to be "annealed", that is, the next time the same agent interacts with z other agents, they are chosen at random anew. Thus, by ignoring spatial correlations, a mean field approximation is achieved.

For $z = N - 1$, that is, when every agent has the information about the entire system, it is easy to see that, in the absence of learning ($\lambda = 0$), the collective decision M follows the evolution equation rule:

$$M^{t+1} = \text{sign}\left[(1 - \mu)M^t - \mu \sum_{\tau=1}^{t-1} M^\tau \right]. \tag{9.32}$$

For $0 < \mu < 1$, the system alternates between the states $M = \pm 1$ (i.e., every agent is a buyer, or every agent is a seller) with a period $\sim 4/\mu$. The residence time at any one state ($\sim 2/\mu$) increases with decreasing μ, and for $\mu = 0$, the system remains fixed at one of the states corresponding to $M = \pm 1$, as expected from RFIM results. At $\mu = 1$, the system remains in the market equilibrium state (i.e., $M = 0$). Therefore, we see a transition from a bimodal distribution of the fractional excess demand, M, with peaks at non-zero values, to an unimodal distribution of M centered about 0, at $\mu_c = 1$. When we introduce learning, so that $\lambda > 0$, the agents try to coordinate with each other and in the limit $\lambda \to \infty$ it is easy to see that $S_i = \text{sign}(M)$ for all i, so that all the agents make identical choice. In the simulations, we note that the bimodal distribution is recovered for $\mu = 1$ when $\lambda \geq 1$.

For finite values of z, the population is no longer "well-mixed" and the mean-field approximation becomes less accurate the lower z is. For $z \ll N$, the critical value of μ at which the transition from a bimodal to a unimodal distribution occurs in the absence of learning, $\mu_c < 1$. For example, $\mu_c = 0$ for $z = 2$, while it is $3/4$ for $z = 4$. As z increases, μ_c quickly converges to the mean-field value, $\mu_c = 1$. On introducing learning ($\lambda > 0$) for $\mu > \mu_c$, we again notice a transition to a state corresponding to all agents being buyers (or all agents being sellers), with more and more agents coordinating their choice.

9.6.2
Agents on a Spatial Lattice

To implement the model when the neighbors are spatially related, we consider d-dimensional lattices ($d = 1, 2, 3$) and study the dynamics numerically. We report results obtained in systems with absorbing boundary conditions; using periodic boundary conditions leads to minor changes, but the overall qualitative results remain the same.

In the absence of learning ($\lambda = 0$), starting from an initial random distribution of choices and beliefs, we observe only very small clusters of similar choice behavior, and the fractional excess demand, M, fluctuates around 0. In other words, at any given time an equal number of agents (on average) make opposite choices so that the demand and supply are balanced. In fact, the most stable state under this condition is one where neighboring agents in the lattice make opposite choices. This manifests itself as a checkerboard pattern in simulations carried out in one- and two-dimensional square lattices (see, for example, Figure 9.20, top left).

The introduction of learning in the model ($\lambda > 0$) gives rise to significant clustering among the choices of neighboring agents (Figure 9.20), as well as a large non-zero value for the fractional excess demand, M. We find that the probability distribution of M evolves from a single peak at 0, to a bimodal distribution (having two peaks at finite values of M, symmetrically located about 0) as λ increases from 0 [47]. The fractional excess demand switches periodically from a positive value to a negative value having an average residence time which increases sharply with λ and with N (Figure 9.21).

For instance, when λ is very high relative to μ, we see that M gets locked into one of two states (depending on the initial condition), corresponding to the majority preferring either one or the other choice. This is reminiscent of *lock-in* for certain economic systems subject to positive feedback [48]. The special case of $\mu = 0, \lambda > 0$ also results in a lock-in of the fractional excess demand, with the time required to get to this state increasing rapidly as $\lambda \to 0$. For $\mu > \lambda > 0$, large clusters of agents with identical choice are observed to form and dissipate throughout the lattice. After sufficiently long times, we observe the emergence of structured patterns having the symmetry of the underlying lattice, with the behavior of agents belonging to a particular structure being highly correlated. Note that these patterns are dynamic, being essentially concentric waves that emerge at the center and travel to the boundary of the region, and continually expand until they meet another such pat-

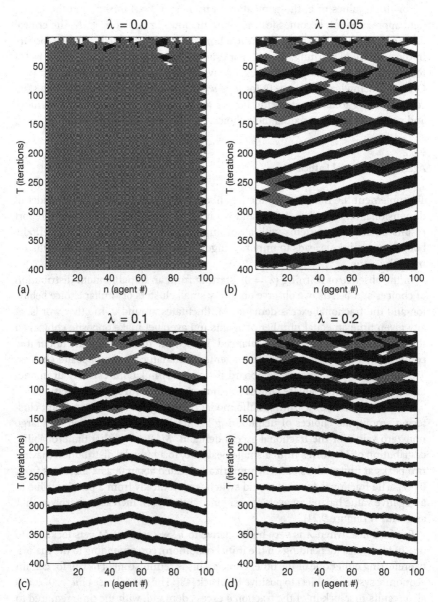

Figure 9.20 The spatiotemporal evolution of choice (*S*) among 100 agents, arranged in a one-dimensional lattice, with a time-evolution of up to 400 iterations starting from a random configuration shown along the vertical axis. The colors (white or black) represent the different choice states (buy or sell) of indi-

vidual agents. The adaptation rate $\mu = 0.1$, and the learning rate λ increases from 0 (a) to 0.2 (d). Note that, as λ increases, one of the two states becomes dominant with the majority of agents at any given time always belonging to this state, although each agent regularly switches between the two states.

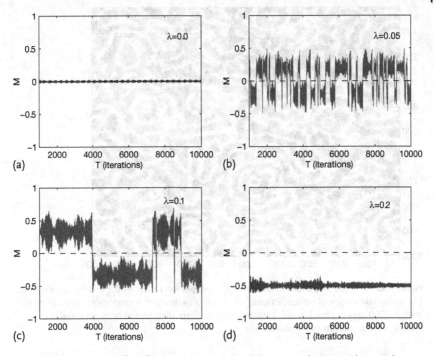

Figure 9.21 Time series of the fractional excess demand M in a two-dimensional square lattice of 100×100 agents. The adaptation rate $\mu = 0.1$ and the learning rate λ is increased from 0 to 0.2 to show the divergence of the residence time of the system in polarized configurations.

tern. Where two patterns meet, their progress is arrested and their common boundary resembles a dislocation line. In the asymptotic limit, several such patterns fill up the entire system. Ordered patterns have previously been observed in the spatial prisoner's dilemma model [49]. However, in the present case, the patterns indicate the growth of clusters with strictly correlated choice behavior. The central site in these clusters act as the "opinion leader" for the entire group. This can be seen as analogous to the formation of "cultural groups" with shared beliefs [50]. It is of interest to note that distributing λ from a random distribution among the agents disrupt the symmetry of the patterns, but we still observe patterns of correlated choice behavior (Figure 9.22). It is the global feedback ($\lambda \neq 0$) which determines the formation of large connected regions of agents having similar choice behavior.

To get a better idea about the distribution of the magnitude of fractional excess demand, we have looked at the rank-ordered plot of M, that is, the curve obtained by putting the highest value of M in position 1, the second highest value of M in position 2, and so on. As explained in [51], this plot is related to the cumulative distribution function of M. The rank-ordering of M shows that with $\lambda = 0$, the distribution varies smoothly over a large range, while for $\lambda > 0$, the largest values are close to each other, and the distribution shows a sudden decrease. In other words, the presence of global feedback results in a high frequency of market events

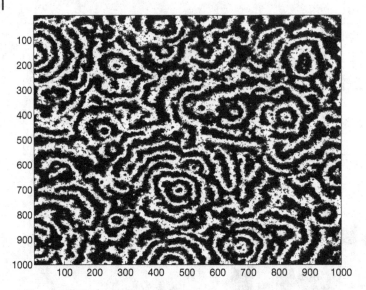

Figure 9.22 The spatial pattern of choice (S) in a two-dimensional square lattice of 1000 × 1000 agents after 2×10^4 iterations starting from a random configuration. The adaptation rate $\mu = 0.1$ and the learning rate λ of each agent is randomly chosen from an uniform distribution between 0 and 0.1.

where the choices of a large number of agents become coordinated, resulting in excess demand or supply. Random distribution of λ among the agents results in only small changes to the curve (Figure 9.23). However, the choice of certain distribution functions for λ elevates the highest values of M beyond the trend of the curve, which reproduces an empirically observed feature in many popularity distributions that has sometimes been referred to as the "king effect" [52, 53].

In summary, the model discussed above shows the emergence of collective action defining market behavior through interactions between agents who make decisions based on personal information that change over time through adaptation and learning. We find that introducing these effects produces market behavior marked by two phases: (a) market equilibrium, where the buyers and sellers (and hence, demand and supply) are balanced, and (b) market polarization, where either the buyers or the sellers dominate (resulting in excess demand or excess supply). There are multiple mechanisms by which the transition to market polarization occurs, for example, (i) keeping the adaptation and learning rate fixed, but switching from an initially regular neighborhood structure (lattice) to a random structure (mean-field) one sees a transition from market equilibrium to market polarization; (ii) in the lattice, by increasing the learning rate λ (keeping μ fixed) one sees a transition from equilibrium to polarization behavior; and (iii) in the case where agents have randomly chosen neighbors, by increasing the adaptation rate μ beyond a critical value (keeping λ fixed) one sees a transition from polarized to equilibrium market state.

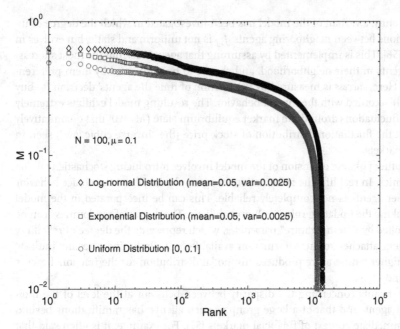

Figure 9.23 Rank-ordered plot of M for a two-dimensional lattice of 100×100 agents. The adaptation rate $\mu = 0.1$ and the learning rate λ of each agent is chosen from three different random distributions: uniform (circle), exponential (square) and log-normal (diamond).

The principal interesting observation seems to be that while, on the one hand, individual agents regularly switch between alternate choices as a result of adapting their beliefs in response to new information, on the other hand, their collective action (and hence, the market) may remain polarized in any one state for a prolonged period. Apart from financial markets, such phenomena has been observed, for example, in voter behavior, where preferences have been observed to change at the individual level which is not reflected in the collective level, so that the same party remains in power for extended periods. Similar behavior possibly underlies the emergence of cooperative behavior in societies. As in our model, each agent can switch regularly between cooperation and defection; however, society as a whole can get trapped in a non-cooperative mode (or a cooperative mode) if there is strong global feedback.

Even with randomly distributed λ we see qualitatively similar results, which underlines their robustness. In contrast to many current models, we have not assumed a priori existence of contrarian and trend-follower strategies among the agents [54]. Rather, such behavior emerges naturally from the micro-dynamics of agents' choice behavior. Further, we have not considered external information shocks, so that all observed fluctuations in market activity is endogenous. This is supported by recent empirical studies which have failed to observe any significant correlation between market movements and exogenous economic variables like investment climate [55].

Recently, a variant of the model has been investigated in which the degree of interactions between neighboring agents J_{ij} is not uniform and static, but evolves in time [56]. This is implemented by assuming that agents seek out the most successful agents in their neighborhood, and choose to be influenced by them preferentially. Here, *success* is measured by the fraction of time the agents' decision (to buy or sell) accorded with the market behavior. The resulting model exhibits extremely large fluctuations around the market equilibrium state ($M = 0$) that quantitatively match the fluctuation distribution of stock price (the "inverse cubic law") seen in real markets.

Another possible extension of the model involves introducing stochasticity in the dynamics. In real life, the information an agent obtains about the choice behavior of other agents is not completely reliable. This can be incorporated in the model by making the updating rule (9.30) probabilistic. The degree of randomness can be controlled by a "temperature" parameter, which represents the degree of reliability an agent attaches to the information available to it. Preliminary results indicate that higher temperature produces unimodal distribution for the fractional excess demand.

The results concerning the disparity between behavior at the level of the individual agent, and that of a large group of such agents, has ramifications beyond the immediate context of financial markets [57]. For example, it is often said that "democracies rarely go to war" because getting a consensus about such a momentous event is difficult in a society where everyone's free opinion counts. This would indeed have been the case had it been true that the decision of each agent is made independently of others, and is based upon all evidence available to it. However, such an argument underestimates how much people are swayed by the collective opinion of those around them, in addition to being aroused by demagoguery and yellow journalism. Studying the harmless example of how market polarizations occur even though individuals may regularly alternate between different choices may help us in understanding how more dangerous mass madness can occur in a society.

References

1 Chakrabarti, A.S. and Chakrabarti, B.K. (2009) *Physica A*, **388**, 4151.

2 Parisi, G. (1999) *Physica A*, **263**, 557.

3 Arthur, W.B. (1999) *Science*, **284**, 107.

4 Lux, T. and Westerhoff, F. (2009) *Nature Physics*, **5**, 2.

5 Samanidou, E., Zschischang, E., Stauffer, D., and Lux, T. (2007) *Reports on Progress in Physics*, **70**, 409–450.

6 Chakraborti, A., Muni Toke, I., Patriarca M., and Abergel, A. (2009) *"Econophysics: Empirical facts and agent-based models"*, available at arXiv:0909.1974.

7 Smith, A. (1776) *An Inquiry into the Nature and Causes of the Wealth of Nations*, Strahan and Caddell, London.

8 Samuelson, P.A. and Nordhaus, W.D. (1998) *Economics*, sixteenth edition, McGraw-Hill Companies, New York, 29–32.

9 Bak, P. (1996) *How Nature Works: The Science of Self-Organised Criticality*, Copernicus Press, New York.

10 Keynes, J.M. (1973) *General Theory of Employment, Interest and Money*, Roy-

al Economic Society, Macmillan Press, London.

11 Chakraborti, A., Pradhan, S., and Chakrabarti, B.K. (2001) *Physica A*, **297**, 253.

12 Dragulescu, A. and Yakovenko, V.M. (2000) *Eur. Phys. J. B*, **17**, 723–729.

13 Chakraborti, A. and Chakrabarti, B.K. (2000) *Eur. Phys. J. B*, **17**, 167–170.

14 Arthur, W.B. (1994) Amer. Econ. Rev. (Papers and Proceedings), **84**, 406, available at http://tuvalu.santafe.edu/~wbarthur/Papers/El_Farol.html

15 Challet, D., Marsili, M., and Zhang, Y.-C. (2005) *Minority Games: Interacting Agents in Financial Markets*, Oxford University Press, Oxford.

16 Challet, D. and Zhang, Y.C. (1997) *Physica A*, **246**, 407.

17 Li, Y., Riolo, R., and Savit, R. (2000) Evolution in minority games. (I). Games with a fixed strategy space. *Physica A*, **276**, 234–264.

18 Li, Y., Riolo, R., and Savit, R. (2000) Evolution in minority games. (II). Games with variable strategy spaces. *Physica A*, **276**, 265–283.

19 Sysi-Aho, M., Chakraborti, A., and Kaski, K. (2003) Adaptation using hybridized genetic crossover strategies. *Physica A*, **322**, 701.

20 Sysi-Aho, M., Chakraborti, A., and Kaski, K. (2003) Biology helps you to win a game. *Physica Scripta T*, **106**, 32.

21 Sysi-Aho, M., Chakraborti, A., and Kaski, K. (2003) Intelligent minority game with genetic crossover strategies. *Eur. Phys. J. B*, **34**, 373.

22 Sysi-Aho, M., Chakraborti, A., and Kaski, K. (2004) Searching for good strategies in adaptive minority games. *Phys. Rev. E*, **69**, 036125.

23 Orléan, A. (1995) Bayesian interactions and collective dynamics of opinion: Herd behavior and mimetic contagion. *J. Econ. Behav. Organ.*, **28**, 257–274.

24 Banerjee, A.V. (1992) A simple model of herd behavior, *Quart. J. Econ.*, **110** (3), 797–817.

25 Chakrabarti, A.S., Chakrabarti, B.K., Chatterjee, A. and Mitra, M. (2009) The Kolkata Paise Restaurant Problem and

26 Kandori, M. (2008) Repeated Games, *The New Palgrave Dictionary of Economics*, second edition, Palgrave Macmillan, New York, Vol. 7, 98–105.

27 Brian Arthur, W. (1994) Inductive reasoning and bounded rationality: El Farol Problem, *American Economics Association Papers & Proceedings*, **84**, 406.

28 Nowak, M. and Sigmund, K. (1993) A strategy of win-stay, lose-shift that outperforms tit-for-tat in the Prisoner's Dilemma game, *Nature*, **364**, 56–58.

29 Freckleton, R.P. and Sutherland, W.J. (2001) Do Power Laws Imply Self-regulation? *Nature*, **413**, 382.

30 Smethurst, D.P. and Williams, H.C. (2001) Power laws: are hospital waiting lists self-regulating? *Nature*, **410**, 652–653.

31 Ghosh, A. and Chakrabarti, B.K. (2009) Wolfram Demonstration of the Kolkata Paise Restaurant (KPR) Problem. http://demonstrations.wolfram.com/.

32 Ghosh, A., Chakrabarti, A.S., and Chakrabarti, B.K. (2010) In *Econophysics & Economis of Games, Social Choices & Quantitative Techniques*, pp. 3–9, Springer, Milano.

33 Ghosh, A., Chatterjee, A., Mitra, M., and Chakrabarti, B.K. (2010) *New J. Phys.*, **12**, 075033.

34 Satinover, J.B. and Sornette, D. (2007) *European Physical Journal B*, **60**, 369–384.

35 Cont, R. and Bouchaud, J.P. (2000) Herd behavior and aggregate fluctuations in financial markets. *Macroecon. Dynam.*, **4**, 170–196.

36 Stauffer, D. (2001) Percolation models of financial market dynamics, *ACS*, **04** (01), 19–27.

37 Kullmann, L. and Kertesz, J. (2001) Crossover to Gaussian Behavior In Herding Market Models. *Int. J. Mod. Phys. C*, **12**, 1211.

38 Lux, T. and Marchesi, M. (1999) *Nature*. **397**, 498.

39 Lux, T. and Marchesi, M. (2000) *Int. J. Theo. Appl. Finance*, **3**, 67.

40 Fama, E.F. (1970) *J. Finance*, **25**, 383.

41 Sinha, S. and Raghavendra, S. (2006) Market polarization in presence of individual choice volatility (ed. C. Bruun), *Advances in Artificial Economics*, Springer, Berlin, pp. 177–190.

42 Schiller, R.J. (2000) *Irrational Exuberance*, Princeton University Press, Princeton.

43 Sinha, S. and Raghavendra, S. (2004) Hollywood Blockbusters And Long-Tailed Distributions: An empirical study of the popularity of movies, *Eur. Phys. J. B*, **42**, 293–296.

44 Keynes, J.M. (1934) *The General Theory of Employment, Interest and Money*, Harcourt, New York.

45 Iori, G. (2002) A microsimulation of traders activity in the stock market: the role of heterogeneity, agents' interaction and trade frictions, *J. Econ. Behav. Organ.*, **49**, 269–285.

46 Weisbuch, G. and Stauffer, D. (2003) Adjustment and social choice, *Physica A*, **323**, 651–662.

47 Sinha, S. and Raghavendra, S. (2005) Emergence of Two-Phase Behavior In Markets Through Interaction and Learning In Agents with Bounded Rationality (ed. H. Takayasu), *Practical Fruits of Econophysics*, Springer, Tokyo, pp. 200–204.

48 Arthur, B.W. (1989) Competing technologies, increasing returns, and lock-in by historical events, *Economic J.*, **99**, 116–131.

49 Nowak, M.A. and May, R.M. (1992) Evolutionary games and spatial chaos, *Nature*, **359**, 826–829.

50 Axelrod, R. (1997) The dissemination of culture: A model with local convergence and global polarization, *J. Conflict Resolution*, 41,203–226.

51 Adamic, L.A. and Huberman B.A. (2002) Zipf's law and the Internet, *Glottometrics*, **3**, 143–150.

52 Laherrere, J. and Sornette, D. (1998) Stretched exponential distributions in nature and economy: "Fat Tails" with characteristic scales, *Eur. Phys. J. B*, **2**, 525–539.

53 Davies, J.A. (2002) The individual success of musicians, like that of physicists, follows a stretched exponential distribution, *Eur. Phys. J. B*, **4**, 445–447.

54 Lux, T. (1995) Herd behaviour, bubbles and crashes, *Economic J.*, **105**, 881–896.

55 Kaizoji, T. (2000) Speculative bubbles and crashes in stock markets: an interacting-agent model of speculative activity. *Physica A*, **287**, 493–506.

56 Sinha, S. (2006) Apparent Madness Of Crowds: Irrational Collective Behavior Emerging From Interactions Among Rational Agents (eds A. Chatterjee, B.K. Chakrabarti), *Econophysics of Stock and Other Markets*, Springer, Milano, pp. 159–162.

57 Sinha, S. and Pan, R.K. (2006) How a "Hit" is born: The emergence of popularity from the dynamics of collective choice (eds A. Chatterjee, A. Chakraborti, B.K. Chakrabarti), *Handbook of Econophysics and Sociophysics*, Wiley-VCH Verlag GmbH, Berlin, pp. 417–447.

10

...and Individuals don't Interact Randomly: Complex Networks

"We are caught in an inescapable network of mutuality, tied in a single garment of destiny. Whatever affects one directly, affects all indirectly."
— Martin Luther King, Jr., *Letter from Birmingham Jail*, April 16, 1963

"Financial markets are the machines in which much of human welfare is decided; yet we know more about how our car engines work than about how our global financial system functions. We lurch from crisis to crisis. In a networked world, mayhem in one market spreads instantaneously to all others – and we have only the vaguest of notions how this happens, or how to regulate it."
— Benoit B. Mandelbrot, *The (Mis)Behavior of Markets*

We have encountered networks earlier in the book, for example the minimal spanning trees and graphs constructed from the cross-correlation information for stock returns that show relations between the different stocks in a financial market, in terms of their price movements (Chapter 4). However, a full chapter is necessary to explain in detail such structures in order to emphasize an oft-forgotten fact: economic exchanges (whether it be trading interactions, ownership, credit-debit relations or strategic alliances) do not necessarily happen between any and every pair of randomly chosen agents (be they nations or firms or banks or individuals). Instead, the actual network according to which these interactions take place can often have a non-trivial structure. For example, certain nodes in such a network can have many more links than others (i.e., they are hubs), there could be correlations between the number of links that a node has and that of the nodes that it connects to (assortativity), or, a high degree of clustering of connections between the nodes of a particular subgroup (cliquishness). Some agents may act as brokers, interacting with many otherwise separated groups, while others confine their exchange to the particular region of the network (module) that they are part of. Moreover, such networks need not have their structure fixed over time. The topology of connections may evolve, growing and shrinking as new nodes enter and old nodes leave the system, as well as, via rearrangement of links between existing nodes.

In this chapter, we will first focus on describing the basic terminology of networks. The theory of complex networks have undergone a renaissance in the past decade, and several models of real-life networks have been proposed during this

Econophysics. Sitabhra Sinha, Arnab Chatterjee, Anirban Chakraborti, and Bikas K. Chakrabarti
Copyright © 2011 WILEY-VCH Verlag GmbH & Co. KGaA, Weinheim
ISBN: 978-3-527-40815-3

time. A few of these will be introduced in this chapter. This will be followed by a discussion of the empirical work done on reconstructing certain economic networks. We will focus in particular on the world trade web, the network of international trade between nations. Networks can be also be defined in terms of the space of all possible products that an economy can produce. We will discuss how a network defined on such a space can show the development of different nations as they have changed stress over time on different types of exports. It is not only between firms and nations that economic networks can exist. They can also occur within an organization, as illustrated by the chart of organizational hierarchy. A hierarchical tree is a specific kind of network that has clearly defined levels. We will use such networks to propose an explanation for the long-tailed distribution of income, expanding on an idea originally proposed by Lydall (based on earlier work by Simon and Mandelbrot). Networks can also significantly affect the kind of dynamics that can happen on the nodes. We will thus look at how the network structure governs the stability of the nodal dynamics. We will illustrate the economic ramifications of this by illustrating how shocks can propagate over banking and financial networks.

10.1
What are Networks?

The different components or interacting units of a complex system, when described as a network, are represented by nodes or vertices, and the interactions or connections between the units are represented by edges or links between pairs of nodes (Figure 10.1). Such networks provide a concise mathematical representation of the topology of interactions between the components. Thus, understanding how economic systems work may often depend partially on understanding their patterns of interactions, that is the underlying networks. The graph theoretical framework has provided the potential synergies among researchers across different multidisciplinary fields to come and work together to solve apparently unrelated problems.

Network architecture may have important functional consequences for the whole system. For example, the topology of the network controls the rate at which information propagates through it [1, 2], its robustness under attack or failure of individual components [3], as well as, adaptation and learning processes on it [4]. Recent work has pointed out the crucial role played by the network structure in determining the emergence of collective dynamical behavior, such as, synchronization of nodal activity [5]. Hence, studying these patterns of interactions between the components of a complex system can lead to a better understanding of its dynamical and functional behavior, in addition to throwing light on the evolutionary mechanism leading to it.

There are several reasons for the emergence and rapid development of the science of networks. Many of the insights and advances in this field are due to the recent availability of large quantities of high-resolution data from different sys-

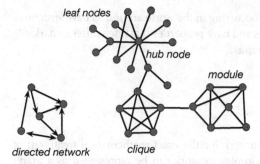

leaf nodes

hub node

module

directed network   *clique*

Figure 10.1 Representations of undirected and directed networks indicating hub and leaf nodes, clique and module.

tems. Obtaining such empirical data has become possible because of technological advances. For example, the network of social interactions among individuals can be constructed from information about the calls they make using their mobile phones [6], leading to better understanding of human social dynamics. There has also been remarkable increase in computational power, and by using this, regularities and patterns in large datasets can be determined. Another reason for the involvement of a large number of physicists in this field is that statistical physics and non-linear dynamics can be used to develop methods and techniques for analyzing and modeling complex networks [7, 8].

The traditional approach in physics for describing an interacting system is to use a lattice embedded in d-dimensional space. Each element of the system is located on a lattice site and interacts with neighboring sites within a range $r (= 1, 2, \cdots)$. On such a regular network, all nodes have the same number of neighbors $(= (2r)^d)$, where r is the range of interaction and d is the dimension of the space on which the lattice is embedded (Figure 10.2a,b). Another commonly used graph in the literature is the Cayley tree or Bathe lattice, where each node has the same number of neighbors but there are no cycles in the structure (Figure 10.2c). At the other extreme, we have the homogeneous random graph (also referred to as Erdős–Rényi or ER graphs) where the edges between any pair of nodes are randomly placed with

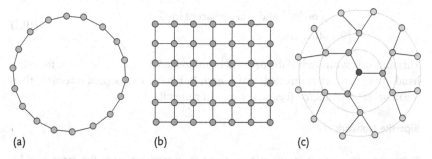

(a)   (b)   (c)

Figure 10.2 Representations of regular graph models: Nodes on (a) a one-dimensional lattice, (b) a two-dimensional lattice, and, (c) a Bethe lattice or Cayley tree with nearest neighbor connections.

probability p. However, networks occurring in the world around us have structures that occur between these extremes and have properties that often differ remarkably from both regular and random graphs.

10.2
Fundamental Network Concepts

Graph theory is the natural framework for the exact mathematical treatment of complex networks. Formally, a complex network can be represented as a graph which is defined in terms of a set of N vertices (or nodes) and E edges (or links). Every edge corresponds to a specific pair of nodes in the graph. We shall use the terms network and graph interchangeably in this chapter.

Depending on the types of its links, a network is classified as directed or undirected (Figure 10.1). A directed network is one whose connections are oriented, that is a direction is associated with the link from one node to the other. In other words, while a link may exist from node i to node j, the connection from j to i may or may not exist. On the other hand, if the links do not have an orientation, that is if i is connected to j then j is also connected to i, then the corresponding network is undirected. Networks in which a number (corresponding to link strength) is associated with each link are known as weighted networks. For most real-world networks, a complex topology is often associated with a large heterogeneity in the strength of the connections (corresponding to their capacity and intensity).

Adjacency Matrix

A pair of nodes that are joined by a link are referred to as being adjacent or neighboring. A complete description of the connection topology of a graph is provided by a tabulation of every connected pair of nodes in it. Alternatively, this information can be gleaned from its adjacency matrix. A matrix $A = \{a_{ij}\}_{N \times N}$ is called the adjacency matrix of a graph G with N nodes, if the elements of A have the following property:

$$a_{ij} = \begin{cases} 1 & \text{if nodes } i \text{ and } j \text{ are adjacent in } G \text{ ,} \\ 0 & \text{otherwise .} \end{cases} \tag{10.1}$$

This matrix is symmetric if the network is undirected (i.e., $a_{ij} = a_{ji}$). On the other hand, if the network is directed, that is each link has an associated direction, then the matrix is asymmetric (i.e., $a_{ij} \neq a_{ji}$, in general).

Bipartite Network

Networks appearing in economic context can sometimes have the form of a bipartite graph (Figure 10.3), which comprise nodes of multiple distinct types and links exist only between nodes of different types. Examples of such graphs are the

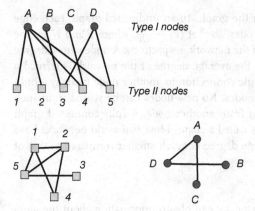

Figure 10.3 A bipartite network comprising two types of nodes can be mapped to two simple networks, one for each class of nodes.

network of companies and their directors. If individual a sits in the board of companies A, B and C, we can put links between the "director" node a and each of the three "company" nodes. Note that, the network of companies and their shareholders may not be strictly a bipartite network, as some of the shareholders may be companies themselves. Therefore, in this latter network, there could be links not only between different classes of nodes (shareholders and companies), but also between the same class of nodes (i.e., between companies) [9]. It is possible to create two simple networks, one for each class of nodes, from the bipartite network, by connecting together all nodes of type I (respectively II) which link to the same node of type II (respectively I) (Figure 10.3).

10.2.1
Measures for Complex Networks

A complex network can be characterized by the various measurable properties of its connection topology. Many local and global measures have been introduced in the literature over the years in order to unveil the organizational principles of networks. Below, we describe some of the most commonly used measures.

Degree
The simplest local characteristic of a node i is its degree, k_i, which is the total number of connections it has to other nodes in the network. It can be calculated from the adjacency matrix as

$$k_i = \sum_{j=1}^{N} A_{ij} .$$

(10.2)

In the case of directed networks, the number of incoming (outgoing) edges of a vertex is called its in-degree (out-degree). The mean degree $\langle k \rangle$ is the average of

k_i over all nodes $i = 1, \ldots, N$ in the graph. In an undirected graph, each edge contributes to the degree of two nodes, so that $\langle k \rangle = \frac{2E}{N}$, where E and N are the total number of links and nodes in the network, respectively. A node whose degree is significantly large compared to the average degree of the network is termed a *hub*. Nodes which only have a single connection to another node (usually a hub) are sometimes referred to as *leaf* nodes. No new nodes can be visited from such a node, once one has arrived at it from another node. A fully connected graph of N nodes with $k_i = N - 1 \forall i$ is called a *clique*. Most real-world networks have *sparse* connectivity and their average degree is much smaller compared to that of the corresponding clique [10, 11].

Degree Distribution

Although degree is a local parameter, we can obtain information about the global topology of the network by looking at its degree distribution, p_k, which is the set of probabilities that a vertex has degree $k = 1, 2, \ldots, N - 1$. A network having a narrow degree distribution with a well-defined mean and a small variance suggests that all its nodes are similar in terms of structural importance, and that the network can be well-described by its average properties. However, many networks occurring in reality are characterized by a degree distribution which decays as a power law

$$p_k \sim k^{-\gamma}, \tag{10.3}$$

with an exponent γ whose value is typically seen to range between 2 and 3. Thus, there is a significantly high probability of observing vertices with large degree relative to the network size [12]. The power-law distribution implies that there is no characteristic scale for the degree of the nodes, so that this class of networks is also termed as *scale-free networks*. In addition to power laws, degree distributions that follow truncated power law or exponential distributions are also observed in many networks occurring in nature and society [13].

Path Length

A global measure of a network can be defined on the basis of the shortest path length or distance between any pair of nodes i and j. This is measured by considering the total number of links that must be traversed to go from node i to node j using the shortest route. The average of shortest path lengths over all pairs of nodes in the graph, also known as the characteristic path length, is an indicator of the compactness of the network. It is defined as

$$\ell = \frac{1}{\frac{1}{2}N(N - 1)} \sum_{i \geq j} d_{ij}, \tag{10.4}$$

where d_{ij} is the shortest path length from node i to node j and N is the total number of nodes in the network. However, if the network consists of disconnected parts, the above definition gives infinite ℓ. To avoid this problem one can define ℓ on such networks to be the harmonic mean of the shortest distance between all pairs

$$E \equiv \ell_{\mathrm{h}}^{-1} \equiv \frac{1}{\frac{1}{2}N(N-1)} \sum_{i>j} \frac{1}{d_{ij}} . \tag{10.5}$$

The inverse of the harmonic mean, is referred to as the *communication efficiency, E,* of the network and is a measure of the speed with which information propagates over it [14].

Most real-world networks have been seen to exhibit the small-world property, which is related to the observation that one can reach a given node from the other nodes in a very small number of steps, on average. In recent years, the term *small-world effect* has taken on a more precise meaning: networks are said to show the small-world effect if ℓ scales logarithmically or slower with network size for fixed average degree, $\langle k \rangle$ [11].

Diameter
Another related measure for compactness of the network is its diameter D, which is defined as the longest of all the shortest paths in the network

$$D = \max\{d_{ij}\}, \forall i - j \text{ pairs of shortest paths} . \tag{10.6}$$

As the diameter and characteristic path length are related properties, sometime these measures are used interchangeably to measure the network compactness.

Clustering
Many real networks have been shown to have a significant transitivity in the pattern of their connections, such that, if the pairs of nodes i, j are connected and the pair j, k are also connected, then so is the pair i, k. This is equivalent to having a significantly high-frequency of triangular structures in the network [15]. In such circumstances, the nodes of the network are said to be clustered. The compactness of the local neighborhood for a node i is measured by the clustering index

$$C_i = \frac{2E_i}{k_i(k_i - 1)} , \tag{10.7}$$

where, E_i is the number of edges among the k_i neighbors of node i. Note that, $C_i = 1$ if the neighbors of node i are fully interlinked, and $C_i = 0$ if none of its neighboring nodes are linked to each other. The average clustering coefficient for the entire network, C, is defined as the average of C_i over all the nodes in the network, that is $C = \frac{1}{N} \sum_{i=1}^{N} C_i$.

This average clustering coefficient is a measure of the "cliquishness" or local compactness of a network. For different real networks, C takes values which are orders of magnitude larger than that of an equivalent random graph with the same number of nodes and edges. If, in addition to the small-world property, a network also possesses a high clustering coefficient C, then it is termed as a *small-world network* (SWN). Many real-world networks are seen to belong to this class [15, 16].

Modularity

Looking beyond microlevel properties such as degree and macroscopic topological features such as efficiency and clustering, it has been observed that at the mesoscopic level, many of the networks in the real world have modular structure [17]. Modules or communities are subnetworks within the network, where connections are much more numerous between nodes belonging to the same subnetwork than between nodes in different subnetworks. The presence of modular structure may also alter the way in which dynamical processes (e.g., spreading of contagion, synchronization, etc.) unfold on the network. Following this realization many recent studies have focused on models of modular networks and the effect of such meso-level organization on the dynamical processes taking place in the nodes of the network [18–20].

Hierarchy

Many networks have also been shown to have *hierarchical* organization, that is they are composed of successive interconnected layers or inter-nested communities [21]. Hierarchy describes the organization of elements in a network: how nodes link to each other to form communities and how communities are joined to form the entire network, for example the metabolic network of several organisms can be organized into highly connected modules that hierarchically combine into larger units [22]. The observed hierarchy also coincides with known metabolic functions, indicating that there may be a functional basis for such meso-level organization.

10.3
Models of Complex Networks

One way of understanding complex networks observed in society is to construct a minimal model that exhibits properties which are similar to those of empirical networks. Such network models can help to explain processes by which these systems evolve and may also shed light on the function of the network. Further, a network model can be used for studying the dynamics on such systems, for example to understand how the processes of synchronization and diffusion are affected by different network topologies [5, 23].

10.3.1
Erdős–Rényi Random Network

The earliest mathematically analyzed nontrivial network model in the literature is that for an ensemble of homogeneous random graphs introduced by Erdős and Rényi (ER) [24]. Starting from a set of N disconnected nodes, each pair of nodes is connected with a probability p. This simple model leads to a surprising list of properties, many of which can be computed exactly in the limit of large N. For a sparse graph, if the average number of edges in the graph is a fraction p of the

$N(N-1)/2$ possible edges, then the average degree

$$\langle k \rangle = \frac{2E}{N} = p(N-1) \,. \tag{10.8}$$

The degree distribution can also be computed, with the probability of a vertex having degree k being

$$p_k = \binom{n}{k} p^k (1-p)^{n-k} \simeq \frac{\langle k \rangle^k \exp^{-\langle k \rangle}}{k!} \,. \tag{10.9}$$

The approximate equality, that is binomial distribution being approximated by a Poisson distribution, becomes exact in the asymptotic limit of large network size. These graphs are therefore also known as *Poisson random graph*.

The expected structure of the random graph varies with the connection probability p. For $p = 0$, there are no edges and the graph is termed an empty graph, whereas for $p = 1$, all possible edges exist and we get a complete graph. As p increases from 0, the edges join nodes together to form several disjoint components, or, subsets of connected nodes. Erdős and Rényi demonstrated that the random graph undergoes a phase transition at a critical value of $p_c = 1/N$, from a low-density state, in which there are few edges and all components are small, to a high-density state, in which an extensive, that is $O(n)$, fraction of all nodes are joined together in a single *giant component*. This component is a set of mutually reachable nodes, whereas the remainder of the nodes occupy smaller components. With increasing p, the giant component captures more and more nodes of the graph. Another important feature is the occurrence of a second connectivity transition at $p_{c1} = \ln N/N$. For $p > p_{c1}$, all sites belong to a single component (in the limit $N \to \infty$), while for $p < p_{c1}$ isolated clusters can exist.

ER graphs have a low clustering coefficient as the probability of connection between two nodes is p regardless of whether they have a common neighbor or not. Hence,

$$C = p = \frac{\langle k \rangle}{N-1} \,, \tag{10.10}$$

which goes to zero as N^{-1} in the limit of large system size [15]. To get an idea of the average path length for the graph, note that the mean number of neighbors at a distance q away from a vertex in a random graph is $\langle k \rangle^q$, so that the value of ℓ needed to span the entire network is given by $\langle k \rangle^\ell \simeq N$. Thus a typical characteristic distance for the network is

$$\ell = \log N/\log\langle k \rangle \,. \tag{10.11}$$

This scaling is much slower than that of a d-dimensional regular lattice where $\ell \sim N^{1/d}$. If the growth of $\ell(N)$ is slower than $N^\alpha (\alpha > 0)$, it is referred to as *small-world effect* [25].

The ease of analysis for random graphs has proven to be very useful in the early development of the field. Although the average path length scales logarithmically with graph size and therefore, shows the small-world effect, in almost all other respects the properties of random graphs do not match those of networks in the real world. Their degree distribution is Poisson, whereas most real-world graphs seem to exhibit broader degree distributions. Also, the random graph lacks clearly defined communities and the clustering coefficient is usually far smaller than that in comparable real-world graphs. The basic Erdős–Rényi model has been extended in several ways, for example to exhibit a power law degree distribution pattern [26, 27]. However, these models do not accurately describe how the properties of complex systems seen around us evolve dynamically, thus making them less useful in understanding the process of network formation in the real world.

10.3.2
Watts–Strogatz Small-World Network

Social networks often show a high tendency of being transitive, that is two people who are friends have a high probability of having one or more *mutual* friends. This kind of clustering is not seen in random graphs, as mentioned previously. In 1998, Watts and Strogatz proposed a mechanism for generating small-world networks with high clustering [15]. This model is often termed as the WS-model and the generative mechanism is as follows: a regular network is first constructed by arranging N nodes on a d-dimensional periodic lattice (a ring for $d = 1$ and torus for $d = 2$). Each node is connected to $k = 2z$ nearest neighbors within the range z, so that all nodes have the same initial degree. Next, one goes through each edge, and with re-wiring probability p, detaches one end of the edge and reconnects it to a randomly chosen vertex (excluding self and multiple connections).

Changing the re-wiring probability p allows us to investigate the transition from a regular graph ($p = 0$) to a random graph ($p = 1$) (Figure 10.4). Let us consider first the limit $p = 0$, where the network is regular and arranged on a ring (i.e., a one-dimensional lattice). The shortest average path length for this system is $\ell_{reg} \sim$

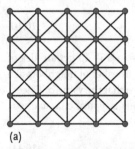 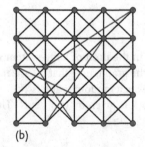 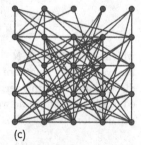

(a) (b) (c)

Figure 10.4 The Watts–Strogatz (WS) small-world network model, constructed on a two-dimensional square lattice substrate. Starting from a regular network (a) where each node is connected to its nearest and next-nearest neighbors, a fraction p of the links are re-wired amongst randomly chosen pairs of nodes. When all the links are rewired, that is $p = 1$, the system is identical to a random network (c). For small p, the resulting network (b) still retains the local properties of the regular network (e.g., high clustering), while exhibiting global properties of a random network (e.g., short average path length).

$N/4z$ for large N, and this grows linearly with N. The clustering coefficient $C_{reg} = (3z - 3)/(4z - 2)$ is constant and tends to 3/4 for large z. This large value indicates the presence of a significant number of triangular structures (i.e., a fully connected triplet of nodes) in the network. On the other hand, when $p = 1$ we have the random graph for which $\ell_{rand} \sim \ln N / \ln z$ and $C \sim 2z/N(\to 0$ as N increases). In the WS model, by changing the re-wiring probability, one finds that there is a broad range of p where $\ell \approx \ell_{rand}$ and $C = C_{reg}$. Thus, globally the network has the small-world property of a random graph, while locally it is clustered like a regular graph. This is because the characteristic length ℓ decreases rapidly when p increases, as adding even a few short-cuts during the re-wiring process reduces the average distance between any pair of nodes significantly. However, the clustering coefficient

$$C = \frac{3(k-1)}{2(2k-1)}(1-p)^3,$$ (10.12)

of the network decreases very slowly with increasing p [28].

The WS model was one of the first models that could explain the co-existence of high clustering and small-world effects. Further, this model introduced the concept of physical distance constraints in network formation. According to this concept, it is easy to form a link between nodes which are geographically close to each other. Although other variations of the WS network have been proposed, in all these models the signature of a physical d-dimensional lattice is still observed. However, the conventional WS model does not exhibit a broad degree distribution, and the discovery of this latter feature in several real-world networks led to the next breakthrough in the physics of complex networks (viz., the existence of scale-free degree distributions) [12, 29].

10.3.3
Modular Small-World Network

In contrast to the implicit notion of a geographical distance (through the use of a lattice) in the WS model, small-world networks can also be defined in terms completely independent of a distance metric. A simple network model that exhibits all the structural characteristics of small-world networks is one that has modular organization. Unlike the WS model, such modular systems show distinct time-scales in their dynamics, corresponding to fast intra-modular and slow inter-modular processes. The simplest model consists of N nodes arranged into m modules [19], where each module contains the same number of randomly connected nodes (Figure 10.5). The connection probability between nodes in a module is ρ_i, and that between different modules is ρ_o. The parameter defining the model is the ratio of inter- to intra-modular connectivity, $\rho_o/\rho_i = r \in [0, 1]$. For $r \to 0$, the network gets fragmented into isolated clusters, while, for $r \to 1$, the network approaches a homogeneous or Erdős–Rényi (ER) random network. The communication efficiency E of the network, defined earlier, is low at small r, when most links are within a module. As the number of inter-modular links increase at higher r, E becomes

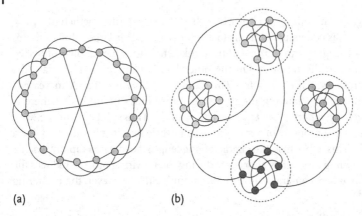

(a)   (b)

Figure 10.5 Schematic diagrams of (a) Watts–Strogatz model (constructed on a one-dimensional ring lattice substrate) and (b) modular network, with modules in the latter indicated by broken circles.

high. For modular networks with large m, clustering C is high at low r and decreases with increasing r. The small-world network property, viz., the coexistence of high E and high C, is observed in the model for an intermediate range of r (Figure 10.6a), exactly as seen in the WS model for intermediate re-wiring probability p (Figure 10.6b).

The extent of modularity in an arbitrary network is difficult to quantify exactly and several measures have been proposed. One of the most popular ones, suggested by Mark Newman, is $Q = \sum_{s=1}^{m}\left[\frac{l_s}{L} - (\frac{d_s}{2L})^2\right]$, where m is the number of modules into which the network is partitioned, L is the total number of links, and l_s and d_s are the links between nodes and the total degree of all nodes belonging to module s, respectively. Note that, this quantity also gives a high value for the WS network at small values of p. Thus, for N nodes with average degree $\langle k \rangle$, the WS model has a maximum Q value of $(1 - p)[1 - \sqrt{(\langle k \rangle + 2)/N}]$, which is very high at low p. For modular random networks, $Q = \frac{(m-1)[N(1-r)-m]}{m[N(1-r+rm)-m]}$, which also yields very high values at low r, where $Q \sim (1 - mr)$ (Figure 10.6c). This implies that community detection algorithms which use Q will be unable to differentiate between these two networks. Other methods, such as, the k-clique percolation cluster technique [30] indicates high local link density relative to the overall connectivity for both the models. Thus, it is difficult to distinguish between the WS model and the modular network model with extant measures that use only structural information.

However, in terms of dynamics, the two network models behave very differently. For example, we can consider diffusive processes on such networks, which can model how information is propagating from node to node. In particular, we look at a discrete random walk on a network, where the walker moves from one node to a randomly chosen neighboring node at each time step. We analyze the time-evolution of the diffusion process by obtaining the distribution of first passage times for random walkers to reach a target node in the modular random network,

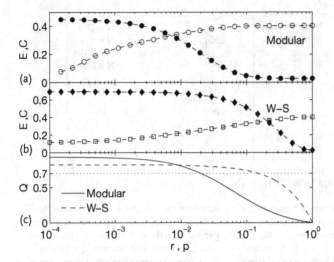

Figure 10.6 Communication efficiency E (empty circle) and clustering coefficient C (filled circle) for (a) modular random network with $m = 16$ modules as a function of r and (b) Watts–Strogatz (WS) network as a function of rewiring probability p ($N = 512$ and $\langle k \rangle = 14$). The data points are obtained by averaging over 100 realizations. Error bars are in all cases smaller than the symbols used. (c) The variation of modularity measure, Q_M, with r for modular random networks (solid line) and with p for WS network (broken line). The dotted line indicates $Q_M = 0.7$ and its intersection with the other two curves gives a pair of r and p values at which we can compare the two model networks.

starting from a source node [31]. Figure 10.7a shows that this distribution differs quite significantly depending on whether the target node belongs to the same module as the source node or in a different module. This suggests two distinct time-scales for the dynamics, with intra-modular diffusion occurring much faster than inter-modular diffusion.

The occurrence of dynamical time-scale separation in modular networks can be understood analytically. The transition probability from node i to j at each step of the random walk is $P_{ij} = A_{ij}/k_i$, where $\mathbf{A}$ is the adjacency matrix and k_i is degree of node i. This relates the transition matrix $\mathbf{P}$ to the normalized Laplacian matrix of the network, $\mathbf{L} = \mathbf{I} - \mathbf{D}^{1/2}\mathbf{P}\mathbf{D}^{-1/2}$, where the Laplacian is defined as $L_{ii} = k_i$ and $L_{ij} = -A_{ij}$, $\mathbf{I}$ is the identity matrix and $\mathbf{D}$ is a diagonal matrix with $D_{ii} = k_i$. The eigenvalues of $\mathbf{P}$ are all real, the largest being 1 while the others are related to the different diffusion time-scales. The eigenvalue spectrum of the Laplacian (and hence that of $\mathbf{P}$, with which it has a one-to-one correspondence) for the modular network model exhibits a gap, which is a manifestation of the difference in diffusion times at different scales. The gap between eigenvalues corresponding to inter- and intra-modular diffusion increases with decreasing value of r (Figure 10.7b). The corresponding spectra for WS networks do not exhibit a gap, indicating that the existence of distinct time-scales for local and global processes has its origin in the modular organization. This dynamical characteristic can be used to identify modules in real networks [18].

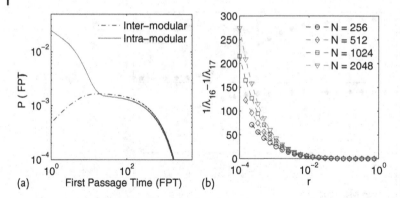

Figure 10.7 (a) Distribution of first passage times (FPT) for diffusion process among nodes in a modular network ($m = 16$, $r = 0.02$, $N = 512$, $\langle k \rangle = 14$). The inter- and intra-modular FPTs indicate two distinct time-scales for random spreading, the process occurring much faster within a module than between modules. (b) The Laplacian spectral gap between the mth and $(m + 1)$th eigen-values increases with decreasing r, shown for different system sizes with the number of modules $m = 16$.

10.3.4
Barabasi–Albert Scale-Free Network

First proposed to explain the degree distribution in citation networks by Derek de Solla Price [32], the idea of preferential attachment (Figure 10.8) has been rediscovered recently by A.-L. Barabasi and R. Albert (BA) in a network model that shows broad degree distributions described by a power law [12]. BA showed that the scale-free nature of these networks can originate from two generic features seen in many real-world networks:

1. *Growth*: Networks are open systems with the number of nodes growing with time (i.e., N increases), and
2. *Preferential attachment*: New nodes in the graph are not connected randomly but preferentially attach to existing nodes which have high degree, thereby making the degree of the latter even higher. This process is sometimes referred to as the *rich getting richer* phenomenon.

If Π, the probability that the new node will be connected to node i, depends linearly on the degree k_i of node i, that is

$$\Pi(k_i) = \frac{k_i}{\sum_j k_j}, \tag{10.13}$$

then, the BA model network evolves into a system with a scale-invariant degree distribution having an exponent $\gamma = 3$.

As the degree distribution of the preferential attachment models match with those occurring in real-world graphs, it suggests that real networks might have been generated by similar processes. However, many networks in nature with a

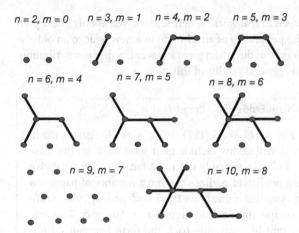

Figure 10.8 The preferential attachment process that results in the construction of a scale-free network. The first 9 steps are shown, starting from a system of 2 unconnected nodes. At each step, a new node with a single link is introduced.

broad degree distribution show deviations from a pure power-law, typically exhibiting an exponential cut-off at high degrees

$$p_k = k^{-\gamma} \phi(k/\xi) \,, \tag{10.14}$$

where $\phi(k/\xi)$ is the cut-off at some scale. In the context of the growing BA model, this phenomenon can be explained due to aging and saturation effects that limit the number of links a node can acquire. Thus, the preferential attachment function, $\Pi(k_i)$ is nonlinear, following $\Pi(k_i) = f(k_i)/\sum_j f(k_j)$, where $f(k)$ is an arbitrary function, resulting in deviations from the power law [33].

The average path length ℓ of the BA network ($\gamma = 3$) grows as

$$\ell(N) \sim \frac{\ln(N)}{\ln \ln(N)} \,, \tag{10.15}$$

with N, which is slower than $\ln N$. This is also termed as the *ultra-small-world effect* [34]. It indicates that the heterogeneous scale-free topology is more efficient in bringing the nodes closer than the homogeneous topology of random network. It is easy to see why the existence of hubs significantly decreases ℓ. If we consider a network which has a single giant hub that all other nodes connect to, then the shortest path length between any pair of nodes is at most 2.

Other scale-free networks with $2 < \gamma < 3$ have a much smaller diameter, with $\ell \sim \ln \ln(N)$, while for networks with $\gamma > 3$, the shortest path length $\ell \sim \ln(N)$ [35]. The clustering coefficient of the BA model decreases with the network size, following approximately a power law, $C \sim N^{-0.75}$. While being slower than the $1/N$ decay observed for C in random graphs, this is still different from the behavior of small-world network models and real-world networks, where C is independent of N [10]. Further there is a strong correlation between age and degree in this model which is rarely seen in real-world systems. Moreover, as only linear preferential attachment (i.e., the probability of attachment of a new node to an old one is a linear function of the degree of the latter) gives a power-law degree distribution, this brings into question the general validity of this process.

Paul Erdős and Alfréd Rényi

Paul Erdős (1913–1996) and *Alfréd Rényi* (1921–1970) were Hungarian mathematicians whose many contributions include their joint work on the theory of random networks. Erdős is arguably the most famous mathematician of the recent past, having published perhaps the most number of papers by any mathematician ever. Awarded a doctorate from Budapest in 1934, Erdős went to Manchester to escape anti-Semitic prejudice in Hungary. This was the beginning of an itinerant life, traveling back and forth between various institutes without staying at any one place for long. Erdős has been very prolific in solving problems arising in a variety of mathematical fields, principally in combinatorics, number theory and graph theory. Erdős was also active in having collaborations with other mathematicians. In 1969, Capser Goffman proposed the concept of Erdős number for a mathematician, that is where a direct collaborator of Erdős has Erdős number 1, any of his/her collaborator has 2, and so on. This is one of the first empirical examples of a collaboration network, and it is said that most of the world's active mathematicians have Erdős number smaller than 8.

Rényi obtained his Ph.D. from the University of Szeged for work on Cauchy–Fourier series. After a brief period spent teaching at Budapest and as Professor at the University of Debrecen, he was appointed as the Director of the Institute of Applied Mathematics of the Hungarian Academy of Sciences. While he has many important contributions to probability, number theory and information theory (including the concept of Rényi entropy), possibly his most famous work is the collaboration with Erdős on two closely related models for random graphs. Often referred to as the Erdős–Rényi model, one is constructed by inserting a link between every pair of nodes in a network with the same probability, independent of all other links. Alternatively, a random network is one that is chosen at random from the ensemble of all possible networks having N nodes and L links. In a 1960 paper, Erdős and Rényi described the behavior of the random network as a function of its probability of connection between each pair, c. In particular, they showed that if the average degree ($k = Nc$) is less than 1, then the network almost surely does not have a connected component of size larger than $O(\log N)$. On the other hand, if it exceeds 1, then the network almost surely contains a unique giant component containing a finite fraction of all nodes. If the average degree is larger than $\ln N$, there will almost surely be no isolated nodes.

Stanley Milgram and Derek de Solla Price

Stanley Milgram (1933–1984) was an American social psychologist who is credited as the source of the six degrees of separation concept (connected to the small-world nature of social contacts), arising out of experiments he conducted while at Harvard. However, he is more famous for his experiments on obedience to authority done at Yale which tested whether normal individuals are willing to inflict any amount of pain on human subjects if instructed to do so by an authority figure. Milgram received his Ph.D. from Harvard and was a faculty member successively at Yale, Harvard and the City University of New York. In 1967 Milgram carried out an experiment to test the "small-world phenomenon", the often observed phenomenon of strangers finding that they are connected to each other by a relatively short chain of acquaintances. The inspiration for this work was a theoretical study by Ithiel de Sola Pool, a political scientist at MIT, and Manfred Kochen, a mathematician with IBM, to see how closely any pair of strangers chosen at random are linked through common acquaintances. In the experiment, several packages were given to randomly selected individuals living in Nebraska, who were asked to forward these to a friend or acquaintance who they thought most likely to eventually bring the packet closer to a target individual in Boston, Massachusetts. Each person was expected to give the package only to someone known to him or her personally, and the details of each person handling the packet was recorded. A number of the packages eventually reached the final destination, and Milgram calculated the average of the number of intermediate people that were required for each packet. Based on this it was stated that on average any two individuals are connected by a chain of six acquaintances ("six degrees of separation"). The original experiment has been criticized for lack of scientific rigor and, in recent times, it has been sought to be replicated, but on the internet using e-mails rather than with physical packages (Dodds, Muhammad and Watts, *Science*, **301**, 827–829, 2003). While Milgram's results were known only in the social network research community until the 1980s, the concept of six degrees became popular knowledge in the 1990s, primarily through the play (and later, its film version) "Six Degrees of Separation". It was also the inspiration for the "Kevin Bacon game" where each player tries to relate an arbitrarily chosen actor to Kevin Bacon through a chain of actor pairs who have been in the same film. The effect investigated by Milgram, that is the ability to span large distances over a network through a small number of steps, is a general feature of "small-world networks", a model for which was proposed in 1998 by Watts and Strogatz.

Derek J. de Solla Price (1922–1983) was an English physicist and pioneer in the field of scientometrics which studies quantitative measures of scientific research and publications. After completing his Ph.D. in experimental physics from the University of London, Price taught at the University of Singapore. Here he developed a theory of how the scientific enterprise grew exponentially from a small, personal scale into the large, organized and institutional form of the present era (and the consequent growth in the number of scientific publications), ideas that would be later put forward in detail in the 1963 book *Little Science, Big Science*. Returning to Cambridge, Price became interested in the study of scientific instruments and obtained his second Ph.D. in the history of science. In 1957 he moved to the United States, becoming a consultant to the Smithsonian Institution, a Fellow of the Institute of Advanced Study and eventually the first Avalon Professor of the History of Science at Yale. Among his many contributions to scientometrics is the first quantitative study (in 1965) of the network of citations between scientific papers. Price discovered that the distributions of both citations to and citations from a paper (i.e., in-degree and out-degree of a node in the network, respectively) exhibit tails that follow power-law scaling. This is therefore the first reported empirical example of a *scale-free network*. In 1976, Price proposed a mathematical model of how a network of citation grows over time based upon a *preferential attachment* process. This mechanism for generating a scale-free network, with nodes selectively connecting to other nodes having higher degree, was rediscovered by Barabasi and Albert in 1999. Among his other interests, Price has also significantly contributed to the analysis of the Antikythera mechanism, an ancient Greek analog computer based on a system of coupled gears and wheels.

10.4
The World Trade Web

The World Trade Web (WTW) (Figure 10.9), also sometimes referred to as the International Trade Network (ITN), is a network defined by the trade relationships and total value traded between the different countries of the world. The nodes are the nations that are involved in trading goods and services with each other and the links correspond to either the existence of a significant amount of trade between the corresponding pairs of nations, or to the total amount exported or imported between the countries within a specified period. Investigating the properties of this network is not only important from the point of view of today's globalized economy, with international trade dominating the economic performance (as measured, for example by the Gross Domestic Product or GDP) of each country, but also because it serves as a conduit for spreading local crises to geographically distant regions. As a result of the tighter coupling between the economies of different countries, even small perturbations in any national or regional economy no longer remains confined to that area but can spread rapidly to become a global crisis. Thus understanding the network structure of international trade can help in locating the most vital nodes of this system.

A paper by Serrano and Boguna [36] is possibly the first study of this network from the point of view of physicists. The database used was the aggregated trade

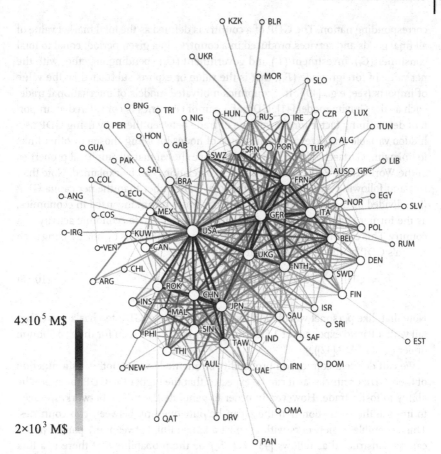

Figure 10.9 A subnetwork of the World Trade Web (2000) showing only 411 connections whose weights (i.e., the annual volume of trade between the two countries in millions of dollars) belong to the largest 4%, and the 80 countries linked by these connections. Node size indicates the strength and link color the connection weight. Source: Bhattacharya *et al.* [47].

statistics tables in the International Trade Centre web-site [37], that are based on COMTRADE database of United Nations Statistics Division. However, these only give a partial list of the trading partners for each country, listing the top forty exchanged goods. A more comprehensive database [39] was used by Garlaschelli and Loffredo [38], which reports the trade between every pair of a large set of countries for all the years between 1948–2000. For example, for 1995, the data reports trade data for 191 countries with trading links between them amounting to a total of 16 225 (out of a possible maximum of 18 145). The early studies of the world trade web considered the network to be undirected, with a link existing between a pair of nations if there is nonzero trade flow in either or both directions.

The 2004 study also looked at the relation between network properties of individual nodes in the world trade web and the gross domestic product (GDP) of the corresponding nation. The GDP of a country is defined as the total market value of all final goods and services produced in a country i in a given period, equal to total consumer (C_i), investment (I_i) and government (G_i) spending, together with the net value of foreign trade (F_i), that is the value of exports subtracted by the value of imports (see, e.g., [40]). In economics-motivated models of international trade, such as the Gravity model [41], GDP of a pair of countries is considered an important determining factor for the volume of trade between them. By using GDP as a hidden variable that determines the *fitness* of nodes for being chosen by other links to link with, Garlaschelli and Loffredo [38] have shown that all empirical properties of the World Trade Web up to third-order correlations can be explained. Note that, this and following studies consider the total GDP rather than the per capita GDP (i.e., divided by the total population) which is more commonly used in economics, as the former is a more accurate indicator of the aggregate economic activity of a country. This can then be used to define the fitness λ of nodes by normalizing each nation's GDP by the total GDP of all countries

$$\lambda_i = \text{GDP}_i / \sum_j \text{GDP}_j . \tag{10.16}$$

Note that, like personal wealth and income, this relative GDP too has a power-law tail with a Pareto exponent of 1 [38] (the power law is also seen for the distribution of per capita GDP [42]).

We can define a connection probability of one node with another as a function of their GDP or fitness, as it can be expected that the higher the GDP the more the ability to foster trade. However, in order to generate the model network, we need to impose the restriction that there can be only one link between two countries. This ensemble of networks with at most a single link between any pair of nodes can be constructed as follows [43]. Let $f_{i,j}$ be the probability that there is a link between the two nodes i, j (correspondingly, the probability that there is no link is $1 - f_{i,j}$). Then the probability of occurrence of any graph G in the ensemble is $\Gamma(G) = \Pi_{(i,j)}(1 - f_{i,j})\Pi_{\text{edges}} \frac{f_{i,j}}{1 - f_{i,j}}$, the first product being over all unique pairs of nodes in the network, while the second is only over those pairs which are linked. Following [43], let us define $P_{i,j} = \frac{f_{i,j}}{1 - f_{i,j}}$, and $\Gamma_0 = \Pi_{(i,j)}(1 - f_{i,j})$, which gives $\Gamma(G) = \Gamma_0 \Pi_{i,j}(P_{i,j})^{\delta_{ij}}$, where $\delta_{ij} = 1$ if i, j are connected and $= 0$, otherwise. We shall now have to choose a functional form for $P_{i,j}$. As all graphs with a given degree sequence should appear in the ensemble with equal probability, a reasonable choice for the function is $P_{i,j} = \beta \lambda_i \lambda_j$. Here, β is a free parameter and λ is the *fugacity* (borrowing a term from statistical mechanics) of each node that determines their expected degree. It is easy to see that

$$\Gamma(G) = \Gamma_0 \beta^L \Pi_i \lambda_i^{k_i} , \tag{10.17}$$

where k_i is the degree of the ith node in a particular graph G and $L = \frac{1}{2}\sum_i k_i$ is the total number of links in the graph.

We next define the grand canonical partition function, $Z = \sum_G \Gamma(G) = \Gamma_0 \sum_{\{\delta_{ij}\}} \Pi_{(i,j)}(P_{i,j})^{\delta_{ij}}$, where the summation is over all possible arrangements of connections between the nodes. Interchanging the sum and the product, and ignoring Γ_0 in all subsequent calculations (as the leading factor cancels out from all observable quantities that are derived from Z), we have $Z = \Pi_{(i,j)} \sum_{\delta_{ij}} (P_{i,j})^{\delta_{ij}} = \Pi_{(i,j)}(1 + P_{i,j}) = \Pi_{(i,j)}(1 + \beta\lambda_i\lambda_j)$.

Now, the free energy is defined as $F = -\log Z = -\sum_{(i,j)} \log(1 + \beta\lambda_i\lambda_j)$. The expected degree is obtained as $\langle k_i \rangle = \lambda_i \frac{\partial}{\partial \lambda_i} \log Z = -\lambda_i \frac{\partial F}{\partial \lambda_i}$. Note that, this is similar to the relation for the average number of particles in a grand canonical ensemble in conventional statistical mechanics, obtained as the product of the fugacity with the partial derivative of the free energy with respect to fugacity. Therefore, the expected value for the degree is $\langle k_i \rangle = \sum_j \frac{\beta\lambda_i\lambda_j}{1 + \beta\lambda_i\lambda_j}$. This gives the probability of a link between a particular pair of vertices as

$$f_{i,j} = \beta\lambda_i\lambda_j(1 + \beta\lambda_i\lambda_j),$$

which lies between 0 and 1. One can see immediately the similarity with the Fermi function used in statistical mechanics.

To fix the free parameter β, we can first calculate the total number of links $L = \frac{1}{2}\sum_{i\neq j} f_{i,j}$ and then tune β so that L equals the empirically observed total number of links in the WTW. Garlaschelli and Loffredo [38] found that $\beta = 80N^2$ fits the observed value of total number of links for a large number of years. Note that, in this model, the knowledge of GDP of the different countries which determines the fitness, indirectly decides the degree sequence, that is the number of trading partners for each country.

To observe how well such a model captures the properties of the World Trade Web, we can consider the relation of the fitness to the expected degree of each node i, $\langle k_i \rangle$. The relation predicted matches very well the empirical data (as expected, wealthier nations have more links), with $\langle k_i \rangle$ being an increasing function of λ and converging to the limiting value of $N-1$ at the asymptotic limit. This immediately follows from the restriction that there can be only one link at most between a pair of nodes. For very low fitness, $\langle k_i \rangle \simeq \beta\lambda_i$. Having obtained the expected degree for each node, we can now determine the degree distribution. A sharp cut-off is observed at large k which results from the limiting value of $N-1$ for the expected degree. Although Serrano and Boguna [36] had claimed the distribution to be scale-free, later analysis has shown that a power law fits only a very small region of the distribution, and the network is not scale-free.

Next, we look at the average nearest neighbor degree $K^{nn}(k)$ which is the average degree of the nodes that are directly linked to a node having degree k (the average is also over all such nodes). Note that, this is a second-order variable, depending on the conditional probability that a node of degree k is connected to another with degree k', $P(k'|k)$. The model predicts the expected value for this quantity for each node to be

$$\langle K_i^{nn} \rangle = \frac{\sum_{j\neq i}\sum_{k\neq j} f_{i,j} f_{j,k}}{\langle k_i \rangle}. \tag{10.18}$$

Again the agreement with empirical data is good, and the decreasing trend of K^{nn} with degree implies that the network is disassortative, that is highly connected nodes tend to preferentially connect to nodes with fewer links.

We can also obtain a measure of the cliquishness of the network by computing the clustering coefficient C_i, which measures the fraction of neighbors of node i which are mutually connected. This quantity depends on the conditional probability that a node with degree k is connected to both a node having degree k' and another node with degree k'', $P(k', k''|k)$. As any realization of the above model is completely determined by the degree sequence, the conditional probabilities are independent and we can simply express the above conditional probability as $P(k', k''|k) = P(k'|k)P(k''|k)$. The expected clustering coefficient for node i is, thus,

$$\langle C_i \rangle = \frac{\sum_{j \neq i} \sum_{k \neq j,i} f_{i,j} \, f_{j,k} \, f_{k,i}}{\langle k_i \rangle [\langle k_i \rangle - 1]}. \tag{10.19}$$

As before, the prediction agrees with observations of the World Trade Web. The decrease of clustering with increasing degree suggests that less wealthy countries trade in closely connected communities, while the interactions of wealthier countries are widely dispersed across the network.

The surprising feature of the above model is that it predicts the higher-order correlated measurables from only a knowledge of the degree sequence. Deviations from the predictions of the model should point out additional features not considered, such as preference for geographically adjacent partners. However, it did not take into account the actual volume of trade and the net balance between exports and imports between any pair of countries. More recent studies have analyzed such a weighted network of world trade [44–47].

The simplest extension of the above formalism to weighted links is to consider the undirected network where each link is proportional to the total volume of trade between the two corresponding countries [47]. For a pair of nodes i, j, this would correspond to the total value of exports by i to j, $\exp_{ij}$ (alternatively, imports by j from i, imp_{ji}) and imports from j by i, imp_{ij} (alternatively, exports by j to i, $\exp_{ji}$): $w_{ij} = \exp_{ij} + \operatorname{imp}_{ij} = \operatorname{imp}_{ji} + \exp_{ji}$. We expect that $\exp_{ij} = \operatorname{imp}_{ji}$ and $\exp_{ji} = \operatorname{imp}_{ij}$. However, because of differences in reporting procedures, there are often minor deviations between these numbers in the empirical data. One possibility is to take an average over these two ways of defining w_{ij}

$$w_{ij} = \frac{(\exp_{ij} + \operatorname{imp}_{ij}) + (\exp_{ji} + \operatorname{imp}_{ji})}{2}. \tag{10.20}$$

These weights have been seen to follow a log-normal distribution [47]

$$P(w) = \frac{1}{w\sqrt{2\pi\sigma^2}} \exp\left(-\frac{[\ln(w/w_0)]^2}{2\sigma^2}\right), \tag{10.21}$$

where $\ln(w_0)$ and σ are the mean and standard deviation of $\ln(w)$. These parameters are seen to have different values in different years. However, it is

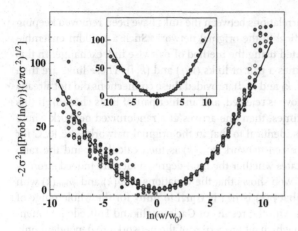

Figure 10.10 The distribution of connection weights of the World Trade Web, showing $2\sigma^2 \ln[\sqrt{2\pi\sigma^2}P(\ln(w))]$ as a function of $\ln(w/w_0)$. The data points fall on a parabola which indicates good fit with the log-normal distribution. Different symbols indicate data aggregated over different 5-year periods. Inset: distribution of all link weights over 1950–2000. Source: Bhattacharya *et al.* [47].

possible to make the distribution independent of these parameters by plotting $-2\sigma^2 \ln[\sqrt{2\pi\sigma^2}P(\ln(w))]$ as a function of $\ln(w/w_0)$. Using the relation $wP(w) = P(\ln(w))$, which follows from the equality $P(\ln(w))d(\ln(w)) = P(w)dw$, we see that the above scaling results in the data corresponding to all years collapsing onto the same parabola of the form $y = x^2$. In [47], this has been done for data from the period 1951–2000 by averaging over every successive 5-year period (viz., 1951–1955, 1956–1960, …) in order to reduce the noise. It is seen that the scaled link weights follow the log-normal distribution form with only slight deviations from the parabola at very low and at very high values (Figure 10.10).

From the link weights, the total volume of trade of any country during a year can be obtained by calculating the node strength $s_i = \sum_j w_{ij}$. It is natural to expect that the total trade (and hence, the nodal strength) of countries with higher GDP will be larger than those with lower GDP. Indeed, a power-law relation between the two is observed such that $s_i \sim \text{GDP}_i^\gamma$. The distribution of the exponent γ for different countries has a peak at 1 with a long tail, having an average value of 1.26 [47].

One can also try to see if the highest weights tend to cluster together. To measure this, first we see how to carry this out in an unweighted network, where the object is to see if the highest degree nodes have relatively dense inter-connections with each other compared to the rest of the network, thus forming a "club". Let us suppose, a club consists of n_k nodes, each of which have at least k links. Then a rich-club-coefficient (RCC) is defined as $\phi(k) = 2E_k/[n_k(n_k-1)]$, where E_k is the actual number of links between members of the "club", and $[n_k(n_k-1)]/2$ is the maximum possible number of such links [48]. However, the value thus obtained is by itself not useful for making any inference unless compared to a equivalent random network, where all correlations between the links have been removed keeping the degree sequence identical to the original network. Such a random ensemble of networks can be generated using the method of pairwise link exchange. In this procedure, one first identifies a pair of links (i, j) and (k, l). These links are then broken and put between (i, k) and (j, l), provided these connections did not already exist (in which case the move is rejected, and another pair of links chosen). If this step is carried out many times, then one arrives at a randomized network where the degree of each node is identical to that in the original network. The RCC for the corresponding randomized network $\phi_{ran}(k)$ is then calculated and the ratio $\rho(k) = \phi(k)/\phi_{ran}(k)$ indicates whether the high degree nodes are indeed strongly coupled. The World Trade Web shows that the variations of $\phi(k)$ and $\phi_{ran}(k)$ with k are nearly the same so that $\rho(k)$ is nearly equal to unity for the whole range of degree values. This agrees with the results of Garlaschelli and Loffredo [38] mentioned earlier, according to which all properties of the network are dependent only on the degree sequence.

Generalizing the concept of RCC to weighted networks, we have

$$R_w(s) = 2 \sum_{i,j} w_{ij}/[n_s(n_s - 1)], \tag{10.22}$$

where n_s is the number of nodes which have strength of at least s. The corresponding random weighted network needs to be generated keeping both the degree $\{k_i\}$ and strength $\{s_i\}$ of each node invariant. In [47], this is done using a self-consistent iteration procedure which preserves the strength sequence $\{s_i\}$ of individual nodes. First, arbitrary random numbers are assigned as weights w_{ij} subject to the restriction that the weights should be symmetric ($w_{ij} = w_{ji}$). Next, the difference from the empirical strength of each node $\delta_i = s_i - \sum_j w_{ij}$ is calculated. The weights of all k_i links for the node i are then updated as

$$w_{ij} \rightarrow w_{ij} + \delta_i \left(\frac{w_{ij}}{\sum_j w_{ij}} \right). \tag{10.23}$$

Note that, the factor $w_{ij}/\sum_j w_{ij}$ is the contribution of the link with each neighboring node of i to the deviation δ_i. Successive iterations of this link weight update process makes the randomized network rapidly converge to the empirical strength sequence $\{s_i\}$. As already seen for degree, for strength, the RCC for the empirical and randomized networks are found to be nearly identical. For high values of strength, $R_w(s) \sim s^{0.85}$ and the ratio $\rho(s) = R_w(s)/R_w^{ran}(s)$ is close to 1 (except close to the highest strength s_{max}).

Why is the rich club coefficient of the World Trade Web so similar to that of the randomized network? To explain this, first observe that the two matrices differ in only about 15% of elements in their respective adjacency matrices. Thus, for a typical node, very few of the links to its neighbors are rearranged as a result of randomization which is a result of the high connection density of the network (e.g., 59% in the year 2000). As this density increases, randomized networks tend to deviate less and less from the empirical system. However, this does not necessarily imply that the coupling among the nodes having high strength is insignificant. If one looks at the fraction of total trade that takes place among members of the club, $f_w(s)$, it is seen to remain close to 1 for a large range of s/s_{max} up to $\sim 10^{-2}$. It decreases gradually to 0.5 only around s/s_{max} before dropping sharply. Thus, a very small number of wealthy countries are responsible for the bulk of the volume of international trade. This heterogeneous nature of trading volume can also be seen in the average pair correlation for strength, $\langle s_i s_j \rangle$ which has a power-law dependence on the link weights: $\langle s_i s_j \rangle \sim w_{ij}^\nu$. It is obvious that links with large weights will connect nodes having high strength ($\langle s_i s_j \rangle \sim s_{max}^2$). On the other hand, for links having weights around unity, $\langle s_i s_j \rangle \sim s_{max}$. Assuming that w_{max} itself is of the order of s_{max} gives an upper bound for the exponent, viz., $\nu = 1$. The empirical value of ν ranges between 0.65 and 0.90 for different years (between 1948–2000). Note that, the functional dependence of the strength of a node as a function of its degree reveals the connection between the distribution of weights and the underlying topological structure of the World Trade Web. This exhibits strong nonlinearity: $\langle s(k) \rangle \propto k^\mu$, where μ varies between 3.4 and 3.7.

So far we have been looking at undirected networks. But in general, in any trading relationship one partner is a net buyer (importer) while the other is a net seller (exporter). So directionality is inherent in the World Trade Web. Serrano et al. [45] have looked at the network of trade imbalances to analyze this asymmetry and its heterogeneity in bilateral trade relations in the worldwide trading system [44, 45]. In this approach, a directed link is drawn between a pair of countries i, j when there is a difference between the quantity imported I_{ij} and quantity exported E_{ij} by i to and from j, respectively, with the arrow pointed along the direction of the net flow of money. It is also weighted by the magnitude of the imbalance $T_{ij} = E_{ij} - I_{ij}$. As $E_{ji} = I_{ij}$ and $I_{ji} = E_{ij}$, it is evident that the trade imbalance matrix T is anti-symmetric. The corresponding adjacency matrix W for the weighted, directed network is constructed from T such that $W_{ij} = |T_{ij}|$ if $T_{ij} < 0$, and $= 0$ if $T_{ij} \geq 0$ (i.e., the direction of the link is towards the country with the positive balance). The sum of all weights W gives the total trade flux, which can be used to calculate the average trade flow that is seen to grow proportionately with the world GDP [44]. Note though that this coupling with GDP is a fairly recent occurrence, seen from the 1960s. By using data that goes back to 1870 [49, 50], it has been seen that prior to the 1960s the average trade imbalance fluctuated around a constant value, while the estimated global GDP was increasing. This has been used to suggest that a transition (related to globalization) took place in the nature of world trade around this period. This has been supported by a study of the distribution of the weights, $P(W_{ij})$. The distribution is seen to have a heavy tail, specifically a log-normal form, in all years that have been considered. Thus, a small fraction of the links in the network carry most of the total flux. This heterogeneity is partly a result of the differences in the size of the economies of different countries (e.g., as measured by their respective GDP). The curves for different years overlap between 1870–1960, but for the period from 1960 onwards, the curves are seen to differ in width from year to year. However, they collapse onto a single curve once the weights are scaled by the global GDP. This suggests that the increasing width of $P(W_{ij})$

over the years is driven solely by the increasing world GDP. In principle, one can then predict the distribution of world trade imbalances in the near future by using the projections of the global GDP for those years. The close relation between world trade imbalance and GDP also points out that from the 1960s, international trade has become a more dominant factor in calculating GDP, possibly affecting the other terms used for its determination: private consumption, business investments in capital and government spending. Garlaschelli and co-workers (2007) have studied further the inter-relation and feedback dynamics between international trade and GDP [46].

Using the above procedure on the data for international trade compiled in [39], one can study the time-evolution of the worldwide trading patterns. For example, a consideration of the number of incoming links, k_{in}, and outgoing links, k_{out}, gives information about the number of trading partner countries which impact positively and negatively to the trade imbalance of a particular nation. By taking into account weights and calculating the incoming and outgoing strengths of a node i, $s_i^{in} = \sum_j W_{ji}$ and $s_i^{out} = \sum_j W_{ij}$, respectively, a country can be classified as net producer or a net consumer by calculating its total trade imbalance $\Delta s_i = s_i^{in} - s_i^{out}$. While the sum of all trade imbalances over the entire network is conserved (global conservation of the trade flux), the local flux for the nodes is not conserved: money flows from consumer nations to producer nations. There is positive correlation between in-degree and out-degree, as well as, between in-strength and out-strength

$$k_{out} \sim k_{in}^{0.5} \quad \text{and} \quad s_{out} \sim s_{in}^{0.6} . \tag{10.24}$$

The data indicates that smaller economies (i.e., those with small s_{in}, s_{out}) tend to be net consumers, that is $\Delta_s < 0$, while bigger economies (i.e., with larger values for s_{in}, s_{out}) are net producers, that is $\Delta_s > 0$.

By looking at the time-evolution of the network, it is seen that as the number of trading countries have increased, there has been a corresponding increase in the average number of trading partners of each node. Also, the total trade flux has been shown to increase proportionately with the world aggregate GDP by Serrano [44]. The macroscopic patterns of the trade flux can be made clearer by extracting the "backbone" of the network. To do this let us first consider the problem of analyzing local heterogeneities in the network. To see whether a few links carry most of the total trade flow for a specific country, we need to calculate the sum of normalized fluxes W_{ji}/s_i^{in} (for incoming flux) and W_{ij}/s_i^{out} (for outgoing flux) for each node. The sum of their products, $Y_i(k) = \sum_{j=1}^k (W_{ji} W_{ij})/(s_i^{in} s_i^{out})$, for a particular node i having k links, can be used to measure the degree of heterogeneity for the trade flux of a country. If the flows are of the same order of magnitude, then $k Y_i(k)$ is independent of k, while dominance of a few links (higher heterogeneity) will show up as a linear dependence on k, and correspondingly a large deviation from constant behavior. As such deviations can also appear at low values of k, the results for the empirical network need to be compared with that of a randomized network keeping the degree sequence unchanged. This equivalent random system is obtained by a simultaneous broken stick process [51], where the [0, 1] interval is broken into

k parts. The random lengths of each of the k sections represent the different values assigned to the k variables p_{ij}. The probability of the length of a section being $x (\in [0, 1])$ is

$$P(x) = (k - 1)(1 - x)^{k-2} . \tag{10.25}$$

Thus, the average and standard deviation of $k Y_{\text{rand}}(k)$ is $2k/(1 + k)$ and $k^2 \left(\frac{20+4k}{(k+1)(k+2)(k+3)} - \frac{4}{(k+1)^2} \right)$, respectively. For a given k, we need to compare each node having a certain in-degree or out-degree with the corresponding random model. As most empirical values of the World Trade Web lie outside the randomized model domain at high k_{in} or k_{out}, the observed heterogeneity cannot be simply an outcome of random fluctuation. Fitting indicates that both in- and out-flux exhibit power-law scaling with degree, $k Y_i(k) \sim k^\beta$, where $\beta_{\text{in}} = 0.6$ and $\beta_{\text{out}} = 0.5$. Thus, it is intermediate between the two extreme cases of perfect homogeneity (independent of k) and extreme heterogeneity (linear dependence on k). The results are consistent with the existence of a dominant subnetwork of high-flux paths for trade imbalance, that is a backbone of the world trade web that carries most of the money flow. To obtain this network by filtering out weaker links we compare the empirical network link fluxes with that of the randomized model used earlier for calculating disparity. For each connection of a country i, the probability α_{ij} that its normalized flux value p_{ij} arises purely through chance is

$$\alpha_{ij} = 1 - (k - 1) \int_0^{p_{ij}} (1 - x)^{k-2} . \tag{10.26}$$

Retaining only those links for which $\alpha_{ij} > \alpha$, a fixed significance level, we identify the subnetwork having high inhomogeneous fluxes. Choosing a global threshold for all countries gives us a homogeneous criteria for comparing inhomogeneities across countries with different numbers of connections k. It also helps us in bringing out links with fluxes that are significantly different from a random distribution. The "backbone" is obtained by preserving all links which exceed this threshold for at least one of the two nodes of a link while not considering the rest. By changing the significance level, progressively stronger heterogeneities and hence, backbones, can be obtained. The algorithm is a systematic procedure to simplify the empirical network by reducing the number of connections without diminishing the number of countries in the world trade web. For the year 2000, at a level of significance 0.05, the backbone consists of only 15% of the original links although accounting for 84% of the total trade imbalance. At higher values of significance, the backbone network starts to break up into disconnected components. A characteristic motif of the resulting network is the occurrence of star-shaped structures (hub-and-spoke arrangements) where a dominant economy (e.g., USA) is seen to be linked with many smaller, dependent economies. By generating similar networks for different years, major changes in the world trade structure can be observed, for example in the development of new hubs such as Japan and China in recent years.

So far we had been only describing the static structure of the network. However, trading over the network is a dynamical phenomenon defined over this structure.

Is it possible to see how such dynamics is governed by the network topology? One way to do this is to observe how diffusion occurs over the given system. For example, in the network of trade imbalances, one can visualize money to be flowing from consumers to producer nations. However, a producer country may not be a complete sink for the incoming flux; a part of the money may travel further on to other producer countries from which it imports goods. Therefore, we can see the flow as a complex process where the net balance of in-flowing and outflowing money is the outcome of the overall diffusion. To see what effect the economy of one nation has on another, we can perform the "dollar experiment" [45]. Here, the network structure is held fixed, as the time-scale at which it changes is much longer than the characteristic diffusion time-scale. First, we imagine that a consumer node ($\Delta s < 0$) injects 1 dollar from its net debit into the network. The dollar travels through the network following fluxes chosen with a probability proportional to the link strengths. The probability of it being trapped at a particular node which is a net producer ($\Delta s > 0$) is $P_{abs} = \Delta s/s_{in}$. Thus, the process is a random walk on a directed network with heterogeneous diffusion probability and sinks. Let the probability that the traveling dollar which originated in the source i is finally absorbed in the sink j be e_{ij}. A corresponding symmetric diffusion process can be considered where each producer node is receiving a dollar and we trace the routes through which this arrives. The dollar travels from node to node following incoming links backward that are chosen with a probability proportional to the link strength. The probability that a fraction of each dollar that a sink country i obtains originated in a particular source country j is g_{ij}. The two probabilities e and g are related by the detailed balance condition

$$|\Delta s_i|e_{ij} = \Delta s_j g_{ji},\tag{10.27}$$

where the matrices e and g satisfy $\sum_{j;sink} e_{ij} = 1$ and $\sum_{i;source} g_{ij} = 1$. Note that, under detailed balance,

$$\Delta s_j = \sum_{i;source} e_{ij}|\Delta s_i|; |\Delta s_i| = \sum_{j;sink} g_{ij}\Delta s_j.\tag{10.28}$$

Under these conditions, the total trade imbalance of a sink or source country can be written as a linear combination of the trade imbalances of the rest of the source or sink countries, respectively. Therefore, by measuring e_{ij}, it is possible to determine the effect that one economy has on another. Similarly, using g_{ij} we can find out which consumer country is contributing most to the income of a producer nation. One of the biggest surprises of this analysis is that a third of all the money Russia gains from trade is coming from the US either directly or indirectly [45].

10.5
The Product Space of World Economy

A different way of looking at international trade is by considering not the network of countries that are trading, but instead, the network of products which they trade.

C.A. Hidalgo and collaborators [52] have constructed a network of relatedness between products, defining a "product space", and they have observed how countries occupy different regions in the product space, with the wealthier nations dominant in the technologically sophisticated products comprising the core of the network, while the poorer nations are dependent on the products needing less sophisticated techniques that are located at the periphery. They also analyze the evolution of a country's economy as it develops by moving through the product space, gradually developing their ability to create products that are close to those they are already making. However, it appears that the poorest economies may never arrive at the economically lucrative core by traveling in network space because of the large topological distances involved.

As there can be many possible factors resulting in the relatedness of products, Hidalgo *et al.* [52] have taken the empirical view that two products are related when a country which exports one has a high probability of also exporting the other. This is based on the idea that related products require similar infrastructure, technical knowledge, availability of resources and institutions, so that the ability of a country to export a product depends on its ability to produce related products. To quantify the strength of this relation between two goods i, j, a measure of proximity $\phi_{i,j}$ is defined as the minimum of the conditional probability that product i (respectively j) forms a particular share of a country's export, given the share of the country's export occupied by product j (respectively i)

$$\phi_{i,j} = \min \{ P(R(x_i)|R(x_j)), P(R(x_j)|R(x_i)) \} . \tag{10.29}$$

Here, R (referred to as the *revealed comparative advantage*) measures whether the share of exports in a certain product from a particular country is more than the share of that product in the global trade

$$R_i^c = \left[x_i^c / \sum_i x_i^c \right] / \left[\sum_c x_i^c / \sum_c \sum_i x_i^c \right] , \tag{10.30}$$

where x is the value of exports of product i by country c. A country is considered to be, in effect, an exporter of product i, when $R_i^c > 1$. The minimum is considered in the definition of proximity in order to symmetrize the matrix, and to avoid problems from cases where a country is the sole exporter of a particular commodity – as this would make the conditional probability for all other goods exported by that country given that one equal to 1, the converse of which is not true.

The dataset used for this analysis is that of bilateral trade by commodity for 1962–2000 available from National Bureau of Economic Research [54] or from the Center for International Data [55] and has been compiled through a NBER project led by Robert Feenstra. The data has been disaggregated according to the four-digit Standardized International Trade Code with country codes similar to the United Nations classification. For each country, information about the value exported to all other countries is available for 775 product classes. Further details about how the data is compiled is given in [56]. Using the data, a 775 × 775 proximity matrix can be constructed showing the relatedness between every pair of products. One of the first things to note is that the product network is not fully connected, with 5% of the matrix elements of Φ being zero, and only 35% of the entries being above 0.2. Thus, it suggests that the relative dominance of weak connections may make a network representation suitable. One possibility of making such a network is to first construct a backbone from the maximal spanning tree. This is the set of $N - 1$ links that connect all the N nodes in the network such that the sum of the proximities is maximized. It is generated by considering the highest value of a non-diagonal entry in Φ and then choosing the strongest link connecting to that pair. Next, the strongest link connecting a new node to the existing triad is chosen, and the process continued by adding further links until all the nodes in the network are included. Next, to consider links which are strong but are not part of the maximal spanning tree, all links above a certain threshold are added to it. The value of the threshold can be chosen based on ease of visualization. In Hidalgo *et al.* [52], all links with proximity value of 0.55 or higher were added to the backbone, resulting in a network having 775 nodes with 1525 links. The corresponding space can be termed as the "product space".

A visual inspection of the network reveals that it has a core-periphery structure, with the densely connected core comprising metal products, machinery and chemicals, while the rest of the products belong to the sparsely connected periphery. There are a few peripheral clusters with products that are strongly related within themselves, corresponding to (a) garments, (b) textiles, (c) animal agriculture, (d) electronics, (e) mining, and (f) forest and paper products. As all products have an associated value (e.g., the average income per capita associated with that product), we can therefore ask if the high-value goods are located in any specific region of the network. Indeed, there appears to be a "rich" region in and around the core of the product space consisting of machinery, electronics and chemicals. To see how the product space is evolving over time, one can compute the Pearson correlation coefficient between the proximity matrices Φ_T at different years T. For example, the correlation between Φ_{1998}, Φ_{1985} is 0.696, indicating that while the network is indeed evolving with time, the overall structure is relatively stable with related products remaining close to each other in the product space while unrelated ones tend to remain far from from each other.

Assuming that the time-scale at which the product space is changing to be far slower than the dynamics of production in various nations as they attain exporter status in new products over time, we can study how the economic development of a country is reflected in its trajectory in the product space. By observing the pattern of specialization of different countries in the product space at any given time (i.e., products i for which $R_i^c > 1$ for a country c), one can observe that industrialized countries such as USA, Canada, Japan and EU member nations tend to occupy the core, while the poorest countries, such as those in sub-Saharan Africa occupy very few nodes, all of which are located at the periphery. East Asian economies have significant presence in the garments, textile and electronics clusters while Latin America's presence is most marked in the mining, agriculture and garment sectors. Thus, the production specialization pattern of each country is brought out by the product space representation. One can take this approach further by looking at how the network can help explain the economic growth of a particular nation, in terms of predicting which products it can branch out to given its existing expertise. For this, the average proximity of a new potential product j to a country's existing export production capabilities is first calculated

$$\omega_j^c = \sum_i X_i \phi_{ij} / \sum_i \phi_{ij} , \qquad (10.31)$$

which is a measure of the occupation density around the jth node of the network, given the export capability of country c ($X_i = 1$ if $R_i^c > 1$, $= 0$ otherwise). Higher values of ω_j would imply that the country can move more easily to producing j for export as it has existing capability to export many other products related to it. In order to test how this lends comparative advantage to a country, one can look at the ω of *transition products* (i.e., products for which a country became an effective exporter over a 5-year period, that is $R^c < 0.5$ in 1990 and $R^c > 1$ in 1995) against *undeveloped* products (i.e., $R^c < 0.5$ in both 1990 and 1995). As expected the distribution of the occupation density, $P(\omega)$ is remarkably different for these two kinds of products, decaying sharply for undeveloped products at low ω while having a slowly decaying long tail at high values of ω for transition products.

An alternative method of finding out whether countries develop export capability for goods that are close to the ones they are already exporting is to calculate the conditional probability of transition to a state of $R > 1$ for a product given that the nearest product with $R > 1$ is at a given proximity ϕ. From the empirical data this appears to be a monotonic relationship, with the probability of transition in the course of 5 years being essentially nil for $\phi = 0.1$ and close to 0.15 for $\phi = 0.8$. A follow-up of this analysis is to ask whether the connectivity of the network of products is sufficient to allow all countries to reach any node (especially the rich core) given enough time. A simple approach to answer this question is to compute the relative size of the largest connected component of the network as a function of the proximity ϕ. One observes that for $\phi \geq 0.6$, the largest connected component is of negligible size, while for $\phi \leq 0.3$, the network is almost fully connected and a path exists from any product to any other product.

As in the case of the World Trade Web, we can try to see how the network affects dynamics by simulating how a specific country can diffuse in the product space starting from its existing export basket of products. By diffusing to related products iteratively, within a few iterations every country can move into the core region containing the highest value exports, provided the network is sufficiently well-connected. This can be simulated by allowing countries to move from a product for which they have $R > 1$ to a product whose proximity is higher than a certain threshold ϕ_0. On reaching a new set of nodes (which now have $R > 1$), in the next iteration the countries again seek out products having proximity $> \phi_0$ and move to those. We can then analyze which countries move to the core within a specific number of iterations (say, 20). When $\phi_0 > 0.65$, there is little change in the distribution of the value exported of countries in the product space from the existing bimodal one that separates the rich from the poor nations. The diffusion process stops quickly after a few iterations when a small number of countries have acquired export capability in a limited range of products. However, by a threshold value of $\phi_0 = 0.55$, most countries are able to diffuse to the richest part of the network, leaving only a few countries behind which belong to the poorest end of the international income distribution. Thus the poor economic performance of certain countries and their lack of development is seen partly to be a result of the difficulty of traversing through a product space where it is harder to reach export capability in products that are sufficiently removed from existing expertise. Thus, international trade does not provide the same opportunities for economic development to all nations, contrary to the extreme proponents of the benefits of globalization. Even among countries with similar levels of economic development (in terms of export production capability), a significant amount of variation is observed in their future growth as suggested by the evolution in product space. Depending on the position of products that are part of their existing expertise, these countries could either continue on to other technologically sophisticated products or could stop at an arrested state of development because other products have significantly lower proximity from the existing capability. Of course, one of the drawbacks of the above approach is that it equates competitive production capability of a country with its share in the export of a particular product. However, it may happen that even though a country produces a large quantity of a particular commodity, the domestic market absorbs this volume almost in its entirety, leaving little to export. Using the above methodology, the country would appear to have little or no expertise in the said product, although in reality this may be far from the case.

10.6
Hierarchical Network within an Organization: Connection to Power-Law Income Distribution

In this section, we shall focus on the network structure within an economic organization. First, we shall look at empirical data of the income of management employees in several Indian companies belonging to the Manufacturing and Information Technology (IT) sectors. As corporate salaries are often quite transparently a manifestation of the hierarchical organization structure of a company, we have analyzed the salary data from 21 manufacturing companies (e.g., Bajaj Auto, Cadbury, Himalaya Drug, etc.) and 16 IT companies (e.g., HCL, Oracle, Satyam, TCS, etc.), the bulk of the data being obtained from an internet-based labor market survey site [53]. In selected cases we verify this data with information from the company's annual report.

Figure 10.11 shows the organizational structure of the management in a typical Indian company. The hierarchy starts at the level of executives, and goes up through managers and divisional heads all the way to the Chief Operating Officer or Vice President. The number of levels need not be same in all divisions,

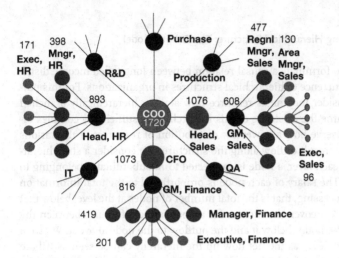

Figure 10.11 Organization of the management in a typical Indian company belonging to the manufacturing sector. The numbers correspond to representative annual salaries (in thousands of Rupees) for the respective positions.

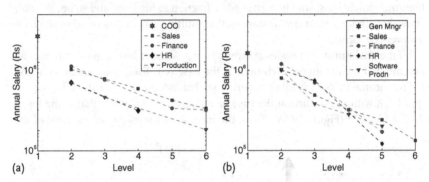

(a) Level

(b) Level

Figure 10.12 The management salary structure of (a) an Indian company belonging to the manufacturing sector, different from the one whose organization structure is shown in Figure 10.11, and (b) that of an Indian company belonging to the Information Technology sector, for the same few divisions in each. Note that salaries are shown on a logarithmic scale, so that exponential increase in salary up the hierarchy will be manifested as a linear trend in the figure.

for example sales and marketing typically has more layers than other divisions, which is reflected in a relatively lower salary ratio between successive levels for people in sales. The salaries at different levels in selected key divisions in companies belonging to the manufacturing and IT sectors are shown in Figure 10.12. The data show exponentially increasing salary along the levels to the top of the hierarchy. Our evidence seems to support that the salary structure in IT companies is relatively more egalitarian than manufacturing companies, but, on the whole, all companies exhibit similarly skewed salary distribution along their hierarchical structure.

10.6.1
Income as Flow along Hierarchical Structure: The Tribute Model

We shall now try to formulate a causal relation between long-tailed income distribution and the occurrence of hierarchical structures in organizations. For this purpose, we shall consider income as resource flow along a hierarchically branched structure. We assume the existence of this hierarchy in a sufficiently complex organization, and observe how a power-law distribution of resources at the various levels can arise as a result of flow along the structure. We consider a strict hierarchy of N levels: at each layer, a node is connected to M subordinates belonging to the level below it. The salary of each node is proportional to the total information arriving at it for processing, that is the total number of nodes at the level below that it is connected to. Moreover, the income for a node is the difference between the total inflow from the nodes below it and the outflow to the node above it. We shall call this the *tribute model*, as the net flow up the hierarchy can be seen as tribute paid by agents belonging in the lower levels of the hierarchy to those above. An obvious realization of this model is in a criminal organization like a drug gang, where the people at the base put a fraction of their earnings for the disposal of their immediate boss, who in turn sends a fraction to his boss, and so on. We now show that, under certain circumstances, the resulting income distribution will have a power-law form.

Let the total number of nodes at layer q be n_q. Without loss of generality, we can assume that the income of each node at the base is 1. Each node sends a fraction f of its income to its immediate superior, so that income of node at level q is $I_q = f M I_{q+1}$, with net income at the base being $I_N = 1 - f$, while that at the top is $I_1 = (f M)^{N-1}$ (Figure 10.13). Thus, the income of a level-q node in terms of the

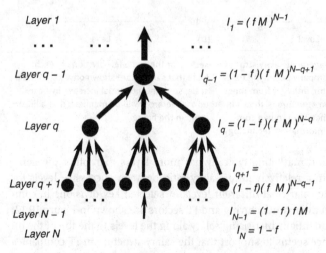

$$I_1 = (f M)^{N-1}$$

Layer 1

Layer $q - 1$ $\qquad I_{q-1} = (1-f)(f\,M)^{N-q+1}$

Layer q $\qquad I_q = (1-f)(f\,M)^{N-q}$

Layer $q + 1$ $\qquad I_{q+1} = (1-f)(f\,M)^{N-q-1}$

Layer $N - 1$ $\qquad I_{N-1} = (1-f)f\,M$

Layer N $\qquad I_N = 1 - f$

Figure 10.13 A schematic diagram of the tribute model representing flow of resources along a hierarchical structure, with the income of nodes at each level q shown on the right.

parameters N, M and f is

$$I_q = (1 - f)(fM)^{N-q} . \tag{10.32}$$

For the income distribution, we obtain n_q in terms of I_q as

$$n_q = n(I) \sim I^{-\log(1-f)/[\log M + \log f]} , \tag{10.33}$$

which has a power-law form. For example, if $f = 1/2$ and $M = 3$, the distribution will be a power law having exponent $\simeq 1.7$. While it may appear that the parameters can be freely chosen to obtain any exponent whatsoever, there are certain restrictions in reality. For example, for a node at the upper layer to benefit from the arrangement, it is necessary that $fM > 1$. Note that, the salary ratio between two consecutive levels is given by $SR = I_q/I_{q+1} = fM$, so that the top-to-bottom income ratio is

$$T_R = \frac{(fM)^N}{1-f} = \frac{(SR)^N}{1-(SR/M)} . \tag{10.34}$$

For example, if $N = 5$, $M = 5$ and $f = 1/2$, we obtain $T_R = 200$. If, as is empirically observed, T_R remains fairly constant for hierarchical organizations with different number of levels, then it follows that as the number of levels, N, increases, the salary ratio between successive levels ($= fM$) decreases. This is indeed observed in the data for management pay structure in Indian companies. So far we had been concerned with inter-level differences in income. Within a level also, incomes for individuals will differ. However, as suggested by Gibrat [57], this is likely to be decided by a large number of independent stochastic factors, such that the intra-level income distribution will most likely be log-normal, the outcome of a multiplicative stochastic process. When seen in conjunction with the multiple hierarchical levels which ensures that the gap between mean income at each level have exponentially increasing separation along with exponentially decreasing population at each level up the hierarchy, this will imply that within a given level the income distribution is log-normal, but a Pareto-like power-law behavior will describe the overall inequality, as inter-level differences will tend to dominate intra-level deviations within an organization.

10.7
The Dynamical Stability of Economic Networks

The economic network is an extremely complex system with each node dynamically interacting with others. Of course, the equations describing the dynamics depends on the process we would like to describe. However, in many cases there are no well-accepted descriptions for these processes so that exact equations cannot be written. Whether the equations are known fully or not, is it possible to say something about whether an equilibrium in such a system will be stable? In particular, general equilibrium theory is a long-established branch of economics which strives to show that given a set of products, and agents who are designing certain allocations to maximize their individual utilities (e.g., profits), there exists an equilibrium state for prices of all goods. While the first attempts to come up with a theory for systems subject to perfect competition can be traced back to Leon Walras in the 1870s, it is only in the twentieth century, first with the work of Paul Samuelson, and later of Kenneth Arrow and Leonid Hurwicz, that mathematically rigorous statements about the stability of such competitive equilibria were made [60, 61]. However, these results, while concerned with a microscopic picture of interaction between agents, do not consider the existence of a network of interactions. How would the stability of a system be affected if agents interact with each other (described by some arbitrarily complex, and in general nonlinear, coupled equations) in a network that restricts the set of possible agents that each agent can deal with?

Questions such as these can also be asked in the context of other systems. In fact, the question of stability of arbitrarily chosen equilibria of N coupled interacting agents has been looked at in great detail in theoretical ecology. Prior to 1972, it used to be taken as a self-evident truth that the more complex a network, the more stable its equilibria. In part, this followed from the intuitive insurance hypothesis: existence of a large number of nodes (redundancy) insures the system against large fluctuations in its overall dynamics because it guarantees that some nodes will continue to function even when others fail. The original concept was based on observations by J.F. Elton that more diverse ecosystems tend to be more robust against foreign species intrusion and parasite outbreaks, and also exhibit low fluctuations in the populations of constituent species. R. MacArthur sought to give a theoretical basis for this idea by suggesting that higher diversity and denser connections per species imply that a small change in the dynamical variables associated with one node will not make any significant impact on the rest of the system, and thus these perturbations (as long as they are not too large) will die out.

A major advance in this area, which ran counter to these earlier ideas, was the suggestion by Robert May that the stability of a network can be inferred from an analysis of the interactions between the network elements [62]. Confining attention only to the local stability of an arbitrary equilibrium of the dynamics, one can ignore explicit dynamics and look at only the leading eigenvalues of the linear stability matrix. Assuming that the network interactions are random, rigorous results on the eigenvalue spectra of random matrices can be applied (see following paragraph). If the stability matrix comprises elements from a normal distribution with zero mean and variance σ^2, then the network is almost certainly stable if $NC\sigma^2 < 1$, and unstable otherwise. Here, N is the number of nodes in the network and C is the network connectivity, that is the probability that any two given elements of the network are coupled to each other, as reflected in the sparsity of the matrix [62, 63]. This result is often referred to as the May–Wigner stability theorem. It is based on the following results for the spectrum of real matrices.

For a random real symmetric matrix (so that all eigenvalues are real) whose elements are chosen from a standard normal distribution, the distribution of eigenvalues follows the Wigner's semicircle law

$$\rho(\lambda) \simeq \frac{1}{\pi}\sqrt{2N - \lambda^2}, \quad \text{if} \quad |\lambda| < \sqrt{2N},$$

$$= 0, \quad \text{otherwise}. \tag{10.35}$$

For nonsymmetric matrices whose entries belong to a standard normal distribution, it is known from Girko's circular law [58] that the (in general) complex eigenvalues for a matrix of order N, is uniformly distributed on a disk of radius $\sqrt{N}$ in the complex plane for large N. At smaller values of N, there is a higher concentration of eigenvalues along the real axis [59].

Thus, it implies that if the complexity of a system is increased by augmenting the number of interacting elements (N), the number of interactions between elements (C) or the strength of interactions (σ), then an arbitrarily chosen equilibrium of the system will be more likely to be unstable. Thus, increasing either the total diversity of the system, or the diversity of connections between the constituent nodes, is more likely to make a system less able to cope with disturbances arising from within or from the external environment. Two of the common charges leveled against the theoretical result of May is that (i) it assumes the interaction network to be random whereas naturally occurring networks are bound to exhibit a certain kind of structure, and (ii) the linear stability analysis assumes the existence of simple steady states, which may not be the case for real systems that may either be having oscillations or be in a chaotic state. However, recent work on networks with complex structural patterns [64, 65], as well as, those considering the full dynamics including periodic oscillations and chaotic behavior [66, 67], show that the result of May is stable with respect to these generalizations.

As such, this approach to understanding the stability of economic systems, in particular, markets that are coupled with each other has been looked at by Hogg *et al.* [68]. Starting out by assuming the existence of an equilibrium within a network of markets, they look at its dynamical stability against small perturbations by using a general model of adjustment processes.

Consider the dynamics of price adjustments for a particular commodity or stock: its price will rise when demand exceeds supply and will fall when supply exceeds demand. If $p(t)$ is the price at any given time, the rate of change is described by the differential equation, $dp/dt = f(p)$, where f is a function expressing the excess demand for the commodity or stock at the instantaneous price p. The equation describes the adjustment of price by a *tatonnement* process (as described by Walras), where a central auctioneer coordinates demand and supply by successively changing the price to satisfy the excess demand. At the equilibrium price p_0 where demand equals supply, $f(p_0) = 0$. On linearizing the dynamics around the equilibrium, the stability of the price p_0 is given by the slope of the function f, $f'(p_0)$ at that point. The equilibrium is stable if the slope has a negative value, with a small perturbation decaying with a time-scale $\tau = 1/|f'(p_0)|$. The rapidity of the adjustment process, which is measured by τ^{-1}, depends on the uncertainty of the agents about the true price.

When there are more than one kind of commodity or stock, or more than one market, we can look at the question of how the coupling between the different elements affects the stability of the prices in the composite system. Let us consider the general case where $\mathbf{p} = \{p^1, p^2, \ldots, p^N\}$, is a price vector containing the individual prices for the N elements (i.e., different markets or different products). The dynamics is now given by a matrix equation $d\mathbf{p}/dt = \mathbf{F}(\mathbf{p})$, which relates the evolution of the different prices as a function of the couplings between them. The stability of the equilibrium $\mathbf{p}_0$ is analyzed by the equation describing the rate of change of an small perturbation $\delta\mathbf{p}$ about the equilibrium

$$\frac{d\delta\mathbf{p}}{dt} = \mathbf{J}_{\mathbf{p}_0}\delta\mathbf{p} , \tag{10.36}$$

where $\mathbf{J} = \{\partial F_i/\partial p_j\}$ is the Jacobian of $\mathbf{F}$ evaluated about the equilibrium. The components of the Jacobian describe how a small increase in the price of one element affects that of another item, which therefore can be of either sign. The diagonal elements indicate the direct effect of a small change in excess demand on the price of a given commodity or stock. We assume that the equilibrium price of each such element in isolation is stable, which implies that the diagonal elements $J_{ii} < 0$, and without loss of generality take to be -1, as this provides a time-scale in terms of the rate at which perturbations decay for the isolated elements.

The effect of small perturbation around the system equilibrium is determined by the eigenvalue of $\mathbf{J}$ having the largest real part, $\lambda_{\max}$. Thus, the long-time behavior of the perturbation is described by $\delta\mathbf{p}(t) \sim \delta\mathbf{p}e^{\lambda_{\max}t}$. If $\lambda_{\max} < 0$, the perturbation will decay within a time-scale $1/\lambda_{\max}$, but if $\lambda_{\max} > 0$, the perturbation will keep growing with time making the system unstable. The value of $\lambda_{\max}$ will of course depend on the particular choice of $\mathbf{J}$, which in turn depends on the adjacency matrix of the system. By assuming a random structure for the interactions between the elements that decide their mutual prices, we can apply the above-mentioned results of May to explain how the stability of markets will change with increasing complexity of interactions. Thus, the condition for stability is seen to be $NC\sigma^2 < 1$. We therefore conclude that, even if every individual element of an economic network tries to restore equilibrium, their interactions in a sufficiently large network with strong, dense connections can make the system as a whole unstable. This lesson has obvious (and sobering) ramifications in today's highly connected economy, implying that crises like the ones we have seen recently in the financial world may become more frequent and severe in the future.

References

1 Boots, M. and Sasaki, A. (1999) 'Small worlds' and the evolution of virulence. *Proc. Roy. Soc. Lond. B*, **266**, 1933–1938.

2 Pastor-Satorras, R. and Vespignani, A. (2001) Epidemic spreading in scale-free networks. *Phys. Rev. Lett.*, **86**, 3200–3203.

3 Albert, R., Jeong, H., and Barabasi, A.-L. (2000) Error and attack tolerance in complex networks. *Nature*, **406**, 378.

4 Araujo, T. and Vilela Mendes, R. (2000) Function and form in networks of interacting agents. *Complex Systems*, **12**, 357–378.

5 Arenas, A., Diaz-Guilera, A., Kurths, J., Moreno, Y., and Zhou, C. (2008) Synchronization in Complex Networks. *Phys. Rep.*, **469**, 93–153.

6 Onnela, J.-P., Saramaki, J., Hyvonen, J., Szabo, G., Lazer, D., Kaski, K., Kertesz, J., and Barabasi, A.-L. (2007) Structure and tie strengths in mobile communication networks. *Proc. Natl. Acad. Sci. USA*, **104**, 7332–7336.

7 Strogatz, S.H. (2001) Exploring complex networks. *Nature*, **410**, 268–276.

8 Mendes, R. (2005) Tools for network dynamics. *Int. J. Bif. Chaos*, **15**, 1185–1213.

9 Garlaschelli, D., Battiston, S., Castri, M., Servedio, V.D.P., and Caldarelli, G. (2005) The scale-free topology of market investments. *Physica A*, **350**, 491–499.

10 Albert, R. and Barabasi, A.-L. (2002) Statistical mechanics of complex networks. *Rev. Mod. Phys.*, **74**, 47–97.

11 Newman, M.E.J. (2003) The structure and function of complex networks. *SIAM Review*, **45**, 167–256.

12 Barabasi, A.-L. and Albert, R. (1999) Emergence of scaling in random networks. *Science*, **286**, 509–512.

13 Amaral, L., Scala, A., Barthilimy, M., and Stanley, H.E. (2000) Classes of small-world networks. *Proc. Natl. Acad. Sci. USA*, **97**, 11149–11152.

14 Latora, V. and Marchiori, M. (2001) Efficient behavior of small-world networks. *Phys. Rev. Lett.*, **87**, 198701.

15 Watts, D.J. and Strogatz, S.H. (1998) Collective dynamics of 'small-world' networks. *Nature*, **393**, 440–442.

16 Dorogovtsev, S.N. and Mendes, J.F.F. (2004) The shortest path to complex networks, http://arxiv.org/abs/cond-mat/0404593.

17 Girvan, M. and Newman, M.E.J. (2002) Community structure in social and biological networks. *Proc. Natl. Acad. Sci. USA*, **99**, 7821–7826.

18 Arenas, A., Díaz-Guilera, A., and Pérez-Vicente, C.J. (2006) Synchronization reveals topological scales in complex networks. *Phys. Rev. Lett.*, **96**, 114102.

19 Pan, R.K. and Sinha, S. (2009) Modularity produces small-world networks with dynamical time-scale separation. *Europhys. Lett.*, **85**, 68006.

20 Dasgupta, S., Pan, R.K., and Sinha, S. (2009) Phase of Ising spins on modular networks analogous to social polarization. *Phys. Rev. E*, **80**, 025101(R).

21 Trusina, A., Maslov, S., Minnhagen, P., and Sneppen, K. (2004) Hierarchy measures in complex networks. *Phys. Rev. Lett.*, **92**, 178702.

22 Ravasz, E., Somera, A.L., Mongru, D.A., Oltvai, Z.N., and Barabasi, A-L. (2002) Hierarchical organization of modularity in metabolic networks. *Science*, **297**, 1551–1555.

23 Dorogovtsev, S.N., Goltsev, A.V., Mendes, J.F.F. (2008) Critical phenomena in complex networks. *Rev. Mod. Phys.*, **80**, 1275–1335.

24 Erdős, P. and Rényi, A. (1960) On the evolution of random graphs. *Publ. Math. Inst. Hungar. Acad. Sci.*, **5**, 17–61.

25 Bollobas, B. (2001) *Random Graphs*, Cambridge University Press, Cambridge.

26 Aiello, W., Chung, F., and Lu, L. (2000) A random graph model for massive graphs. *Proc. 32nd Annual ACM Symposium on Theory of Computing*, p. 171–180.

27 Newman, M.E.J. (2001) Scientific collaboration networks I: Network construction and fundamental results. *Phys. Rev. E*, **64**, 016131.

28 Barratt, A. and Weigt, M. (2000) On the properties of small-world network models. *Eur. Phys. J. B*, **13**, 547–560.

29 Caldarelli, G., Capocci, A., De Los Rios, P., and Muñoz, M.A. (2002) Scale-free networks from varying vertex intrinsic fitness. *Phys. Rev. Lett.*, **89**, 258702.

30 Derényi, I., Palla, G., and Vicsek, T. (2005) Clique percolation in random networks. *Phys. Rev. Lett.*, **94**, 160202.

31 Baronchelli, A. and Loreto, V. (2006) Ring structures and mean first pas-

sage time in networks. *Phys. Rev. E*, **73**, 026103.

32 de Solla Price, D.J. (1976) A general theory of bibliometric and other cumulative advantage processes. *J. Amer. Soc. Inform. Sci.*, **27**, 292–306.

33 Krapivsky, P.L., Redner, S., and Leyvraz, F. (2000) Connectivity of growing random networks. *Phys. Rev. Lett.*, **85**, p. 4629–4632.

34 Cohen, R. and Havlin, S. (2003) Scale-free networks are ultrasmall. *Phys. Rev. Lett.*, **90**, 058701.

35 Bollobas, B. and Riordan, O. (2004) The diameter of a scale-free random graph. *Combinatorica*, **24**, 5–34.

36 Serrano, M.A. and Boguna, M. (2003) Topology of the world trade web. *Phys. Rev. E*, **68**, 015101(R).

37 http://www.intracen.org

38 Garlaschelli, D. and Loffredo, M.I. (2004) Fitness-dependent topological properties of the world trade web. *Phys. Rev. Lett.*, **93**, 188701.

39 Gleditsch, K.S. (2002) Expanded trade and GDP data. *J. Conflict Resolution*, **46**, 712–724.

40 http://www.investorwords.com

41 Tinbergen, J. (1962) An analysis of world trade flows, (ed. J. Tinbergen), *Shaping The World Economy*, The Twentieth Century Fund, New York.

42 Iwahashi, R. and Machikata, T. (2004) A new empirical regularity in world income distribution dynamics 1960–2001. *Economics Bulletin*, **6**, 1–15.

43 Park, J. and Newman, M.E.J. (2003) The origin of degree correlations in the Internet and other networks. *Phys. Rev. E*, **68**, 026112.

44 Serrano, M.A. (2007) Phase transition in the globalization of trade. *J. Stat. Mech.*, L01002.

45 Serrano, M.A., Boguna, M., and Vespignani, A. (2007) Patterns of dominant flows in the world trade web. *J. Econ. Interac. Coord.*, **2**, 111–124.

46 Garlaschelli, D., Di Matteo, T., Aste, T., Caldarelli, G., and Loffredo, M.I. (2004) Interplay between topology and dynamics in the World Trade Web. *Eur. Phys. J. B*, **57**, 159–164.

47 Bhattacharya, K., Mukherjee, G., Saramäki, J., Kaski, K., and Manna, S.S. (2008) The international trade network: Weighted network analysis and modeling. *J. Stat. Mech.*, P02002.

48 Colizza, V., Flammini, A., Serrano, M. A., and Vespignani, A. (2006) Detecting rich-club ordering in complex networks. *Nature Physics*, **2**, 110–115.

49 Maddison, A. (2001) *The World Economy: A Millenial Perspective*, OECD Development Centre, Paris.

50 Maddison, A. (2003) *The World Economy: Historical Statistics*, OECD Development Centre, Paris.

51 MacArthur, R.H. (1957) On the relative abundance of bird species. *Proc. Natl. Acad. Sci. USA*, **43**, 293–295.

52 Hidalgo, C.A., Klinger, B., Barabasi, A.-L., and Hausmann, R. (2007) The product space conditions the development of nations. *Science*, **317**, 482–487.

53 http://www.paycheck.in

54 http://www.nber.org/data

55 http://cid.econ.ucdavis.edu/data/undata/undata.html

56 Feenstra, R.C., Lipsey, R.E., Deng, H., Ma, A.C., and Mo, H. (2004) World trade flows: 1962–2000. *NBER Working Paper*, 11040.

57 Gibrat, R. (1931) *Les Inégalites Économiques*, Librairie du Recueil Sirey, Paris.

58 Girko, V.L. (1990) *Theory of Random Determinants*, Kluwer, Boston.

59 Reichl, L.E. (2004) *The Transition to Chaos: Conservative classical systems and quantum manifestations*, 2nd edition, Springer, New York.

60 Samuelson, P.A. (1941) The stability of equilibrium: Comparative statics and dynamics. *Econometrica*, **9**, 97–120.

61 Arrow, K.J., Hurwicz, L. (1958) On the stability of the competitive equilibrium I. *Econometrica*, **26**, 522–552.

62 May, R.M. (1972) Will a large complex system be stable? *Nature*, **238**, 413–414.

63 May, R.M. (1973) *The Stability of Complex Ecosystems*, Princeton, Princeton University Press.

64 Sinha, S. (2005) Complexity vs. stability in small-world networks. *Physica A*, **346**, 147–153.

65 Pan, R.K. and Sinha, S. (2007) Modular networks emerge from multi-constraint optimization. *Phys. Rev. E*, **76**, 045103(R).

66 Sinha, S. and Sinha, S. (2005) Evidence of universality for the May–Wigner stability theorem for random networks with local dynamics. *Phys. Rev. E*, **71**, 020902(R).

67 Sinha, S. and Sinha, S. (2006) Robust emergent activity in dynamical networks. *Phys. Rev. E*, **74**, 066117.

68 Hogg, T., Huberman, B. A., and Youssefmir, M. (1995) The instability of markets, http://arxiv.org/abs/adap-org/9507002.

11
Outlook and Concluding Thoughts

"One of the most fateful errors of our age is the belief that the problem of production has been solved. This illusion ... is mainly due to our inability to recognize that the modern industrial system, with all its intellectual sophistication, consumes the very basis on which it has been erected. (...) it lives on irreplaceable capital which it cheerfully treats as income."

– E.F. Schumacher [1]

"Development, to be sure, is a must for us. But what is the aim of development and what should be its programme? Will the Euro-American model be our model also? Industrialisation is necessary both in socialism and in capitalism. But let us not take a superficial view. The sky-high chimney of a steel plant belching smoke may not necessarily be the symbol of construction of socialism, even if the plant is State-owned. The ultimate touchstone is the life-style we want to create."

– P. Dasgupta [2]

The discussions in Chapter 10 brings us to one of the most poignant questions that is facing our society today: Is the current state of growth of our global civilization sustainable? Or is it just about ready to collapse? We know numerous examples in history of advanced and complex societies that reached a peak in terms of cultural achievements as well as material prosperity, and then collapsed, often within a very short span of time [3]. It now appears that a large number of these collapses were caused by unsustainable growth resulting in enormous environmental damage to the surrounding areas (such as, massive deforestation around the main urban centers of a civilization). The immediate trigger of the collapse was often a sudden climatic change such as a drought. Although this change by itself may not have been significant, such that simpler societies were able to survive it with relative ease, the enormous complexity of advanced societies also made them fragile, so that they collapsed as a result of an apparently small perturbation. Needless to say, given the unprecedented complexity and energy intensiveness of our present global civilization, questions about its sustainability are being asked more and more often [4–7].

Econophysics. Sitabhra Sinha, Arnab Chatterjee, Anirban Chakraborti, and Bikas K. Chakrabarti
Copyright © 2011 WILEY-VCH Verlag GmbH & Co. KGaA, Weinheim
ISBN: 978-3-527-40815-3

11.1
The Promise and Perils of Economic Growth

Possibly the biggest impact that econophysics can have on the way traditional economics is done is by making it possible to perform a scientifically rigorous reappraisal of the consequences of economic growth, and even whether growth is desirable under all circumstances (Figure 11.1). This will be of immense consequence in view of the current search for sustainable development, that is "development that meets the needs of the present without compromising the ability of future generations to meet their own needs" (as defined in the Brundtland Report [8]). There have always been voices in the periphery, such as that of Fritz Schumachar, calling for smaller economies. This has been echoed by other people in other places, such as in India by the Gandhian Marxist social activist, Pannalal Dasgupta. However, the mainstream economics community has always considered these views to be heretical. The anarchist utopia of society as a system of self-sufficient villages, where affluence is rejected in favor of stability and social justice, has never been considered to be a practical alternative. Indeed the "development at all costs" mindset has permeated through society at large, as reflected by the obsession of the mass media with economic growth and gross domestic product (GDP). Newspapers and television channels are always worried about whether the rate of growth is slowing down (rumors of "an economic downturn", whether justified or not, always seem to appear in the media every few months) and headlines announce by what percentage the economy has grown in a quarter. However, whether economic growth is a panacea for all social ills, or whether growth itself is the cause for most problems, is not as settled a question as it may seem. This is especially so in view of the

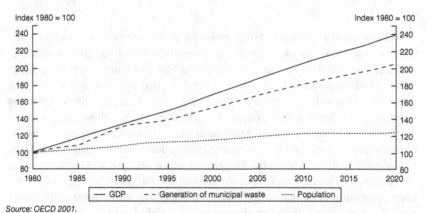

Source: OECD 2001.

Figure 11.1 The ecologically catastrophic consequences of economic growth is underlined by the increase in waste generated by the OECD countries (a group of 30 economically developed nations) along with their GDP, even though the population has not increased significantly. In 1997, OECD countries produced waste corresponding to 500 kg per person in a year and it is estimated that by 2020 this will increase to 640 kg per capita. Source: E. Geyer-Allely, *Towards Sustainable Household Consumption*, OECD 2002.

social and ecological consequences of growth. As the linear models of mainstream economics are inadequate for tackling such questions, the systems dynamics approach developed by Jay Forrester at MIT and his students in the 1960s was one of the first scientifically rigorous approaches towards this complex problem.

11.2
Jay Forrester's World Model

The Club of Rome in the late 1960s wanted to know how major global problems such as poverty and hunger, unemployment, depletion of natural resources, environmental degradation, etc., were related and if there were ways to solve them. Forrester's computer model suggested that the leverage point (i.e., the factor in a complex system where a small change eventually results in a large overall change in the entire system) was economic growth [9]. The problem was that growth has negative consequences, although traditionally economic development is seen as an unalloyed boon. From a modeling exercise it becomes clear that many problems that are sought to be solved by growth, such as poverty and hunger, can in fact be exacerbated by it. Sometimes the solution might be to slow the growth of the economy, or even to turn it back.

The model developed by Forrester's students, Meadows and collaborators [10] has taken this original work forward by looking at several resource stocks and their flows, and trying to predict resource availability at future times. One of the striking observations of this model is that it is not so much the depletion of resources that is the key problem but the increasing cost of capital (e.g., as a result of environmental pollution, among several reasons). Thus, according to Meadows et al., the solution lies not in unconditional economic growth, but in the efficient use of resources.

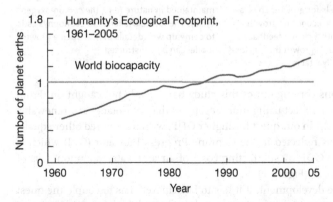

Figure 11.2 Is growth sustainable? The ecological footprint of humanity, measured in terms of the area of biologically productive land and water needed for providing the resources and services to maintain the human population at the present standard of living, has overshot the biocapacity of the planet (i.e., the amount of biologically productive area available). Source: WWF, [11].

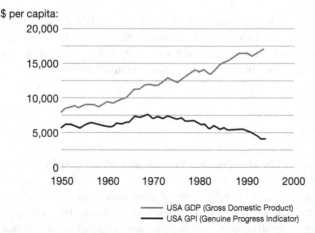

Figure 11.3 Economic growth or decay? The fall by a factor of half in the Genuine Progress Indicator (GPI), an alternative measure of economic growth suggested by Cobb, Halstead and Rowe in 1995 that takes into account the cost of environmental degradation accompanying economic activity, even as the Gross Domestic Product of the US has increased three-fold over the latter half of the twentieth century. Source: [12]

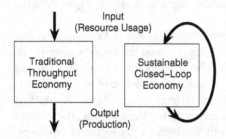

Figure 11.4 Two models for growth. The traditional non-sustainable economic growth model is contrasted with a closed feedback loop model of sustainable growth that is closer to the way biologically vital resources are maintained in nature (e.g., the carbon-oxygen cycle). The future challenge to econophysics is to come up with details of how such a growth model can be constructed.

However, the lessons coming out of this study have clearly not caught on. From 1985 onwards, we are in fact using more resources than our planet can renewably produce (Figure 11.2). In our quest for higher GDP, we have ignored other equally important factors, as reflected in the Genuine Progress Indicator (GPI) which is measured from the GDP by subtracting costs of air and water pollution, loss of farm and wetlands, etc. (Figure 11.3).

Thus, sustainable development, if it has to be achieved, has to couple the quest for growth with conservation of nature and the achievement of acceptable social conditions. Sustainability will be achieved when we stop depleting not only economic capital, but also the social and environmental capital bequeathed to us so that they can be carried over to future generations. Econophysics, by being able to view the problem in the light of insights gleaned from looking at other sustainable systems, such as in the biological world, is in a unique position to develop simple models that can suggest possible solutions. If in place of the traditional throughput economy (wasteful of resources) we want a sustainable model that re-uses products in a closed-loop cycle (Figure 11.4), would using the concepts learnt from the study of ecological food webs help us? Can a network economy give a more sustainable alternative to development? These questions are hard to answer, but possibly the most important that the econophysics community will have to tackle in the near future.

References

1 Schumacher, E.F. (1973) *Small Is Beautiful: A Study of Economics as if People Mattered*, Harper & Row, New York.

2 Dasgupta, P. (1983) *Production by the Masses and the Philosophy of Charka*, Sribhumi, Calcutta.

3 Tainter, J.A. (1988) *The Collapse of Complex Societies*, Cambridge University Press, Cambridge.

4 von Foerster, H., Mora, P., and Amiot, L. (1960) Doomsday: Friday, 13 November, A.D. 2026. *Science*, **132**, 1291–1295.

5 Cohen, M.N. (1977) *The Food Crisis in Prehistory: Overpopulation and the Origins of Agriculture*, Yale University Press, New Haven, Conn.

6 Cohen, J.E. (1995) *How Many People Can the Earth Support?* W.W. Norton, New York.

7 Beddoe, R., Costanza, R., Farley, J., Garzaa, E., Kent, J., Kubiszewski, I., Martinez, L., McCowen, T., Murphy, K., Myers, N., Ogden, Z., Stapleton, K., and Woodward, J. (2009) Overcoming systemic roadblocks to sustainability: The evolutionary redesign of worldviews, institutions, and technologies, *Proc. Natl. Acad. Sci. USA*, 106, 2483–2489.

8 Bruntland, G. (ed.) (1987) *Our Common Future: Report of the World Commission on Environment and Development* Oxford University Press, Oxford.

9 Forrester, J.W. (1971) *World Dynamics*, Productivity Press, Portland, Oregon.

10 Meadows, D., Randers, J., and Meadows, D. (2004) *Limits to Growth: The 30-Year Update*, Chelsea Green, White River, Vermont.

11 http://www.panda.org

12 http://www.sustainabilitydictionary.com

Appendix A
Thermodynamics and Free Particle Statistics

A.1
A Brief Introduction to Thermodynamics and Statistical Mechanics [1, 2][1]

Thermodynamics refers to the changes in the thermal state of macroscopic systems (consisting of many, typically $O(10^{23})$, basic dynamical constituents like atoms, molecules, spins, etc.). The macroscopic states of systems like fluids, for example, are represented by pressure, volume, temperature, etc. (or field, magnetization and temperature respectively, for magnets) and the relations satisfied by these quantities are called equations of state. This phenomenological study of macroscopic objects is called thermodynamics.

Knowing the laws of mechanics, it is in principle possible to derive all the properties of a material by solving the equations of motion. But this approach is highly constrained due to the fact that even if we use computers, it is impossible to go beyond a few hundred particles using this approach, while a macroscopic system contains about Avogadro's number ($O(10^{23})$) of particles. Therefore, this approach is absolutely unfit for such systems, unless averages are performed to eliminate minute details, which are in any case unnecessary to derive thermodynamic properties, and statistical mechanics deals with such studies of "average" thermodynamic systems.

A.1.1
Preliminary Concepts of Thermodynamics

Let us first discuss very briefly some preliminary concepts which are useful in the subsequent discussions.

1. A *thermodynamic system* is a macroscopic system which is distinct from the rest of the universe.
2. A *boundary* separates a system from its surroundings and may be real or imaginary.

1) Please find a list of notations for all Appendices proceeding Appendix E.

Econophysics. Sitabhra Sinha, Arnab Chatterjee, Anirban Chakraborti, and Bikas K. Chakrabarti
Copyright © 2011 WILEY-VCH Verlag GmbH & Co. KGaA, Weinheim
ISBN: 978-3-527-40815-3

3. The *surroundings* of a system is that part of the universe which is outside the system and can influence a change in the parameters of the system.

4. The *thermodynamic variables* or *parameters* of a system are the quantities that specify the macroscopic properties of the system, like pressure (P), volume (V), number of particles (N) for a gaseous system, magnetic field (h), magnetization (m) or the number (N) of spin moments for a magnetic system, etc.

5. All of the above mentioned variables are not independent. The relations that connect them are called *equations of state*.

6. A thermodynamic variable is called *extensive* if its value is proportional to the system size, for example volume (V), internal energy (U), etc. For an *intensive* variable the value does not depend upon system size, for example pressure (P), temperature (T), etc.

7. By saying that a system is in *thermodynamic equilibrium* we imply three conditions: *mechanical equilibrium*, *chemical equilibrium* and *thermal equilibrium*. As the names suggest, mechanical equilibrium is said to exist when there is no unbalanced force or torque in the system and surroundings. For chemical equilibrium, there should be no unbalanced chemical reaction rate or diffusion present in the system. Finally, thermal equilibrium is attained when there are no spontaneous changes in the system when it is separated from the surroundings by a diathermal boundary (which allows exchange of energy). A system satisfying all these conditions is said to be in thermodynamic equilibrium.

8. The concept of thermal equilibrium is utilized to derive the concept of temperature. The following statement is assumed to be true: If two systems are in thermal equilibrium with a third system, then they are in thermal equilibrium with themselves. This is sometimes called the *zeroth law of thermodynamics*. The concept of temperature arises by saying that the parameter which is common between two systems in thermal equilibrium is called *temperature*. The zeroth law forms the basis of thermometric measurements.

9. The process of a system going from one state of thermodynamic equilibrium (characterized by a set of values of the thermodynamic variables) to another is called a *thermodynamic process*. A process is *reversible* if the system can be traced back to the initial state in the same path that led to the original process. A process is *quasi-static* if it departs from a thermodynamic equilibrium only infinitesimally. A reversible process is necessarily quasi-static. Again, a process is *isothermal* if temperature does not change, *adiabatic* if energy does not change, *isochoric* if volume does not change and *isobaric* if pressure does not change.

10. When the surroundings of a system exert force on it and the system undergoes a displacement, *work* is said to be done *on* the system. Again, if the surroundings undergo a displacement due to force exerted by the system, then work is said to be done *by* the system. Work is said to be a *path variable* in the sense that its value depends upon the path which is followed in going from one state to another. On the other hand, a *state* variable only depends upon the initial and final state of the system, for example internal energy of the system.

11. When all the constituents of the (thermodynamic) system have a strictly correlated motion (typical example is a rigid body), the total energy of the system is

mechanical energy. Heat energy corresponds to the case when there is no correlation between the motion of the constituents (like in the case of gases, the
molecules have kinetic energy due to motions in different directions, although
the system as a whole remains motionless).

A.1.2
Laws of Thermodynamics

We now discuss the laws of thermodynamics, which are very useful axioms with
far reaching consequences. But it must be remembered that these are valid only
for macroscopic systems in thermodynamic equilibrium.

First Law of Thermodynamics   In thermodynamic processes, forms of energy can
be converted or changed. In an infinitesimal process, the mathematical statement
of the first law is

$$\bar{d}Q = dU + \bar{d}W ,$$ (A1)

where $\bar{d}Q$ is the heat supplied to the system, dU is the change in internal energy of the system and $\bar{d}W$ is the work done by the system. The differentials with
bars denote that they are *inexact* in the sense that their value depends upon path
and not on the initial and final states of the system. The internal energy U on the
other hand, is a state function (dU does not depend upon path). The first law of
thermodynamics is essentially a statement of conservation of energy.

Second Law of Thermodynamics   This law gives the natural direction of the thermodynamic (energy) changes. There are many equivalent statements of this law.
The Clausius statement reads: there is no cyclic process, the only effect of which is
to transfer heat from a colder to a hotter body.

  As we shall see shortly, the second law helps in formulating the concept of entropy (S), the growth of which in turn determines the direction of natural processes.
It also helps in giving an absolute scale of temperature (T). Specifically, the second
law says that the inexact differential $\bar{d}Q$ can be transformed to an exact differential
dS by defining $dS = \bar{d}Q/T$, where the entropy (S), is a state function.

Heat Engines: Carnot Efficiency   A heat engine is a device which operates in a cycle
and converts heat into work and Carnot efficiency (which depends, as we shall show
shortly, only on the temperature of the source and sink) is the maximum allowed
efficiency of a heat engine by the laws of thermodynamics. Suppose, in a cyclic
process,

  Q_1 = amount of heat absorbed by the system (form source).
  Q_2 = amount of heat rejected by the system (to the sink).
  W = amount of work done by the system.

The thermal efficiency (ϑ) of a heat engine is defined as,

$$\vartheta = \frac{\text{Work done by the system}}{\text{Heat supplied to the system}} . \tag{A2}$$

Since the system operates in a cycle, it must reach the initial configuration after each cycle. Hence there is no net change in internal energy of the system ($dU = 0$) as it is a state function. Therefore from the first law of thermodynamics (A1) we have

$$W = Q_1 - Q_2 .$$

For the thermal efficiency therefore, we get

$$\vartheta = \frac{W}{Q_1} = \frac{Q_1 - Q_2}{Q_1} = 1 - \frac{Q_2}{Q_1} . \tag{A3}$$

As the second law says that $Q_2 > 0$ the thermal efficiency of a heat engine can never be unity. However, the question of *maximum efficiency* is an intriguing one. Let us consider an ideal gas having the equations of state $PV - Nk_BT$ for isothermal process and $PV^{\phi} = \text{const.} = k$ for adiabatic process and evaluate the efficiency of an imaginary idealized reversible engine, called *Carnot's engine*; the efficiency will be independent of the choice (ideal gas).
The Carnot cycle consists of the following reversible operations:

a. An isothermal expansion (AB) at temperature T_1 where the volume changes from V_1 to V_2. The heat absorbed is Q_1 and work done by the system is W_1.
b. An adiabatic expansion (BC) where the volume changes from V_2 to V_3 and by definition there is no heat exchange. The work done is W_2. Temperature drops from T_1 to T_2.
c. An isothermal compression (CD) where the the volume changes from V_3 to V_4 at constant temperature T_2. The heat generated by the system is Q_2 and the work done on the system is W_3.
d. Finally, an adiabatic compression (DA) in which the volume changes from V_4 to V_1 and the temperature increases from T_2 to T_1 and the work done by the system is W_4.

The net amount of work done by the system is

$$W = W_1 + W_2 + W_3 + W_4 .$$

We shall calculate these quantities individually.
To begin with,

$$W_1 = Q_1 = \int_{V_1}^{V_2} P\,dV = nRT_1 \int_{V_1}^{V_2} \frac{dV}{V} = Nk_B T_1 \ln \frac{V_2}{V_1} ,$$

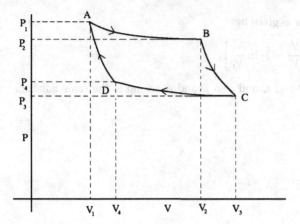

Figure A.1 P–V diagram of the Carnot cycle.

for the isothermal path, and

$$W_2 = \int_{V_2}^{V_3} P\,dV = k \int_{V_2}^{V_3} \frac{dV}{V^\phi}$$

$$= \frac{k}{1-\phi}\left(\frac{1}{V_3^{\phi-1}} - \frac{1}{V_2^{\phi-1}}\right)$$

$$= \frac{1}{\phi-1}\left[\frac{k\,P_2\,V_2}{P_2\,V_2^\phi} - \frac{k\,P_3\,V_3}{P_3\,V_3^\phi}\right]$$

$$= \frac{Nk_B}{\phi-1}(T_1 - T_2).$$

Also,

$$W_3 = \int_{V_3}^{V_4} P\,dV = Nk_B T_2 \ln\frac{V_4}{V_3}$$

$$= -Nk_B T_2 \ln\frac{V_3}{V_4}.$$

Finally,

$$W_4 = \int_{V_4}^{V_1} P\,dV = k \int_{V_4}^{V_1} \frac{dV}{V^\phi}$$

$$= \frac{Nk_B}{\phi-1}(T_2 - T_1)$$

$$= -\frac{Nk_B}{\phi-1}(T_1 - T_2).$$

Hence the total work done is given by

$$W = N k_B \left[T_1 \ln \frac{V_2}{V_1} - T_2 \ln \frac{V_3}{V_4} \right].$$

We further notice that B and C and also D and A belong to the same adiabatic curves. So,

$$T_1 V_2^{\phi-1} = T_2 V_3^{\phi-1},$$

and

$$T_1 V_1^{\phi-1} = T_2 V_4^{\phi-1}.$$

Using these two we obtain

$$\frac{V_2}{V_1} = \frac{V_3}{V_4}.$$

Hence,

$$W = N k_B (T_1 - T_2) \ln \frac{V_2}{V_1}.$$

Now, the thermal efficiency will be

$$\vartheta = \frac{W}{Q_1}$$

$$= \frac{N k_B (T_1 - T_2) \ln(V_2/V_1)}{N k_B T_1 \ln(V_2/V_1)}$$

$$= 1 - \frac{T_2}{T_1}.$$

Clearly, the efficiency only depends upon the temperature between which the Carnot cycle operates. There is no property of the substance that is finally left in this estimate of the efficiency.

A.2
Free Particle Statistics [1, 2]

Let us consider an ideal gas containing a large ($O(10^{23})$)number of particles. In equilibrium, one can ask the question: what is the expected average number of particles in the gas having energy ϵ? In other words, we seek the form of the distribution function $n(\epsilon)$ or $n_q(\epsilon_q)$, where we have denoted the momentum states by the subscript q.

In deriving this statistics, we use the principle that *entropy* of the system will be maximized in equilibrium. The entropy of a system is defined as follows:

$$S = \ln \Omega (E, N),$$

where $\Omega\,(E, N)$ is the number of microstates of the system when some macrostate is already specified by the variables like the total energy E and the total number of particles N, which are kept constant

$$E = \sum_q n_q \epsilon_q$$

$$N = \sum_q n_q \,.$$

Now, the principle of *ergodicity* states that if we have a system developing its dynamics, then it will evolve in such a way that all the microstates will be equally probable. If we could track the system for long enough time, we would find that it has gone through all the microstates. Hence, if we take the average over an ensemble of all possible microstates at a given time, then the averages over the ensemble and over the dynamics would match. This hypothesis may not be valid in some cases, for example in *spin glasses* (see Chapter 11). Let,

n_q = number of particles with energy ϵ_q
g_q = degeneracy number for energy ϵ_q
Ω = total number of microscopic configurations.

The degeneracy number for a given eigenvalue (in this case, energy) refers to the number of states that have the same eigenvalue. Therefore, entropy can be written as

$$S = \ln \Omega \,. \tag{A4}$$

We want to maximize this subject to the conditions of constancy of E and N.

A.2.1
Classical Ideal Gas: Maxwell–Boltzmann Distribution and Equation of State

In Maxwell–Boltzmann statistics, we take the particles to be distinguishable objects, as opposed to the indistinguishable objects considered in the quantum cases. There is no limit in the number of particles that can stay in a given state. Now if n_q is the number of particles in the energy state ϵ_q then from particle and energy conservations, we have as before

$$\sum_i n_i = N \,, \quad \sum_q n_q \epsilon_q = E \,. \tag{A5}$$

Here, as the particles are distinguishable, to find the total number of configurations, we need to consider first what is the number of ways in which n_q particles can be selected from N particles, for all q values. It is given by

$$\Omega_1 = \frac{N!}{\prod_q n_q!} \,.$$

Now, in finding the number of ways in which n_q particles can be put into g_q boxes, where g_q is the degeneracy factor, we see that all the n_q particles can be put into any of the g_q boxes. Hence the number of ways is given by

$$\Omega_2 = \prod_q g_q^{n_q} .$$

The total number of configurations is obtained by multiplying these two contributions

$$\Omega^{MB} = \Omega_1 \Omega_2 = \frac{N!}{\prod_q n_q!} \prod_{q'} g_{q'}^{n_{q'}} .$$

Hence entropy is given by

$$S = \ln \Omega^{MB}$$
$$= \ln(N!) + \sum_q n_q \ln g_q - \sum_q \ln(n_q!)$$
$$= N \ln N + \sum_q n_q \ln g_q - \sum_q n_q \ln n_q .$$

For the most probable distribution, as before, entropy is to be maximized. Hence,

$$\delta \ln \Omega^{MB} = 0$$
$$\Rightarrow \sum_q (\ln g_q - \ln n_q) \, \delta n_q = 0 .$$

We need to incorporate the following conditions

$$\sum_q \delta n_q = 0 , \quad \sum_q \epsilon_q \delta n_q = 0 .$$

Using Lagrange's multipliers, we get

$$\sum_q (\ln g_q - \ln n_q + \theta \mu - \theta \epsilon_q) \, \delta n_q = 0 , \tag{A6}$$

where $\theta \mu$ and μ are the undetermined multipliers. This equation must now hold for any δn_q, giving

$$n_q = g_q e^{-\theta(\epsilon_q - \mu)} . \tag{A7}$$

This is the Maxwell–Boltzmann distribution.

A.2.1.1 Ideal Gas: Equation of State

With the identification $\left(\frac{\partial S}{\partial E}\right)_V = \frac{1}{T}$, we see that $\theta = \frac{1}{k_B T}$. Now we can evaluate $\theta \mu$ as well. If the energy levels are densely spaced, we can replace the sum by an integral and write

$$N = \int_0^\infty n(\epsilon) d\epsilon = e^{-\theta \mu} \int_0^\infty g(\epsilon) e^{-\theta \epsilon} d\epsilon. \tag{A8}$$

Let us replace, $g(\epsilon) = \frac{2\pi V}{(2\pi\hbar)^3}(2M)^{3/2}\epsilon^{1/2}$, where M is the mass of each particle. Then we get

$$N = e^{\theta\mu}\frac{2\pi V}{(2\pi\hbar)^3}(2M)^{3/2}\int_0^\infty \epsilon^{1/2}e^{-\theta\epsilon}\,d\epsilon$$

$$= e^{\theta\mu}\left(\frac{2M\pi k_B T}{(2\pi\hbar)^2}\right)^{3/2}V .$$

Then we invert

$$e^{\theta\mu} = \frac{N}{V}\left(\frac{(2\pi\hbar)^2}{2M\pi k_B T}\right)^{3/2} .$$

With these in hand, we may now go on to derive the *equipartition theorem*. It says that each particle has energy $\frac{k_B T}{2}$ for each of its degrees of freedom. Let us now calculate the total energy:

$$E = \int_0^\infty n(\epsilon)\epsilon\,d\epsilon$$

$$= \int_0^\infty \epsilon g(\epsilon)e^{-\theta(\epsilon-\mu)}\,d\epsilon$$

$$= \frac{2\pi V}{(2\pi\hbar)^3}(2M)^{3/2}e^{\theta\mu}\int_0^\infty \epsilon^{3/2}e^{-\theta\epsilon}\,d\epsilon$$

$$= \frac{3}{2}N k_B T .$$

So, energy per particle will be

$$\frac{E}{N} = \frac{3}{2}k_B T . \tag{A9}$$

Hence, corresponding to each degrees of freedom (in the case above, each mass point has three translational degrees of freedom), there is an associated energy $\frac{1}{2}k_B T$. This is called the equipartition law.

We can also derive the *equation of state of an ideal gas* from here. To do that let us first evaluate the partition function:

$$Z = \int_0^\infty g(\epsilon)e^{-\epsilon/k_B T}\,d\epsilon$$

$$= \int_0^\infty \frac{2\pi V(2M)^{3/2}}{(2\pi\hbar)^3}\epsilon^{1/2}e^{-\epsilon/k_B T}\,d\epsilon$$

$$= \frac{V}{(2\pi\hbar)^3}(2\pi M k_B T)^{3/2} .$$

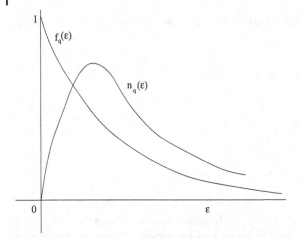

Figure A.2 Plot of MB distribution and Gibb's distribution.

Now, from thermodynamics, we know that pressure is given by

$$P = - \left(\frac{\partial F}{\partial V} \right)_T .$$

Here,

$$F = -N k_B T \ln Z$$
$$= -N k_B T \left[\ln V - \ln(2\pi\hbar)^3 + \frac{3}{2} \ln (2\pi M k_B T) \right].$$

So,

$$\left(\frac{\partial F}{\partial V} \right)_T = - \frac{N k_B T}{V} ,$$

implying,

$$PV = N k_B T , \tag{A10}$$

which is the familiar equation of state for an ideal gas in equilibrium.

The plots of $n_q = f_q g_q$, as given in (A7), and of the Gibbs distribution function ($f_q = e^{-\epsilon/k_B T}$) are shown in Figure A.2.

A.2.2
Quantum Ideal Gas

If the particles are quantum mechanical, they will either obey Bose–Einstein (any number of particles can stay in a given energy state) or Fermi–Dirac (at most one particle can occupy an energy state) statistics. Let us separately consider both the cases.

A.2.2.1 Bose Gas: Bose–Einstein (BE) Distribution

In BE distribution we can think of the problem of finding $\Omega^{\text{BE}}_{n_q, g_q}$ as distributing n_q particles in g_q boxes. Suppose we place the boxes and particles in a row with the

understanding that the particles placed to the right of a box, belong to the box. All the configurations can be found by rearranging the boxes and particles among themselves with the constraint that one box must be placed at the left most place. Also, as the particles are identical, interchanging the positions of the boxes and the particles, keeping the order of appearance unchanged, does not give new configurations. Now we can write the total number of microstates for all q as

$$\Omega^{\text{BE}} = \prod_q \frac{(n_q + g_q - 1)!}{n_q!\,(g_q - 1)!} . \tag{A11}$$

Then the entropy will be

$$
\begin{aligned}
S^{\text{BE}} &= \ln \Omega^{\text{BE}} \\
&= \sum_q \{\ln (n_q + g_q - 1)! - \ln n_q! - \ln (g_q - 1)!\} \\
&= \sum_q \{(n_q + g_q - 1) \ln (n_q + g_q - 1) - (n_q + g_q - 1) \\
&\quad - n_q \ln n_q + n_q - (g_q - 1) \ln (g_q - 1) + (g_q - 1)\} \\
&= \sum_q \{(n_q + g_q) \ln(n_q + g_q) - n_q \ln n_q - g_q \ln g_q\} ,
\end{aligned}
$$

where we have neglected 1 with respect to n_q and g_q in the last step. We now need to find its variation with respect to n_q (since g_q is fixed) and set it to zero in extremum

$$dS^{\text{BE}} = \sum_q dn_q \left\{\ln \frac{n_q + g_q}{n_q}\right\} .$$

We also need to incorporate the conditions

$$dE = \sum_q dn_q \epsilon_q = 0 , \tag{A12}$$

$$dN = \sum_q dn_q = 0 . \tag{A13}$$

As before, we use the Lagrange undetermined multiplier method and write

$$dS^{\text{BE}} - \theta\,dE + \theta\mu\,dN = 0 . \tag{A14}$$

Therefore,

$$\sum_q dn_q \left\{\left[\ln \left(\frac{n_q + g_q}{n_q}\right)\right] - \theta \epsilon_q + \theta \mu\right\} = 0 ,$$

implying

$$\ln\left(\frac{n_q + g_q}{n_q}\right) = \theta\left(\epsilon_q - \mu\right).$$

Rearranging, the number distribution for BE statistics comes out to be

$$n_q^{BE} = \frac{g_q}{e^{\theta(\epsilon_q - \mu)} - 1}. \tag{A15}$$

From BE distribution a very interesting phenomenon at low temperature can be predicted, which is called the *Bose–Einstein condensation*. Recently, this has been experimentally shown as well.

BE distribution is valid for the energy range $0 \leq \epsilon_q < \infty$, because this is the energy range of a free particle. Now, suppose $\mu > 0$, then for minimum energy $\epsilon_q = 0$ we will have

$$n_q = \frac{g_q}{e^{-\theta\mu} - 1}. \tag{A16}$$

However, if $e^{-\theta\mu} < 1$, n_q will be negative and that is absurd. This leads us to conclude that $\mu \leq 0$ in BE distribution. Now, one can further show that for BE distribution, $\frac{d\mu}{dT} < 0$, that is as temperature is decreased, μ increases. It implies that μ starts from a large negative value and approaches zero as temperature is decreased. There would have been no problem if $\mu \to 0$ as $T \to 0$. But this is not the case. μ becomes zero at a *finite* value of temperature. Let us determine that value first. We have the following identity

$$\sum_q n_q = \sum_q \frac{g_q}{e^{(\epsilon_q - \mu)/k_B T} - 1} = N.$$

Let the temperature at which $\mu = 0$ be T_B. Then we will have

$$N = \sum_q \frac{g_q}{e^{\epsilon_q / k_B T_B} - 1}$$

$$= \int_0^\infty \frac{g(\epsilon)\, d\epsilon}{e^{\epsilon / k_B T_B} - 1}$$

$$\sim \int_0^\infty \frac{\epsilon^{1/2}\, d\epsilon}{e^{\epsilon / k_B T_B} - 1}$$

$$\sim (k_B T_B)^{3/2} \int_0^\infty \frac{\left(\frac{\epsilon}{k_B T_B}\right)^{1/2} d\left(\frac{\epsilon}{k_B T_B}\right)}{e^{\epsilon / k_B T_B} - 1}$$

$$\sim (k_B T_B)^{3/2} \int_0^\infty \frac{x^{1/2}\, dx}{e^x - 1}.$$

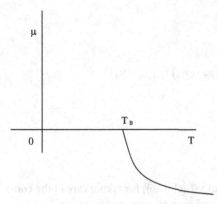

Figure A.3 Plot of chemical potential with temperature for Bose gas.

Let the value of the integral be C. Therefore, $N = C\,T_B^{3/2}$. In terms of $\rho(= N/V)$

$$k_B T_B = C'\rho^{2/3} \,. \tag{A17}$$

Clearly, unless $\rho = 0$, T_B is not zero as C' is always finite. Therefore to save the situation, we keep $\mu = 0$ even below T_B. But this is a nontrivial statement, because then for $T < T_B$

$$N' = T^{3/2}C \,. \tag{A18}$$

Here, C is constant. So, the above equation implies, N' decreases with temperature for $T < T_B$. This means that as the temperature is lowered below T_B some of the particles are condensed into a state from which they no longer take part in the BE distribution, that is $N' < N$. The condensed fraction is, of course, proportional to

$$N - N' \sim \left(T_B^{3/2} - T^{3/2}\right). \tag{A19}$$

A.2.2.2 Fermi Gas: Fermi–Dirac Distribution

In this case, at most one particle can occupy an energy state. If we go back to the picture of placing particles in boxes, we see that g_q boxes will be placed, and to the right of each box there is a place for one particle. n_q particles will be placed in those g_q places:

the number of microstates is then simply the number of ways in which n_q boxes can be chosen out of g_q boxes. Remembering that the particles are identical, the expression will be (for all q)

$$\Omega^{FD} = \prod_q \frac{g_q!}{n_q!\,(g_q - n_q)!} \,.$$

Hence the entropy is given by

$$S^{\mathrm{FD}} = \ln \Omega^{\mathrm{FD}}$$

$$= \sum_q \left\{ g_q \ln g_q - n_q \ln n_q - (g_q - n_q) \ln (g_q - n_q) \right\}.$$

The variation, as before, will be

$$d S^{\mathrm{FD}} = \sum_q d n_q \left(\ln \frac{g_q - n_q}{n_q} \right).$$

And using the Lagrange multipliers $\theta \mu$ and θ (cf. (A6)) for taking care of the conservation of number of particles and the total energy, respectively, we get

$$\sum_q d n_q \left[\ln \frac{g_q - n_q}{n_q} - \theta \epsilon_q + \theta \mu \right] = 0 ,$$

finally leading to

$$n_q^{\mathrm{FD}} = \frac{g_q}{e^{\theta (\epsilon_q - \mu)/k_{\mathrm{B}} T} + 1} \tag{A20}$$

with the identification $\theta = \frac{1}{k_{\mathrm{B}} T}$.

Appendix B
Interacting Systems: Mean Field Models, Fluctuations and Scaling Theories

B.1
Interacting Systems: Magnetism [1–4]

B.1.1
Heisenberg and Ising Models

So far we have dealt with noninteracting systems. Now we shall consider interacting systems. Consider the Heisenberg Hamiltonian, representing a set of spins (S_i; $i = 1, 2, \ldots, N$) on lattice sites i, interacting through the exchange interaction J_{ij}

$$H = -\sum_{\langle i,j \rangle} J_{ij} S_i \cdot S_j . \tag{B1}$$

If $J_{ij} > 0 \forall i, j$, at zero temperature, the spins are all aligned without any external magnetic field (spontaneous magnetization). This explains ferromagnetism at low temperature. At higher temperatures, excited or higher energy configurations (not satisfying the minimum energy criterion for the above Hamiltonian) will be populated and spontaneous magnetization will disappear at $T > T_c$, the transition point.

For classical treatments, it was assumed that the only excitations are provided by continuous spin orientations. In general for the quantum Heisenberg Hamiltonian, an additional problem arises due to quantum noncommutative relations. Ising's proposal was to consider the interaction only in one (e.g., $\hat{z}$) direction, thereby making the cooperative Hamiltonian classical

$$H = -\sum_{\langle i,j \rangle} J_{ij} S_i^z S_j^z , \tag{B2}$$

where J_{ij} denotes the interaction between the nearest neighbor spins. We assume $S_i^z = \pm 1$, that is minimum number of degrees of freedom is considered. In absence of any thermal noise, nearest neighbor spins will be parallel to minimize the energy. Let us consider the simple case when $J_{ij} = J \forall i, j$. In Figure B.1, we note that the spin at A has four neighboring spins, all of which are aligned in parallel. If the spin at A is oriented in the same direction as its neighbors, then there is an energy cost of $\Delta E = 8J$ if its wants to flip to the opposite orientation. Therefore

Econophysics. Sitabhra Sinha, Arnab Chatterjee, Anirban Chakraborti, and Bikas K. Chakrabarti
Copyright © 2011 WILEY-VCH Verlag GmbH & Co. KGaA, Weinheim
ISBN: 978-3-527-40815-3

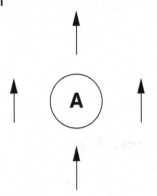

Figure B.1 A site in the environment of all parallel spins.

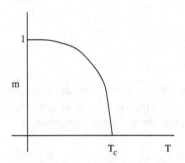

Figure B.2 Variation of magnetization with temperature.

the probability to flip will be $e^{-8J/k_B T}$, which is very small unless T is very large (k_B is the Boltzmann constant). If m ($= \langle S^z \rangle$) denotes the average magnetization, then the temperature at which m vanishes is called the critical temperature (T_c). The plot of m versus T is shown in Figure B.2. If there is an external field, denoting S^z by S only, we can express the Hamiltonian of the system as

$$H = -J \sum_{\langle ij \rangle} S_i S_j - h \sum_i S_i .$$
(B3)

The partition function reads, $Z = \mathrm{Tr}\left(e^{-H/k_B T}\right)$. The exact solution is possible in one dimension. In two dimensions Onsager solved this problem for $h = 0$. In three and higher dimensions, exact solutions have not yet been completed.

B.1.2
Mean Field Approximation (MFA)

We shall now see how analytical calculations can be done for the Ising system under the mean field assumption. Suppose

$$H = -h^{\mathrm{ext}} \sum_i S_i .$$
(B4)

Then, the partition function is given by

$$
\begin{aligned}
Z &= \mathrm{Tr}\left[\exp\left\{\frac{h^{\mathrm{ext}}}{k_{\mathrm{B}}T}\sum_i S_i\right\}\right] \\
&= \mathrm{Tr}\prod_i \exp\left\{\frac{h^{\mathrm{ext}}}{k_{\mathrm{B}}T}S_i\right\} \\
&= \prod_i \mathrm{Tr}\exp\left\{\frac{h^{\mathrm{ext}}}{k_{\mathrm{B}}T}S_i\right\} \\
&= \prod_i 2\cosh\left(\frac{h^{\mathrm{ext}}}{k_{\mathrm{B}}T}\right) \\
&= \left[2\cosh\left(\frac{h^{\mathrm{ext}}}{k_{\mathrm{B}}T}\right)\right]^N ,
\end{aligned}
$$

where N is the total number of spins. As the free energy $F = -k_{\mathrm{B}}T\ln Z$, one obtains

$$
F = -Nk_{\mathrm{B}}T\ln\left[2\cosh h^{\mathrm{ext}}/k_{\mathrm{B}}T\right]. \tag{B5}
$$

This brings out the extensive property of the free energy.

However, for a cooperative interacting system like (B1) we cannot neglect the interaction term. The idea of mean field approximation is to consider an average effect of the states of the nearest neighbors and deal with an average or effective field

$$
h^{\mathrm{eff}} = Jz\langle S_i\rangle + h^{\mathrm{ext}} ,
$$

where z, called the coordination number, denotes the number of nearest neighbors of any spin. Note that we have replaced the average over spin states by its thermodynamic average. Hence,

$$
H \approx -h^{\mathrm{eff}}\sum_{i=1}^{N} S_i . \tag{B6}
$$

Now consider the spin at a particular site, for example, S_0. Then its average will give the magnetization. Hence,

$$
m = \langle S_0 \rangle = \frac{\mathrm{Tr}\left(S_0 e^{-\frac{H}{k_B T}}\right)}{\mathrm{Tr}\left(e^{-\frac{H}{k_B T}}\right)}
$$

$$
= \frac{\mathrm{Tr}\left(S_0 e^{\frac{h^{\mathrm{eff}}}{k_B T} \sum_i S_i}\right)}{\mathrm{Tr}\left(e^{\frac{h^{\mathrm{eff}}}{k_B T} \sum_i S_i}\right)}
$$

$$
= \frac{\mathrm{Tr}\left(S_0 e^{\frac{h^{\mathrm{eff}}}{k_B T} S_0} e^{\frac{h^{\mathrm{eff}}}{k_B T} \sum_i' S_i}\right)}{\mathrm{Tr}\left(e^{\frac{h^{\mathrm{eff}}}{k_B T} S_0} e^{\frac{h^{\mathrm{eff}}}{k_B T} \sum_i' S_i}\right)}
$$

$$
= \frac{\mathrm{Tr}\left[S_0 e^{\frac{h^{\mathrm{eff}}}{k_B T} S_0}\right]}{\mathrm{Tr}\left[e^{\frac{h^{\mathrm{eff}}}{k_B T} S_0}\right]}
$$

$$
= \frac{e^{\frac{h^{\mathrm{eff}}}{k_B T}} - e^{-\frac{h^{\mathrm{eff}}}{k_B T}}}{e^{\frac{h^{\mathrm{eff}}}{k_B T}} + e^{-\frac{h^{\mathrm{eff}}}{k_B T}}}
$$

$$
= \tanh\left(\frac{h^{\mathrm{eff}}}{k_B T}\right). \tag{B7}
$$

Therefore,

$$
\langle S \rangle \equiv \langle S_0 \rangle = m = \tanh\left(h^{\mathrm{eff}}/k_B T\right)
$$
$$
= \tanh\left[(Jz\langle S \rangle + h)/k_B T\right]
$$
$$
= \tanh\left[(Jzm + h)/k_B T\right].
$$

For $h = 0$, (zero external field) the (spontaneous) magnetization is given by

$$
m = \tanh\left(\frac{Jzm}{k_B T}\right). \tag{B8}
$$

This is a self-consistent equation (m appearing on both sides) and can be solved in a graphical way. Consider the curves $y_1 = m$ and $y_2 = \tanh\left(\frac{Jzm}{k_B T}\right)$. When $Jz < k_B T$ the two curves intersect only at one point $m = 0$, that is paramagnetic state (no spontaneous magnetization). When $Jz > k_B T$ there are two solutions, one corresponds to $m = 0$ and the other corresponds to a finite nonzero m (spontaneous magnetization) indicating a ferromagnetic state. So, the point at which a finite magnetization first appears is called the critical point and it is determined from the condition

$$k_B T_c = Jz .$$

(B9)

It may be noted that as the mean field theory does not consider fluctuation effects, it wrongly predicts a finite critical temperature even for the one-dimensional Ising model with finite range interactions.

B.1.2.1 Critical Exponents in MFA

When $T \approx T_c$, then m is very small. Hence we can expand $\tanh\left(Jzm/k_B T\right)$ in a series as

$$m \approx Jzm/k_B T - \frac{\left(Jzm/k_B T\right)^3}{3} + \cdots$$

When $T < T_c$, $m \neq 0$, so we can divide both sides by m

$$1 = Jzm/k_B T - \frac{\left(Jz/k_B T\right)^3}{3} m^2$$

$$1 = \left(\frac{T}{T_c}\right) - \frac{\left(T_c/T\right)^3}{3} m^2 .$$

Rearranging, we get

$$m^2 = \frac{\left[3\frac{T_c - T}{T_c}\right]}{\left(T_c/T\right)^2} .$$

Implying,

$$m \sim \left(\frac{T_c - T}{T_c}\right)^{1/2} .$$

(B10)

Defining the order parameter exponent β as

$$m \sim \left(\frac{T_c - T}{T_c}\right)^{\beta} ,$$

(B11)

we get $\beta = \frac{1}{2}$ under MFA.

When an external field is present, we have

$$m = \tanh\left(Jzm + h\right)/(k_B T) .$$

Again expanding for small m near T_c we get

$$m \approx \left(Jzm + h\right)/(k_B T) - \frac{\left(Jzm + h\right)^3}{3 k_B^3 T^3} + \cdots$$

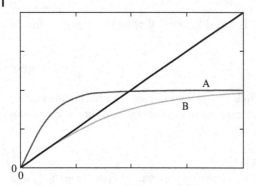

Figure B.3 Graphical solution of mean field (B8).

Differentiating the above equation with respect to h we get

$$\frac{\partial m}{\partial h} = \left[\frac{Jz}{k_B T} \frac{\partial m}{\partial h} + \frac{1}{k_B T} - \frac{(Jzm + h)^2}{k_B^2 T^2} \left(\frac{Jz}{k_B T} \frac{\partial m}{\partial h} + \frac{1}{k_B T} \right) \right].$$

Again, the definition of susceptibility reads

$$\chi = \left. \frac{\partial m}{\partial h} \right|_{h \to 0}.$$

Hence,

$$\chi = \frac{Jz}{k_B T} \chi + \frac{1}{k_B T} - \frac{(Jzm)^2}{k_B^2 T^2} \left[\left(\frac{Jz}{k_B T} \chi \right) + \frac{1}{k_B T} \right].$$

After some algebra we get

$$\chi \left[1 \frac{T_c}{T} + \left(\frac{T_c}{T} \right)^3 m^2 \right] = \frac{1}{k_B T} \left[1 - \left(\frac{T_c}{T} \right)^2 m^2 \right].$$

When $T > T_c$, we can neglect higher powers of (T_c/T),

$$\chi \left(1 - \frac{T_c}{T} \right) \approx \frac{1}{k_B T}.$$

Hence we obtain

$$\chi \sim \left(\frac{T_c - T}{T_c} \right)^{-1}. \tag{B12}$$

But the definition of the critical exponent γ is

$$\chi \sim \left(\frac{T_c - T}{T_c} \right)^{-\gamma}. \tag{B13}$$

Comparing we get $\gamma = 1$ under MFA.

Let us now look into a more general case of the Ising model where the system is not homogeneous, that is J_{ij} is not constant.

$$H = -\sum_{i,j} J_{ij} S_i S_j - h \sum_i S_i$$

$$= \sum_{i=1}^{N} h^{\text{eff}} S_i$$

with $h^{\text{eff}} = h + \sum_j J_{ij} \langle S_j \rangle$. Now, we can expand J_{ij} in a Fourier series

$$J(q) = \sum_{\langle ij \rangle} J_{ij} e^{iqi(R_i - R_j)} \, ,$$

where R_i denotes the position vector for site i. We also have as before, $\langle S_i \rangle = \tanh(\beta h^{\text{eff}})$. If we replace $\langle S_i \rangle$ by m_i and expand in a series keeping up to the first term, then we obtain

$$m_i = \beta h^{\text{eff}} = \frac{h}{k_B T} + \frac{1}{k_B T} \sum_j J_{ij} m_j \, .$$

Differentiating with respect to h

$$\frac{\partial m_i}{\partial h} = \chi_i = \frac{1}{k_B T} \left[1 + \sum_j J_{ij} \chi_j \right].$$

Taking the Fourier transform of both sides gives

$$\sum_i e^{iqR_i} \chi_i = \frac{1}{k_B T} \left[\sum_i e^{iqR_i} + \sum_{i,j} e^{iq(R_i - R_j)} J_{ij} e^{iqR_j} \chi_j \right], \tag{B14}$$

implying,

$$\chi_q = \frac{1}{k_B T} \delta(q) + \frac{1}{k_B T} J(q) \chi_q \, . \tag{B15}$$

Rearranging, we have

$$\chi_q (k_B T - J(q)) = \delta(q) \, . \tag{B16}$$

For nearest neighbor interactions one can expand

$$J(q) = J(0) - C q^2 + \dots \, , \tag{B17}$$

where $J(0) = J z = k_B T_c$ (z denoting the coordination number), $C = J(0) a^2$ and the odd order terms vanish. Using this, we get

$$\chi_q = \frac{1}{(T - T_c) + q^2}$$

$$= \frac{1}{(T - T_c)(1 + q^2 \xi^2)} \, ,$$

where $\xi \sim (T - T_c)^{-1/2}$. Note that this ξ has the dimension of length and this length diverges as $(T - T_c)^{-1/2}$ near $T = T_c$. When ξ is expressed as

$$\xi = \left| \frac{T - T_c}{T_c} \right|^{-\nu}, \tag{B18}$$

and we get $\nu = 1/2$ for the correlation length exponent in MFA.

B.1.2.2 Free Energy in MFA

We have the Ising Hamiltonian as $H = \sum_i h^{\text{eff}} S_i$, where $h^{\text{eff}} = \sum_{\langle ij \rangle} J_{ij} \langle S_j \rangle + h$. The free energy of the system is given by, $F = E - TS$, where $E = \langle H \rangle$, is the internal energy of the system. Now

$$\langle H \rangle = - \left\langle \sum_i h_i^{\text{eff}} S_i \right\rangle$$

$$= - \sum_i h^{\text{eff}} \langle S_i \rangle$$

$$= - N h^{\text{eff}} m .$$

Implying,

$$E = \langle H \rangle = -N \left[Jzm^2 + hm \right].$$

To find S, we look for the number of spin configurations $\Omega(m)$ for fixed m. Suppose there are N sites. If N_+ is the number of up-spins and N_- the number of down-spins, we shall have

$$N_+ + N_- = N , \quad N_+ - N_- = mN .$$

Therefore,

$$\Omega = \frac{N!}{N_+! N_-!} = \frac{N!}{\left\{ \frac{N(1+m)}{2} \right\}! \left\{ \frac{N(1-m)}{2} \right\}!} .$$

Using Stirling's approximation we have

$$S = -N \left[\left(\frac{1+m}{2} \right) \ln \left(\frac{1+m}{2} \right) + \left(\frac{1-m}{2} \right) \ln \left(\frac{1-m}{2} \right) \right].$$

Recalling that $\ln \left(\frac{1+m}{2} \right) \Big|_{m \to 0} = m - \ln 2$, we get on simplification

$$-TS = NTm^2 + NT \ln 2 .$$

So, the free energy per site f will be

$$f = F/N = -Jzm^2 + Tm^2 + T \ln 2 - hm + \dots , \tag{B19}$$

where we can add higher order terms of even power, if needed. Hence, generally one expects

$$f = (T - zJ)\, m^2 + C_1 T m^4 + C_2 T m^6 - hm + \ldots$$
$$= (T - T_c)\, m^2 + A m^4 + B m^6 + \cdots - hm \,, \tag{B20}$$

where $A > 0$ and $B > 0$. Suppose, $h = 0$ and consider up to A term. Then for $T > T_c$, the free energy minimum occurs at $m = 0$. Note that the highest order term must have a positive coefficient. This is to ensure a lower bound in the free energy. For $T < T_c$, there will be an initial decrease in free energy, then with increase in m, the second term will dominate and free energy will increase again. For minima

$$\frac{\partial F}{\partial m} = 2(T - T_c) m + 4 A m^3 = 0 \,. \tag{B21}$$

So, for $T < T_c$, we will again have, $m \sim (T - T_c)^{1/2}$.

B.1.3
Landau Theory of Phase Transition

The order parameter in a phase transition will have nonzero value in the ordered phase and will vanish in the disordered phase. Landau theory performs a series expansion of the free energy in terms of the order parameter. Consider the case of second-order phase transition, when $F(\tilde{m}) = F(-\tilde{m})$, with $\tilde{m}$ being the order parameter. This theory says that if there is no external field in the Hamiltonian, then for $\tilde{m}$ and $-\tilde{m}$ we will have the same Hamiltonian. Let us write the general series as

$$F(\tilde{m}) = A + B \tilde{m} + C \tilde{m}^2 + D \tilde{m}^3 + E \tilde{m}^4 + \ldots \tag{B22}$$

We intend to keep the minimum number of terms, but also we must ensure that the last term is positive. For the moment, we truncate the series at $E \tilde{m}^4$ with $E > 0$. From the symmetry $F(\tilde{m}) = F(-\tilde{m})$, the coefficient of the odd terms will vanish, that is $B = 0 = D$. We can always choose the zero of energy arbitrarily, so we put $A = 0$. Therefore,

$$F(\tilde{m}) = C \tilde{m}^2 + E \tilde{m}^4 \,. \tag{B23}$$

If we want to keep up to the $C \tilde{m}^2$ term, then $C > 0$. Then the form of the free energy is shown in Figure B.4. Clearly, to minimize f, $\tilde{m} = 0$, that is the disordered phase. So, we must consider the $E \tilde{m}^4$ term with $E > 0$. Then, the form of the free energy for $C > 0$ and $C < 0$ are shown. Clearly, for $C < 0$ and $E > 0$, we get free energy minima for nonzero value of the order parameter. Obviously, there is an order-disorder transition if C changes sign. Let us assume

$$C(T) = c(T - T_c) \,, \qquad c > 0 \,.$$

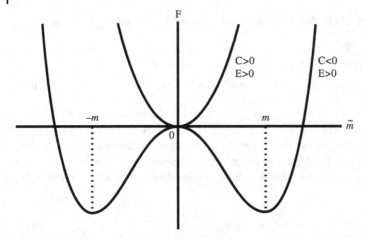

Figure B.4 Landau free energy for different signs of the coefficients.

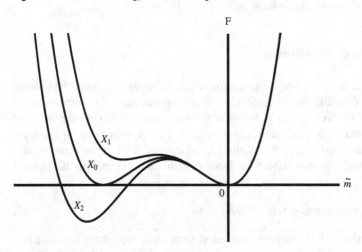

Figure B.5 Landau free energy for first-order transition.

Now, for minima, $\frac{\partial F}{\partial \tilde{m}}\big|_{\tilde{m}=m} = 0$. This implies

$$2Cm + 4Em^3 = 0.$$

For, $C < 0$ we can write

$$|C|m = 2Em^3.$$

Then, putting in the form of C we obtain

$$m \sim (T - T_c)^{1/2}. \tag{B24}$$

Thus we get the behavior of the order parameter near T_c with very simplifying assumptions.

Now let us discuss the first-order phase transition. Here, $F(\tilde{m}) \neq F(-\tilde{m})$. So we write the free energy as

$$F(\tilde{m}) = C\tilde{m}^2 + D\tilde{m}^3 + E\tilde{m}^4 + \ldots . \qquad (B25)$$

We cannot stop at the $D\tilde{m}^3$ term to ensure the lower bound for large negative $\tilde{m}$ values. The form of the free energy is shown in Figure B.5 for $C > 0$, $D > 0$, $E > 0$. We see that for curve X_0, F is zero when $\tilde{m} = 0$ and also at a nonvanishing value of $\tilde{m}$. As the coefficients can change, for example, with temperature, the global minima can discontinuously shift from zero to a nonzero value of the order parameter. Consequently, there is a jump is free energy at the transition point. This implies the existence of *latent heat* in the case of the first-order phase transition.

B.1.4
When is MFA Exact?

So far we have seen that MFA gives approximate results which are sometimes even qualitatively wrong. But one may ask, what is the limit when MFT is exact? To answer this, let us consider the following Hamiltonian:

$$H = -\frac{J}{2N} \sum_{i=1}^{N} \sum_{j=1}^{N} S_i S_j - h \sum_j S_j . \qquad (B26)$$

Here every spin interacts with every other spin as opposed to the nearest neighbor interactions taken before. The two sums give N^2 dependence of E, so we normalize by dividing J by N and to avoid double counting also multiply by 1/2. This Hamiltonian can be rewritten as

$$H = -\left[\left(\frac{J}{2N}\right)^{1/2} \sum_i S_i \right]^2 - h \sum_j S_j . \qquad (B27)$$

Therefore, the partition function can be written as

$$Z = \mathrm{Tr}\left(e^{-\frac{H}{k_B T}} \right)$$

$$= \mathrm{Tr}\left\{ \exp\left\{ \frac{1}{k_B T} \left[\left(\frac{J}{2N}\right)^{1/2} \sum_i S_i \right]^2 + \frac{1}{k_B T} h \sum_j S_j \right\} \right\} .$$

One can further simplify

$$Z = \mathrm{Tr}\left\{ e^{\frac{1}{k_B T} h \sum_i S_i} \left(\frac{1}{\sqrt{2\pi}}\right)^N \right.$$

$$\left. \times \int_{-\infty}^{+\infty} \exp\left(-\frac{1}{2}\sum_i x_i^2 + \sqrt{2\frac{1}{k_B T}} \left(\frac{J}{2N}\right)^{1/2} \sum_i S_i x_i \right) dx_i \right\}$$

$$= \left(\frac{1}{\sqrt{2\pi}}\right)^N \prod_i \int_{-\infty}^{+\infty} e^{-\frac{1}{2}x_i^2} \mathrm{Tr}\left(e^{\left(\frac{1}{k_B T} h + A x_i\right) S_i} \right) dx_i ,$$

where $A = \sqrt{2\frac{1}{k_B T}} \left(\frac{J}{2N}\right)^{1/2}$. We have used the transformation

$$e^{a^2} = \frac{1}{\sqrt{2\pi}} \int\limits_{-\infty}^{+\infty} e^{-\frac{1}{2}x^2 + \sqrt{2}ax}\, dx\, . \tag{B28}$$

Now,

$$\mathrm{Tr}\left[e^{\left(\frac{1}{k_B T}h + Ax_i\right)S_i}\right] = e^{\frac{1}{k_B T}h + Ax_i} + e^{-\frac{1}{k_B T}h + Ax_i}$$

$$= 2\cosh\left(\frac{1}{k_B T}h + Ax_i\right).$$

Therefore,

$$Z = \left(\frac{1}{2\pi}\right)^{N/2} \int\limits_{-\infty}^{+\infty} e^{-f(x)}\, dx\, ,$$

where $f(x) = \frac{x^2}{2} - \ln\left(2\cosh\left[\frac{1}{k_B T}h + \left(\frac{k_B T J}{N}\right)^{1/2}x\right]\right)$. One can now use the identity

$$\int\limits_{-\infty}^{+\infty} e^{-f(x)}\, dx = C e^{-f(x_0)}\, ,$$

where C is a constant and $\left.\frac{\partial f}{\partial x}\right|_{x=x_0} = 0$. Using the expression for $f(x)$ we get

$$x_0 = \left(\frac{\frac{1}{k_B T}J}{N}\right)^{1/2} \tanh\left[\frac{1}{k_B T}h + \left(\frac{\frac{1}{k_B T}J}{N}\right)^{1/2}x_0\right]. \tag{B29}$$

In the absence of an external field, we have

$$x_0 = \left(\frac{\frac{1}{k_B T}J}{N}\right)^{1/2} \tanh\left[\left(\frac{\frac{1}{k_B T}J}{N}\right)^{1/2}x_0\right]. \tag{B30}$$

This again is a transcendental equation and can be solved graphically. One can easily see that the critical point is at $T_c = \frac{J}{k_B N}$. In deriving the solution, we have not made any approximation. Therefore we conclude that mean field theory becomes exact when interactions are long-range type or the dimensionality goes to infinity.

B.1.5
Transverse Ising Model (TIM)

In many physical systems, cooperative interactions between spin-like (two state) degrees of freedom tend to establish some kind of order in the system, while the presence of some noise effect (due to temperature, external transverse field, etc.) tend to destroy it. The transverse Ising model can quite successfully be employed to study the order-disorder transitions in many of such systems.

An example above is the study of ferroelectric ordering in potassium di-hydrogen phosphate (DP) type systems. To understand such ordering, the basic structure can be viewed as a lattice, where in each lattice point there is a double-well potential created by an oxygen atom or a hydrogen atom, or a proton resides within any of the two wells. In the corresponding Ising (or pseudo-spin) picture the state of a double-well with a proton at the left well and that with one at the right well are represented by, say, $|\uparrow\rangle$ and $|\downarrow\rangle$, respectively. The protons at neighboring sites have mutual di-polar interactions. Hence, if the proton had been a classical particle, then there would be the zero temperature configuration of the system, one with either all the protons residing at their respective left well or, with all protons at their respective right well (corresponding to the all-up or all-down configuration of the spin system in presence of cooperative interactions alone, at zero temperature). Considering no fluctuation in zero temperature, the Hamiltonian for the system in the corresponding pseudo-spin picture will be identical to the classical Ising Hamiltonian (without any transverse term). However, since the proton is a quantum particle, there is always a finite probability for it to tunnel through the finite barrier between two wells even at zero temperature due to quantum fluctuations. To formulate the term for the tunneling in the corresponding spin-picture, we notice that S^x is the right operator. This is because

$$S^x |\uparrow\rangle = |\downarrow\rangle \quad \text{and} \quad S^x |\downarrow\rangle = |\uparrow\rangle . \tag{B31}$$

Hence the tunneling term will exactly be represented by the transverse field term in the transverse Ising Hamiltonian. here the transverse field coefficient Γ will represent the tunneling integral, which depends on the width and height of the barrier, mass of the particle, etc.

Such a system as discussed above, can be represented by a quantum Ising system, having Hamiltonian

$$H = -\sum_{\langle i,j \rangle} J_{ij} S_i^z S_j^z - \Gamma \sum_i S_i^x . \tag{B32}$$

Where the spins (S^α's) have the following representation:

$$S^x = \begin{pmatrix} 0 & 1 \\ 1 & 0 \end{pmatrix}, \quad S^y = \begin{pmatrix} 0 & -i \\ i & 0 \end{pmatrix}, \quad S^z = \begin{pmatrix} 1 & 0 \\ 0 & -1 \end{pmatrix}$$

The order parameter for such a system is generally taken to be the expectation value of the z-component of the spin, that is $\langle S^z \rangle$. In such a system absolute ordering (complete alignment along z-direction) is not possible even at zero-temperature, that is $\langle S^z \rangle_{T=0} \neq 1$, when $\Gamma \neq 0$. In general, therefore, the order ($\langle S^z \rangle \neq 0$) to disorder ($\langle S^z \rangle = 0$) transition can be brought about by tuning the field Γ and/or temperature T.

B.1.5.1 MFA for TIM

For $T = 0$: Let

$$S_i^z = |S| \cos \theta , \quad \text{and} \quad S_i^x = |S| \sin \theta ,$$

where θ is the angle between $\mathbf{S}$ and z-axis. This renders the two mutually non-commuting part of the Hamiltonian (B32) commuting, since both are expressed in terms of $|S|$ operator only. If S is the magnitude of $|S|$, then the energy per site of the semiclassical system is given by

$$E = -S\Gamma \sin \theta - \frac{1}{2} S^2 J(0) \cos^2 \theta , \tag{B33}$$

$J(0) = J_i(0) = \sum_{\langle i,j \rangle} J_{ij}$, where j indicates the jth nearest neighbor of the ith site. The average of the spin components are given by

$$\langle S^z \rangle = \cos \theta$$
$$\langle S^x \rangle = \sin \theta .$$

The energy in (B33) is minimized for

$$\sin \theta = \frac{\Gamma}{S J(0)} \quad \text{or,} \quad \cos \theta = 0 .$$

Thus we see that if $\Gamma = 0$, then $\langle S^x \rangle = 0$ and the order parameter $\langle S^z \rangle = 1$, indicating perfect order.

On the other hand, if $\Gamma < S J(0)$, then the ground state is partially polarized, since none of $\langle S^z \rangle$ or $\langle S^z \rangle$ is zero. However, if $\Gamma \geq S J(0)$, then we must have $\cos \theta = 0$ for the ground state energy, which means $\langle S^z \rangle = 0$, that is the state is a completely disordered one. Thus, as Γ increases from 0 to $J(0)$, the system undergoes a transition from ordered (ferro)-phase with order parameter $\langle S^z \rangle = 1$ to disordered (para)-phase with order parameter $\langle S^z \rangle = 0$.

For $T \neq 0$: The mean field method can be extended to obtain the behavior of this model at nonzero temperature. In this case we define a mean field $\mathbf{h}_i$ at each site i, which is, in some sense, a resultant of the average cooperation enforcement in the z-direction and the applied transverse field in x-direction. Precisely, we take, for the general random case

$$\mathbf{h}_i = \Gamma \hat{x} + \left(\frac{1}{2} \sum_j J_{ij} \langle S_j^z \rangle \right) \hat{z} , \tag{B34}$$

and the spin-vector at the ith site follows $\mathbf{h}_i$ ($\hat{x}$ and $\hat{z}$ represent unit vectors in the x and z-direction, respectively). The spin-vector at the ith site is given by

$$\mathbf{S}_i = S_i^x \hat{x} + S_i^z \hat{z} ,$$

and the Hamiltonian thus reads

$$H = -\sum_i \mathbf{h}_i \cdot \mathbf{S}_i \,. \tag{B35}$$

For nonrandom case, all the sites have identical ambiance, hence $\mathbf{h}_i$ is replaced by $\mathbf{h} = \Gamma \hat{x} + \langle S^z \rangle J(0)\hat{z}$. And the resulting Hamiltonian takes the form (see (B6))

$$H = -\mathbf{h} \cdot \sum_i \mathbf{S}_i \,. \tag{B36}$$

The spontaneous magnetization can readily be written down as (cf. (B7))

$$S = \tanh\left(\frac{|\mathbf{h}|}{k_B T}\right) \cdot \frac{\mathbf{h}}{|\mathbf{h}|}$$

$$|\mathbf{h}| = \sqrt{\Gamma^2 + (J(0)\langle S^z \rangle)^2} \,.$$

Now if $\mathbf{h}$ makes an angle θ with the z-axis, then $\cos\theta = J(0)\langle S^z \rangle / |\mathbf{h}|$ and $\sin\theta = \Gamma / |\mathbf{h}|$ and hence we have

$$\langle S^z \rangle = \left[\tanh\left(\frac{|\mathbf{h}|}{k_B T}\right)\right]\left(\frac{J(0)\langle S^z \rangle}{|\mathbf{h}|}\right), \tag{B37}$$

and

$$\langle S^x \rangle = \left[\tanh\left(\frac{|\mathbf{h}|}{k_B T}\right)\right]\frac{\Gamma}{|\mathbf{h}|} \,. \tag{B38}$$

These self-consistent equations can be solved graphically or otherwise, to obtain the order parameter $\langle S^z \rangle$ at any temperature T and transverse field Γ. Clearly, the order-disorder transition is tuned both by Γ and T.

For $\Gamma = 0$ (Transition Driven by T): Here,

$$\langle S^z \rangle = \tanh\left(\frac{J(0)\langle S^z \rangle}{k_B T}\right),$$

and

$$\langle S^x \rangle = 0 \,. \tag{B39}$$

One can easily see graphically, that the above equation has a nontrivial solution only if $k_B T < J(0)$, that is

$$\langle S^z \rangle \neq 0 \quad \text{for} \quad k_B T < J(0)$$
$$\langle S^z \rangle = 0 \quad \text{for} \quad k_B T > J(0) \,.$$

This shows that there is a critical temperature $T_c = J(0)$ above which there is no order.

For $k_B T \to 0$ (Transition Driven by Γ): Here,

$$\langle S^z \rangle = \frac{J(0)\langle S^z \rangle}{\sqrt{\Gamma^2 + (J(0)\langle S^z \rangle)^2}} \qquad (\text{since, } \tanh x|_{x \to \infty} = 1) .$$

From this equation we easily see that there is a critical transverse field $\Gamma_c = J(0)$ such that for any $\Gamma \gg \Gamma_c$ there is no order even at zero temperature. In general one sees that at any temperature $T < T_c$, there exists some transverse field Γ_c, at which the transition from the ordered state ($\langle S^z \rangle \neq 0$) to the disordered state ($\langle S^z \rangle = 0$) occurs.

For $T \neq 0$: Phase Boundary The equation for the phase boundary in the ($\Gamma - T$) plane is obtained by putting $\langle S^z \rangle \to 0$ in (B37). This gives the relation between Γ_c and T_c as follows:

$$\tanh \left(\frac{\Gamma_c}{k_B T} \right) = \frac{\Gamma_c}{J(0)} . \tag{B40}$$

One may note that for the ordered phase, since $\langle S^z \rangle = 0$

$$\langle S^x \rangle = \tanh \left(\frac{\Gamma}{k_B T} \right) . \tag{B41}$$

Using magnetic mapping, mean field theory of this type was indeed applied to the BCS theory of superconductivity.

B.1.5.2 Dynamical Mode-Softening Picture

The elementary excitations in such a system as described above are known as spin waves and they can be studied using Heisenberg equation of motion for S^z using the Hamiltonian. The equation of motion is then given by

$$\dot{S}_i^z = (i\hbar)^{-1} \left[S_i^z, H \right]$$

or

$$\dot{S}_i^z = 2\Gamma S_i^y \quad (\text{with } \hbar = 1) .$$

Hence,

$$\ddot{S}_i^z = 2\Gamma \dot{S}_i^y = 4\Gamma \sum_j J_{ij} S_i^z S_i^x - 4\Gamma^2 S_i^z .$$

With Fourier transforms and random phase approximation ($S_i^x S_j^z = S_i^x \langle S_j^z \rangle + \langle S_i^x \rangle S_j^z$, with $\langle S^z \rangle = 0$ in para phase), we get

$$\omega_q^2 = 4\Gamma (\Gamma - J(q) \langle S^x \rangle) , \tag{B42}$$

for elementary excitations (where $J(q)$ is the Fourier transform of J_{ij}). The mode corresponding to $q = 0$ softens, that is ω_0 vanishes at the same phase boundary given by (B42).

B.2
Quantum Systems with Interactions [3, 4]

B.2.1
Superfluidity and Superconductivity

He^4 when cooled below 4.2 K, becomes liquid. If liquid He^4 is placed in a container, with the liquid filled almost to the edge, then there is constant dripping out of the liquid from the container. This indicates the disappearance of friction or viscosity of He^4 liquid (in the surface layers) to hold the liquid in the container. There were attempts to explain this from the concept of Bose–Einstein (BE) condensation. However, this is not a system of free particles but an interacting system. Therefore, it is not a Bose gas but a Bose liquid. Hence, superfluidity cannot be explained through BE condensation. Landau suggested the following scenario.

For free particles, the dispersion relation reads as $\epsilon_q \sim q^2$. However, such a free particle dispersion cannot show superfluidity. Landau assumed that for interacting boson systems the dissipation relation may become, $\epsilon_q \sim cq$ (for electron systems $\epsilon_q \sim \Delta + cq^2$ is possible) and that might lead to frictionless flow. He considered the superfluid system consisting of the (quantum) fluid placed in contact with a (classical) massive wall or surface. If some friction is encountered in the flow, then energy of the fluid and hence the momentum has to change. This change in momentum has to be taken up by the classical mass in contact with the liquid. If it can be shown that this heavy mass cannot accept every excitation satisfying the energy and momentum conservations, then the fluid has no other choice but to flow without friction.

If v and v' denote velocities of the classical mass before and after energy exchange, then from energy conservation

$$\frac{1}{2}Mv'^2 = \frac{1}{2}Mv^2 + \epsilon_q , \tag{B43}$$

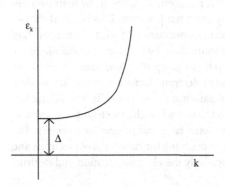

Figure B.6 Dispersion relation for interacting systems.

and for momentum conservation

$$Mv' = Mv + \hbar q .$$

Squaring and dividing the above equation by $2M$ we get

$$\frac{1}{2M} M^2 v'^2 = \frac{1}{2} M v^2 + \frac{1}{2M} \hbar^2 q^2 + \hbar v q .$$

Subtracting this from (B43), we get

$$\epsilon_q = \frac{1}{2M} \hbar^2 q^2 + \hbar v q .$$

Since M is very large, the above energy-momentum conservation is satisfied for

$$\frac{\epsilon_q}{q} \bigg|_{min} = \hbar v . \tag{B44}$$

If $\epsilon = cq^2$, then the above condition is satisfied for all nonvanishing velocities, indicating the possibility of exchange of energy and momentum. Hence no superfluidity. But if $\epsilon_q = cq$ (as in the cases of phonon and roton excitations) then only for $v > c$ is the condition satisfied. For lower values of v, one gets flow without resistance or superfluidity. The other case of interest is $\epsilon_q = \Delta + q^2$ where one also gets superfluidity for $\Delta \neq 0$ and this can occur for the BCS theory of superconductivity discussed next.

B.2.2
MFA: BCS Theory of Superconductivity

The resistivity in some materials practically vanishes when the temperature is reduced below a critical value. These materials then also become completely diamagnetic. This is called superconductivity. Superconductivity cannot be a property of free electrons. In a lattice, electrons are of course not free. Apart from the Coulomb repulsion among them they face a fluctuating periodic potential (electron–phonon scattering). It is observed that replacement of atoms by their isotopes induces changes in the superconductivity onset temperature. This is called the isotope effect, which often relates the critical temperature (T_c) with the mass (M) of the atoms by the relation $M^{1/2} T_c = $ constant. When we change the atomic mass, we basically do nothing to the electron but only change the phonon modes. This led to the belief that superconductivity is due to electron-electron interactions, mediated by phonons. Cooper showed that the maximum value of the effective interaction between the electron pairs can become attractive when the electron pairs are near the Fermi surface and they have opposite spin states and momentum vectors. The BCS Hamiltonian can be written, with q, q' denoting the momenta of electrons and $\tilde{q}$ for phonons and $C^\dagger$, C denoting respectively the electron creation and destruction operators, as (see, e.g., [3, 4])

$$H = \sum_q \epsilon_q^0 C_q^\dagger C_q + \sum_{q,q',\tilde{q}} V_{qq'\tilde{q}} C_{q+\tilde{q}}^\dagger C_{q'-\tilde{q}}^\dagger C_q C_{q'} . \tag{B45}$$

We assume here $V_{qq'\bar{q}} \equiv -V$, the maximum value of the effective attractive interaction when q and q' $(= -q)$ pairs (with opposite spin states) are formed and the q are near Fermi vector. Then the Hamiltonian can be written as

$$H = \sum_q \epsilon_q^0 C_q^\dagger C_q - V \sum_{qq'} C_{q'}^\dagger C_{-q'}^\dagger C_q C_{-q} . \tag{B46}$$

Because of the pairing of electrons, we have $(q, -q)$ and $(q', -q')$ pairs, both having the same energy ϵ_q^0. Hence the Hamiltonian takes the form

$$H = \sum_q \epsilon_q^0 \left(C_q^\dagger C_q + C_{-q}^\dagger C_{-q} \right) - V \sum_{qq'} C_{q'}^\dagger C_{-q'}^\dagger C_q C_{-q} .$$

We choose $\sum_q \epsilon_q^0 = 0$. Therefore,

$$H = -\sum_q \epsilon_q^0 \left(1 - C_q^\dagger C_q - C_{-q}^\dagger C_{-q} \right) - V \sum_{qq'} C_{q'}^\dagger C_{-q'}^\dagger C_q C_{-q} .$$

The last term is still not in a diagonal form. We intend to map this Hamiltonian (in the lowest energy states) to a pseudo-spin Hamiltonian.

Let us consider only the low lying states of this Hamiltonian, namely the electron pair occupied ($|-\rangle$) and pair unoccupied ($|+\rangle$)

$$|+\rangle \equiv |0_q, 0_{-q}\rangle \quad |-\rangle \equiv |1_q, 1_{-q}\rangle .$$

Hence,

$$\left(1 - C_q^\dagger C_q - C_{-q}^\dagger C_{-q} \right) |+\rangle = |+\rangle$$

$$\left(1 - C_q^\dagger C_q - C_{-q}^\dagger C_{-q} \right) |-\rangle = -|-\rangle .$$

We therefore make the correspondence $(1 - C_q^\dagger C_q - C_{-q}^\dagger C_{-q}) \equiv S_q^z$. Since

$$C_q^\dagger C_{-q}^\dagger |+\rangle = |-\rangle , \quad C_q^\dagger C_{-q}^\dagger |-\rangle = 0 ,$$

$$C_{-q} C_q |-\rangle = |+\rangle \quad \text{and} \quad C_{-q} C_q |+\rangle = 0 ,$$

we immediately identify its correspondence with raising and lowering operators S^+/S^-:

$$S^\pm = S^x \pm i S^y = \begin{pmatrix} 0 & 0 \\ 2 & 0 \end{pmatrix} \quad \text{or} \quad \begin{pmatrix} 0 & 2 \\ 0 & 0 \end{pmatrix} ,$$

and therefore

$$C_q^\dagger C_{-q}^\dagger = \frac{1}{2} S_q^- , \quad C_{-q} C_q = \frac{1}{2} S_q^+ . \tag{B47}$$

In terms of these spin operators we finally arrive at

$$H = -\sum_q \epsilon_q^0 S_q^z - \frac{V}{4} \sum_{qq'} \left(S_q^x S_{q'}^x + S_q^y S_{q'}^y \right)$$
(B48)

or

$$H = -\sum_q h_q \cdot S_q \,,$$
(B49)

where the effective field h_q components are:

$$h_q^x = \frac{V}{4} \sum_{q'} \langle S_{q'}^x \rangle \,; \quad h_q^y = \frac{V}{4} \sum_{q'} \langle S_{q'}^y \rangle \,; \quad \text{and} \quad h_q^z = \epsilon_q^0 \,.$$
(B50)

As x and y components are symmetric we consider only one component (with proper counting) and hence $|h_q| = \sqrt{\epsilon_q^{0^2} + \frac{V^2}{4} \left(\sum_{q'} \langle S_{q'}^x \rangle \right)^2}$. The pseudo-spin S_q will therefore be in the direction of the field h_q and its magnitude will depend on the temperature.

At $T = 0$:

Here the spin magnitude will be its maximum and hence

$$\langle S_q \rangle = \frac{h_q}{|h_q|} \,.$$
(B51)

Let

$$\langle S_q^x \rangle = \sin \theta_q = \frac{\langle h_q^x \rangle}{|h_q|} \quad \text{and} \quad \langle S_q^z \rangle = \cos \theta_q = \frac{\langle h_q^z \rangle}{|h_q|} \,.$$

Hence,

$$\langle S_q^x \rangle = \frac{\sum_{q'} \frac{V}{2} \langle S_{q'}^x \rangle}{\sqrt{\epsilon_q^{0^2} + \frac{V^2}{4} \left(\sum_{q'} \langle S_{q'}^x \rangle \right)^2}} \quad \text{and} \quad \langle S_q^z \rangle = \frac{\epsilon_q^0}{\sqrt{\epsilon_q^{0^2} + \frac{V^2}{4} \left(\sum_{q'} \langle S_{q'}^x \rangle \right)^2}} \,.$$

If we define

$$\Delta = \frac{V}{2} \sum_{q'} \langle S_{q'}^x \rangle \,,$$
(B52)

then we can write

$$\langle S_q^x \rangle = \sin \theta_q = \frac{\Delta}{\sqrt{\epsilon_q^{0^2} + \Delta^2}} \quad \text{and} \quad \langle S_q^z \rangle = \cos \theta_q = \frac{\epsilon_q^0}{\sqrt{\epsilon_q^{0^2} + \Delta^2}} \,.$$
(B53)

Putting (B53) into (B52) we get the self-consistent equation for Δ

$$\Delta = \frac{V}{2} \sum_{q'} \frac{\Delta}{\sqrt{\epsilon_{q'}^{0\,2} + \Delta^2}} \, . \tag{B54}$$

We can express this gap in (B53) as an integral in the following form:

$$\Delta = \frac{V}{2} \int \frac{\rho(\epsilon_F) \Delta \, d\epsilon}{\sqrt{\epsilon^2 + \Delta^2}} \, ,$$

where $\rho(\epsilon_F)$ is the density of states at the Fermi level. Therefore, the so-called *gap equation* takes the form

$$\frac{V}{2} \int \frac{\rho(\epsilon_F) \, d\epsilon}{\sqrt{\epsilon^2 + \Delta^2}} = 1 \, . \tag{B55}$$

The ground state of the Hamiltonian (B49) occurs when S_q is aligned in the direction of h_q. The excited state will be when the alignment of S_q is opposite to h_q. Hence the change in energy is $2\,|h_q|$. Therefore,

$$\epsilon_q = 2\,|h_q| = 2\sqrt{\epsilon_q^{0\,2} + \Delta^2} \, .$$

When $q \to 0$, $\epsilon_q^0 \sim q^2 \to 0$, but $\epsilon_q \neq 0$. Hence Δ represents the zero temperature energy gap, independent of q. As discussed in the previous section, the nonvanishing value of this gap Δ ensures a "superfluid-like" flow of the (charged) electrons and hence superconductivity.

At $T \neq 0$:

Here, unlike in (B51), the average spin magnitude will be given by $\tanh\left(\frac{|h_q|}{k_B T}\right)$. Hence,

$$\langle S_q \rangle = \frac{h_q}{|h_q|} \tanh\left(|h_q| / k_B T\right), \quad |h_q| = \sqrt{\epsilon_q^{0\,2} + \Delta^2} \, ,$$

and consequently the generalization of (B54) would be

$$\Delta = \frac{V}{2} \sum_{q'} \tanh\left(\frac{|h_{q'}|}{k_B T}\right) \sin\theta_{q'} \, ; \quad \sin\theta_q = \frac{\Delta}{|h_q|} \, . \tag{B56}$$

The gap Δ is now a function of temperature T and if we define the critical temperature T_c as the temperature where the gap Δ vanishes (see previous section), then

$$\frac{V}{2} \sum_{q'} \frac{1}{\epsilon_{q'}^0} \tanh\frac{\epsilon_{q'}^0}{k_B T_c} = 1 \, . \tag{B57}$$

Numerically solving (B56) and (B57), we get

$$2\Delta\,(T = 0) = 3.5\,T_c\,. \tag{B58}$$

For most conductors this relation is seen to be fairly accurate, like

$$2\Delta \approx 3.5\,T_c \quad \text{for Sn}$$
$$\approx 3.4\,T_c \quad \text{for Al}$$
$$\approx 4.1\,T_c \quad \text{for Pt}\,.$$

B.3
Effect of Fluctuations: Peierls' Argument [5–7]

B.3.1
For Discrete Excitations

Let us introduce the lower critical dimension d_L and upper critical dimension d_U. Although we are considering cooperative phenomena, fluctuations may be such that $T_c = 0$ until $d \geq d_L$. In our case (Ising model) $d_L = 1$. Mean field theory gives approximate, even qualitatively wrong results near the lower critical dimension. As the dimensionality is increased, MFT results approach the exact value.

Let us now come to Peierl's argument, which attempts to explain why there is no finite T_c for the Ising model in one-dimension and why there is one in two-dimensions. Consider the situation (in the one-dimensional Ising model) when all the spins are up as compared to the situation when there is one domain wall in the system. There are N ways to place a domain wall in the system. Hence, the entropy gain will be $k_B T \ln N$. Again, change in internal energy will be $2J$. Therefore the change in free energy can be written as

$$\Delta F = \Delta E - T\Delta S$$
$$= 2J - k_B T \ln N\,. \tag{B59}$$

In the thermodynamic limit, the second term always wins. Therefore at any finite temperature, $\Delta F < 0$. So, a domain wall is always favored. Now, as all the possible configurations of the domain wall are equally probable, the net magnetization will vanish.

In the two-dimension, however, the situation is different. Here, $\Delta E = 2J L$, where L is the size of the boundary of the domain having different spin orientation. Keeping L constant, there can be many configurations of the boundary. Remembering that the boundary is like a path of a self-avoiding walker, the number of ways can be written as

$$G_L \sim \bar{z}^L L^{\gamma-1}\,, \tag{B60}$$

where $\bar{z}$ is the self-avoiding connectivity constant. Therefore, the change in entropy will be $\Delta S = \ln G_L \approx \ln \bar{z}^L = L \ln \bar{z}$. The $\ln L$ term can be neglected in

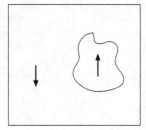

Figure B.7 Domain of up-spins in a down-spin environment.

comparison with unity. So, the change in free energy will be

$$\Delta F = L(2J - k_B T \ln \bar{z}). \tag{B61}$$

Now, ΔF can be both positive and negative depending upon the value of T. So, at the transition temperature T_c

$$2J - k_B T_c \ln \bar{z} = 0. \tag{B62}$$

Therefore we get

$$k_B T_c = \frac{2J}{\ln \bar{z}}. \tag{B63}$$

An Estimate of $\bar{z}$ (by Temperly–Fisher–Sykes)

One can consider the domain wall contours as solid-on-solid. One can move its contour as a path of a self-avoiding walker, which is not allowed to take back-steps in one particular lattice direction. This is called a directed self-avoiding walk. We want to find the number G_L of distinct configurations for a fixed length of the walk. We divide G_L into two parts: $(G_L)^h$ gives the number of walks ending horizontally and $(G_L)^v$ gives that ending vertically. Therefore,

$$G_L = (G_L)^v + (G_L)^h. \tag{B64}$$

If we now increase the length by unity

$$G_{L+1} = 3(G_L)^v + 2(G_L)^h. \tag{B65}$$

That is because a walk ending vertically has three possible ways to take the next step, as opposed to a walk that ends horizontally, which has two possible ways. We can further write

$$G_{L+1} = 2G_L + G_L^v$$
$$= 2G_L + G_{L-1}. \tag{B66}$$

The last step follows from the fact that each configuration of step $L-1$ will contribute one configuration with a vertical step for the next one. So,

$$G_{L+1} = 2G_L + G_{L-1}. \tag{B67}$$

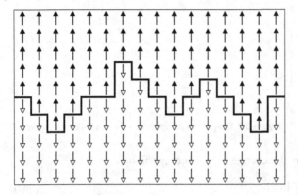

Figure B.8 Schematic diagram of a solid-on-solid type domain boundary.

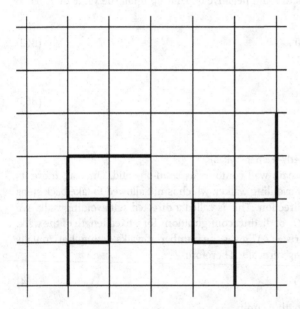

Figure B.9 Two equal length directed walks (directed upwards); one ending horizontally and the other vertically.

For the self-avoiding walker (see Chapter 7) $G_L = \bar{z}^L L^{\gamma-1}$ and if one assumes $\gamma = 1$ for such walks, we get

$$\bar{z}^{L+1} = 2\bar{z}^L + \bar{z}^{L-1} \, . \tag{B68}$$

Solving it and neglecting the negative solution we get, $\bar{z} = 1 + \sqrt{2}$. So,

$$G_L = \left(1 + \sqrt{2}\right)^L \, . \tag{B69}$$

B.3.2
For Continuous Excitations

Preserving order in a solid means that when the particles are excited from their equilibrium positions, they come back to those positions. When melting starts, this order is destroyed. We want to find the temperature at which melting starts and if that temperature vanishes then there can be no order at any finite temperature. That is the case in one and two dimensions. The coupled equations of motion of the atoms are of the form

$$\frac{d^2 u_i}{dt^2} = -\kappa \sum_{i,\delta} (u_i - u_{i-\delta})^2 , \tag{B70}$$

where u_i denotes the displacement of the ith lattice site and δ denotes the nearest neighbor. We then diagonalize the Hamiltonian using the Fourier transform ($u_i = \sum_q u_q e^{i(\omega_q t - q \cdot r_i)}$; r_i denoting the equilibrium lattice position vector of site i) and get the total energy of the lattice

$$E = \sum_q \omega_q^2 u_q^2 , \tag{B71}$$

where $\omega_q^2 = \kappa (1 - \cos qa)$ are the square of the frequencies of the phonon modes (a denotes lattice constant). Instead of individual degrees of freedom (which are coupled as in (B70)) we go over to the degrees of freedom of the individual decoupled phonon modes and apply the equipartition theorem to these modes to get

$$\omega_q^2 \left\langle \left| u_q^2 \right| \right\rangle \sim k_B T .$$

Fluctuation at a lattice site at temperature T is, therefore, given by

$$\left\langle \left| u_i^2 \right| \right\rangle \sim \sum_q \left\langle \left| u_q^2 \right| \right\rangle$$

$$= k_B T \sum_q \frac{1}{\omega_q^2}$$

$$= \frac{k_B T}{C^2} \int \frac{q^{d-1} dq}{q^2} , \tag{B72}$$

using the fact that (for phonons) $\omega_q \sim Cq$ for $q \to 0$, as obtained above ($C = \kappa a$) and d denotes the lattice dimension. The lower limit is determined from the system size, that is the minimum wave vector will be π/L. The upper cut-off comes from the finiteness of the lattice constant, that is the maximum wave vector will be π/a, where a is the lattice constant. Therefore,

$$\left\langle \left| u_i^2 \right| \right\rangle = \frac{k_B T}{C^2} \int_{\pi/L}^{\pi/a} \frac{q^{d-1} dq}{q^2} .$$

For an infinite lattice, the lower limit will be zero. Then in the one dimension we will have

$$
\langle |u_i^2| \rangle = \frac{k_B T}{C^2} \int_0^{\pi/a} \frac{dq}{q^2}
$$

$$
= \frac{k_B T}{C^2} \left(-\frac{1}{q} \right)_0^{\pi/a} .
$$

But it diverges due to the lower limit. For $d = 2$ we have

$$
\langle |u_i^2| \rangle = \frac{k_B T}{C^2} [\ln q]_{\pi/L}^{\pi/a} \sim \frac{k_B T}{C^2} \ln L .
$$

This also diverges with the system size L. Even if the elastic constant κ (or C) is strong, the divergence cannot be competed with. Hence, we see that fluctuations diverge in one and two dimensions, so there can be no ordering in those dimensions when the excitation is continuous. In three dimensions

$$
\langle |u_i^2| \rangle = \frac{k_B T}{C^2} \int_{\pi/L}^{\pi/a} dq = \frac{k_B T}{C^2} [q]_{\pi/L}^{\pi/a} . \tag{B73}
$$

This is a finite number. Only if T is large enough will the value of $\langle |u_i| \rangle$ grow beyond the lattice constant value a, and we thus see a melting transition in three dimensions. So, the lower critical dimension in this case is two.

The solid order (lattice correlations) therefore is lost (due to the divergence of $\langle |u_i^2| \rangle$) for $d \leq 2$, when the excitations (over the ordered phase) are continuous as in (B73); both phonons and magnons can cause similar instabilities in $d \leq 2$.

B.4
Effect of Disorder [8, 9]

Harris and Fisher gave heuristic arguments which suggested that the critical behavior of a system would be affected by the presence of disorder (quenched or annealed, respectively), if the internal energy fluctuation or specific heat of the (pure) system diverges (with specific heat exponent α). For quenched disorder, the arguments by Fisher also give the nature of the new (Fisher renormalized) critical behavior. These findings, using these arguments, have later been supported using renormalization group techniques.

B.4.1
Annealed Disorder: Fisher Renormalization

Since the impurities here are in the same thermal bath as the magnetic system, one can consider a total Hamiltonian of the system including the impurity. Let the impurity coupling be denoted by λ. For such annealed systems, Fisher assumed

$$F(\Delta T, \lambda) \approx F\left(\Delta T^*\left(\Delta T, \lambda\right)\right), \tag{B74}$$

where F denotes (the singular part of the) free energy of the system with $F(\Delta T^*) \sim |\Delta T^*|^{2-\alpha}$, α denotes the specific heat of the pure system, and the renormalized temperature interval ΔT^* is an analytic function of its arguments ΔT and λ. The impurity concentration p is then given by

$$p \sim \frac{\partial F}{\partial \lambda} \sim \left(\frac{\partial F}{\partial T^*}\right)\left(\frac{\partial T^*}{\partial \lambda}\right) \sim |\Delta T^*|^{1-\alpha} \quad \text{for } \alpha > 0$$

$$\sim |\Delta T^*| \quad \text{for } \alpha < 0. \tag{B75}$$

Assuming now analytic constraints over p, both sides of the above expression can be expanded and the lower-order terms in ΔT and ΔT^* on both sides may be compared to give (since both ΔT and ΔT^* are much less than unity)

$$|\Delta T| \sim |\Delta T^*|^{1-\alpha} \quad \text{or} \quad |\Delta T^*| \sim |\Delta T|^{1/(1-\alpha)} \quad \text{for} \quad \alpha > 0, \tag{B76}$$

and

$$|\Delta T^*| \sim |\Delta T| \quad \text{for} \quad \alpha < 0. \tag{B77}$$

Usual critical behavior of the annealed system, expressed in terms of ΔT^*, gives in the effect the Fisher renormalized exponents $\alpha_R = -\alpha/(1-\alpha)$, $\beta_R = \beta/(1-\alpha)$, $\gamma_R = \gamma/(1-\alpha)$ and $\nu_R = \nu/(1-\alpha)$ for the impure system when $\alpha > 0$ for the pure system. These renormalizations of exponents keeps the scaling and hyperscaling relations $\alpha + 2\beta + \gamma = 2$ and $\alpha = 2 - d\nu$ unchanged.

B.4.2
Quenched Disorder: Harris Criterion

Let the concentration of the quenched disorder (which is not in thermal equilibrium with the other degrees of freedom) be denoted by p. The mean square fluctuation in the concentration in a typical volume element ξ^d (ξ denoting the thermal correlation length of the system) will then be given by

$$(\Delta p)^2 \equiv \sum_i \left[\langle p_i^2 \rangle - \langle p_i \rangle^2\right] \sim \xi^d p(1-p). \tag{B78}$$

Here p is not critical. Since the change in internal energy due to this fluctuation in the interaction strength corresponds to a change in local transition temperature T_c, then

$$\xi^d \Delta T_c \sim \Delta p \sim \xi^{d/2}, \tag{B79}$$

or $\Delta T_c \approx \xi^{-d/2}$. However, since $\xi \sim |\Delta T|^{-\nu}$ where $\Delta T \equiv |T - T_c|/T_c$, we can express ΔT as $\sim \xi^{-1/\nu}$. For a sharp transition, we require $\Delta T_c \ll \Delta T$, or

$d/2 > v^{-1}$, or $(2 - dv) < 0$. Using the hyperscaling relation (see Section 10 for details) $\alpha = 2 - dv$, the above condition becomes $\alpha < 0$, indicating nontrivial effect (a new different sharp transition with negative value of α, or a smeared transition) due to quenched disorder for (pure) systems with $\alpha > 0$.

B.5
Flory Theory for Self-Avoiding Walk (SAW) Statistics [10]

B.5.1
Random Walk Statistics

In a random walk, we are interested in the following question: what is the distance $G_N(r)$ between the initial and final points after N steps are taken? We may also be interested in different moments of the distribution, that is $\sum_r r^m G_N(r)$. The first moment is zero, because there is no preference in direction for an unbiased walker and therefore one can show that $G_N(r) \sim e^{-r^2/N}$. Let us consider the second moment $\langle R_N^2 \rangle$, where, $R_N = \sum_{i=1}^{N} r_i$. Therefore,

$$
\langle R_N^2 \rangle = \sum_{i=1}^{N} \sum_{j=1}^{N} \langle r_i \cdot r_j \rangle
$$
$$
= \sum_{j=1}^{N} r_j^2
$$
$$
= Na^2 , \tag{B80}
$$

where a is the length of each step. Therefore the average end-to-end distance grows as $\sqrt{N}$. Initially, it was thought that the configuration of a polymer can be mapped to a random walk, each monomer describing one step direction. But it was observed that the radius of gyration of a polymer does not go as $N^{1/2}$ but goes as N^v, where $v > 1/2$. In three dimensions, $v \approx 0.6$.

B.5.2
SAW Statistics

For random walk or Brownian motion, the ensemble of walk configurations on a lattice grows with the wall length N as z^N, where z is the lattice coordination number. Let us now consider only those configurations, which do not intersect themselves, that is we consider a self-avoiding walk (SAW). Using this ensemble, if we calculate the average R_N^2, we will have, $v = 0.75$ for $d = 2$ and $v = 1$ for $d = 1$.

When we are stretching or compressing the polymer we are changing the entropy. On either side of the equilibrium point, entropy will decrease. Therefore, an elastic force will be felt. This will be on the order $\left(\frac{R - R_0}{R}\right)^2$ and will be a loss in free energy. Again, the gain in the free energy will be due to repulsion between the monomers. If $\frac{N}{R^d}$ is the monomer density, then this interaction energy is approximated by $R^d \left(\frac{N}{R^d}\right)^2$. We can now write the free energy as

$$F(R) = \left(\frac{R - R_0}{R_0}\right)^2 - R^d \frac{N^2}{R^{2d}} \sim \frac{R^2}{N} - \frac{N^2}{R^d}, \tag{B81}$$

where we have used $R_0 \sim N^{1/2}$ and kept only the leading order term. Now we have

$$F(R) = \frac{R^2}{N} - \frac{N^2}{R^d}.$$

Minimizing with respect to R we have

$$\frac{2R}{N} = \frac{d N^2}{R^{d+1}}. \tag{B82}$$

Implying that, $R \sim N^{\nu_F}$, where $\nu_F = 3/(2 + d)$. This result, though derived from an approximate theory, gives very accurate results for ν_F. A SAW subset for dimensions above four forms statistics identical to those of RW.

B.6
Percolation Theory [11]

Consider a nonconducting plate, on which one may spray uniformly some conducting dye. If one applies a potential difference across any two opposite ends of the plate, with an ammeter in series, there will be no current initially when no dye is spread. Also, if the entire area of the plate is covered by the dye then obviously the plate is conducting. But, is it necessary to cover the entire plate to get a nonvanishing current? The answer is, no. Current starts to flow when there is a marginally connected path of the overlapping clusters of the dye grains across the plate. The point at which the conduction first takes place is called the percolation threshold. In order to make the discussion more quantitative and precise, let us consider now the lattice percolation model. There are two versions of the model: site percolation and bond percolation. In the site percolation problem, each site of a large lattice is randomly occupied with probability p. Clusters are defined as a graph of neighboring lattice sites. In bond percolation, each bond of a lattice is occupied randomly, with a probability p. A cluster is defined as a graph of overlapping bonds, sharing a common site. Most of the physical properties of such random systems depend on the geometric properties of these random clusters, and in particular, on the existence of an infinite, connected cluster which spans the system. Percolation theory deals with the statistics of the clusters formed.

Let us define some quantities of interest in percolation theory. Let $n_s(p)$ denote the number of clusters (per lattice site) of size s. A detailed knowledge of $n_s(p)$ would give a lot of information about the percolation statistics, as most of the

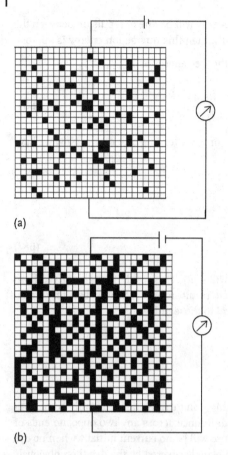

(a)

(b)

Figure B.10 Sample below (a) and above (b) the percolation threshold.

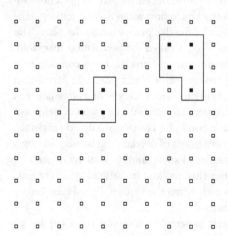

Figure B.11 Clusters in site percolation.

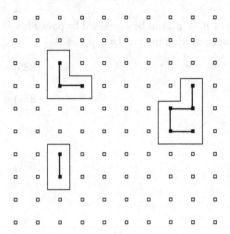

Figure B.12 Clusters in bond percolation.

quantities of interest can be extracted from the various moments of the cluster size distribution n_s.

The probability that a given site (bond) is occupied and is a part of an s-size cluster is $sn_s(p)$. Let, $P(p)$ denote the probability that any occupied site (bond) belongs to the infinite (lattice spanning) cluster. Then we have the obvious relation

$$\sum_s sn_s + P = 1 ,\tag{B83}$$

where the summation extends over all finite clusters. Clearly, at $p = 1$, $P(p) = 1$ and $P(p) = 0$ for $p < p_c$ as the infinite cluster does not exist for $p < p_c$. $P(p)$ can therefore be taken as the order parameter of the percolation phase transition. Another quantity of interest is the mean size of the finite clusters, denoted by $S(p)$, which is related to $n_s(p)$ through the relation

$$S(p) = \frac{\sum_s s^2 n_s(p)}{\sum_s sn_s(p)} ,\tag{B84}$$

where the summation is again over all finite clusters. One can also define a pair connectedness (or two point correlation) function $C(p, r)$ as the probability that two occupied sites (bonds) at a distance r are members of the same cluster. The sum over the pair connectedness over all distances gives the mean cluster size:
$S(p) = \sum_r C(p, r)$.

B.6.1
Critical Exponents

Most of the quantities defined above, have power-law variations near p_c. For example, the variation of the total number of clusters per site $G(p) = \sum_s n_s(p)$ (sum extends over all finite clusters), the decay of the order parameter $P(p)$ and the divergence of the mean cluster size: $S(p)$, as $p \to p_c$, can be expressed by power-law variations of these quantities with the concentration interval $|p - p_c|$ as

$$G(p) \equiv \sum_s n_s(p) \sim |p - p_c|^{2-\alpha} \tag{B85}$$

$$P(p) \sim (p - p_c)^\beta \tag{B86}$$

$$S(p) \sim |p - p_c|^{-\gamma} \tag{B87}$$

$$C(p, r) \sim \frac{\exp(-r/\xi(p))}{r^{d-2+\eta}}, \tag{B88}$$

where the correlation length

$$\xi(p) \sim |p - p_c|^{-\nu} \tag{B89}$$

diverges at $p = p_c$.

These powers $\alpha, \beta, \gamma, \eta$ and ν are called the critical exponents. These exponents are observed to be universal in the sense that although p_c depends on the details of the model under study, these exponents depend only on the lattice dimensionality.

B.6.2
Scaling Theory

Scaling theory assumes that the cluster distribution function $n_s(p)$ is a homogeneous function near $p = p_c$. Thus $n_s(p)$ is basically a function of the single scaled variable $s/S_\xi(p)$, where S_ξ denotes the typical cluster size

$$n_s(p) \sim s^{-\tau} f\left(\frac{s}{S_\xi(p)}\right), \tag{B90}$$

with $S_\xi(p) \sim |p - p_c|^{-1/\sigma}$. Here, τ and σ are two independent exponents and the scaling theory intends to relate all the above exponents to these exponents through the scaling relations. The function, $f(x)$ is assumed to have the asymptotic behavior, $f(x) \to 1$ as $x \to 0$ and $f(x) \to 0$ as $x \to \infty$. Further details of this function is unspecified in the theory. It may be noted that the above form implies, $n_s(p_c) \sim s^{-\tau}$. This has been checked through Monte Carlo simulations.

Assuming the above scaling form for $n_s(p)$, the mth moment of $n_s(p)$ can be expressed as

$$\sum_s s^m n_s \sim \sum_s s^{m-\tau} f\left(\frac{s}{|p - p_c|^{1/\sigma}}\right)$$

$$\sim |p - p_c|^{\frac{\tau-m-1}{\sigma}} \int x^{m-\tau} f(x) dx$$

$$\sim |p - p_c|^{\frac{\tau-m-1}{\sigma}},$$

assuming the integral over $x \, (= s/|p - p_c|^{1/\sigma})$ converges because of the asymptotic behavior of $f(x)$. Noting that $G(p), P(p)$ and $S(p)$ correspond to the zeroth, first and second moments of n_s, respectively, we get $\alpha = 2 - \frac{\tau-1}{\sigma}$, $\beta = \frac{\tau-2}{\sigma}$ and $\gamma = -\frac{\tau-3}{\sigma}$. This gives the scaling relation

$$\alpha + 2\beta + \gamma = 2 . \tag{B91}$$

This is satisfied by the observed values of the exponents. Also, since $S(p) = \sum_r C(r, p) = \int r^{d-1} C(r, p) dr$, one immediately gets

$$S(p) \sim |p - p_c|^{-\gamma}$$

$$= \int r^{d-1} \frac{e^{-r/\xi}}{r^{d-2+\eta}} dr$$

$$\sim \xi^{2-\eta} \int y^{1-\eta} e^{-y} dy$$

$$\sim |p - p_c|^{-\nu(2-\eta)} , \tag{B92}$$

giving the scaling relation

$$\gamma = \nu(2 - \eta) . \tag{B93}$$

Assuming that near p_c, the typical density $sn_s \sim S_\xi^{1-\tau}$ where $S_\xi \sim |p - p_c|^{-1/\sigma}$, one gets $sn_s \sim |p - p_c|^{(\tau-1)/\sigma}$ for the density. Assuming further that this density scales with the inverse of the typical volume element $\xi^d \sim |p - p_c|^{-d\nu}$, we get the hyperscaling relation

$$d\nu = (\tau - 1) / \sigma = 2 - \alpha . \tag{B94}$$

These scaling relations are satisfied by the numerically estimated values and in special cases the exact values of the critical exponents.

B.7
Fractals [12]

We are familiar with Euclidean dimensions of space, where the number of independent variables to specify the dynamics of the system gives the dimension. But we can also define dimensions in a different way. Suppose we decrease the linear size of a system by a factor $1/b$. Then it is obvious that a quantity $Q(L_1)$, measured

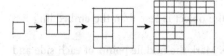

Figure B.13 Construction of a fractal object starting from a square (shown up to the third generation). The fractal (self-similar at all length scales) is formed as the generation number goes to infinity.

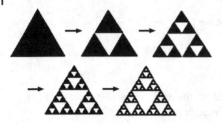

Figure B.14 The steps of formation of the Sierpinski triangle; the fractal is formed at infinite step or generation number.

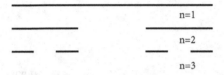

Figure B.15 Steps of forming the Cantor set; the fractal set is formed for $n \to \infty$.

with the initial length scale L_1, is related to the same quantity $Q(L_2)$ measured with the changed length scale L_2 ($= L_1/b$), by the following relation

$$\frac{Q(L_1)}{Q(L_2)} = b^{-D} , \tag{B95}$$

where D is the dimension of the system. Defined in this way, D need not be an integer. We shall consider a few examples when this is so. These fractional dimensional objects are called *fractals*.

Example 1: Take a square and divide it into four equal parts. Now drop the fourth quadrant and do the same for the remaining three. Continuing in this way up to infinite steps one can generate an object which is a fractal. This statement is justified from the fact that if we reduce the length to half of the original one, then the mass of the system reduces to one third. Hence,

$$\left(\frac{1}{2}\right)^D = \frac{1}{3} .$$

Giving,

$$D = \frac{\ln 3}{\ln 2} = 1.5849\ldots \tag{B96}$$

Note that the fractal dimension of the object is less than the embedding Euclidean dimension.

Example 2: Take an equilateral triangle. Mark the middle points of each side and join them. We now have three triangles where each side is halved. One can continue the same process with these three triangles. Eventually one would again reach a fractal object. In this object too, if we reduce the length to half of its original value, the mass of the object reduces to one third of the original value. So, as before

$$D = \frac{\ln 3}{\ln 2},$$ (B97)

implying that we again have the fractal dimension less than the Euclidean dimension. This particular structure, which is an example of fractal, is called a *Sierpinski triangle*.

Example 3: Take a line segment of unit length. Cut it into three pieces and remove the middle piece. One can continue to do this indefinitely. In this case, the length is scaled by a factor 1/3 and the number of line segment increases by a factor 2. Hence, we have again

$$\left(\frac{1}{2}\right)^D = \left(\frac{1/3}{1}\right),$$

implying,

$$D = \frac{\ln 2}{\ln 3},$$ (B98)

which is also not an integer. This example of a fractal object is called a *Cantor set*.

Appendix C
Renormalization Group Technique

C.1
Renormalization Group Technique [13]

C.1.1
Widom Scaling

Let us use the following definitions for the various critical exponents:

$$C \sim |T - T_c|^{-\alpha} , \tag{C1}$$

$$m \sim |T - T_c|^{\beta} , \tag{C2}$$

$$\chi \sim |T - T_c|^{-\gamma} , \tag{C3}$$

where C denotes the specific heat, m the magnetization and χ the susceptibility. It may be mentioned that the predictions of mean field approximation and the experimentally observed values show that there exists a relation $\alpha + 2\beta + \gamma = 2$, satisfied as an equality. We need to investigate this scaling relation. We have the Rushbrook inequality as follows:

$$\alpha + 2\beta + \gamma \geq 2 . \tag{C4}$$

We can prove it in the following way. We have

$$C_P - C_V = \frac{TV\lambda^2}{\kappa_T} ,$$

where C_P and C_V are the specific heat at constant pressure and volume, respectively, and $\lambda = \frac{1}{V} \left(\frac{\partial V}{\partial T} \right)_P$ and $\kappa_T = -\frac{1}{V} \left(\frac{\partial V}{\partial P} \right)_T$. Using it

$$C_P - C_V = -\frac{T \left(\frac{\partial V}{\partial T} \right)_P^2}{\left(\frac{\partial V}{\partial P} \right)_T} .$$

Econophysics. Sitabhra Sinha, Arnab Chatterjee, Anirban Chakraborti, and Bikas K. Chakrabarti
Copyright © 2011 WILEY-VCH Verlag GmbH & Co. KGaA, Weinheim
ISBN: 978-3-527-40815-3

For magnets, pressure (P) is to be replaced by external field (h) and volume (V) is to be replaced by magnetization (m). Hence,

$$C_h - C_m = \frac{T\left(\frac{\partial m}{\partial T}\right)_h^2}{\left(\frac{\partial m}{\partial h}\right)_T}.$$

The negative sign is not here because, we had to put negative sign to make κ positive. Since κ is an elastic constant, it should be positive for stability and the volume decreases when pressure increases, so the derivative is negative. But here magnetization increases as the field increases, so there is no need for the negative sign. Now, as the specific heat must be positive, we can write

$$C_h \geq \frac{T\left(\frac{\partial m}{\partial T}\right)_h^2}{\left(\frac{\partial m}{\partial h}\right)_T}, \tag{C5}$$

and the equality applies when $C_m = 0$. Hence,

$$C_h \sim \left|\frac{T - T_c}{T_c}\right|^{-\alpha} \sim |\Delta T|^{-\alpha}, \tag{C6}$$

and

$$m \sim \left|\frac{T - T_c}{T_c}\right|^{\beta}, \tag{C7}$$

where $\Delta T = \frac{T - T_c}{T_c}$. Therefore, we have

$$\frac{\partial m}{\partial T} \sim \left|\frac{T - T_c}{T_c}\right|^{\beta-1}, \tag{C8}$$

implying,

$$\left(\frac{\partial m}{\partial T}\right)^2 \sim \left|\frac{T - T_c}{T_c}\right|^{2\beta-2} \sim (\Delta T)^{2\beta-2}. \tag{C9}$$

Also, we know

$$\chi = \frac{\partial m}{\partial h} \sim \left|\frac{T - T_c}{T_c}\right|^{-\gamma} \sim (\Delta T)^{-\gamma}. \tag{C10}$$

Now using the inequality we can write

$$(\Delta T)^{-\alpha} \geq \frac{(\Delta T)^{2\beta-2}}{(\Delta T)^{-\gamma}}$$

$$\Rightarrow (\Delta T)^{-\alpha} \geq (\Delta T)^{2\beta-2+\gamma}.$$

Now, since $\Delta T < 1$, we have the inequality

$$-\alpha \leq 2\beta - 2 + \gamma$$

$$\Rightarrow \alpha + 2\beta + \gamma \geq 2, \tag{C11}$$

which, however, is observed to be an equality.

Scaling Hypothesis   The Widom scaling hypothesis states that the free energy per site can be written as a sum of a regular part and a singular part. The singular part is of the following scaling form:

$$f_{\text{sing}} \sim |\Delta T|^{2-\alpha} \, Y \left[\frac{\Delta h}{(\Delta T)^{\Delta}} \right],$$ (C12)

where $\Delta T \equiv T - T_c$ and $\Delta h \equiv h - h_c = h$ (for ferromagnets $h_c = 0$). We then assume that the function Y is defined in the asymptotic limits of its arguments. Then we can write the specific heat as

$$C = \frac{\partial^2 f}{\partial T^2}\bigg|_{h\to0} \sim (\Delta T)^{-\alpha} \, Y \left(\frac{h}{(\Delta T)^{\Delta}} \right)\bigg|_{h\to0},$$

and with $Y \to 1$ as $h \to 0$, one gets $C \sim |\Delta T|^{-\alpha}$.

Now for the magnetization

$$m = \frac{\partial f}{\partial h}\bigg|_{h\to0} \sim |\Delta T|^{2-\alpha-\Delta} \sim |\Delta T|^{\beta} .$$

Similarly,

$$\chi \sim \frac{\partial^2 f}{\partial h^2}\bigg|_{h\to0} \sim |\Delta T|^{2-\alpha-2\Delta} \sim |\Delta T|^{-\gamma} .$$

Hence, $2-\alpha-\Delta = \beta$ and $2-\alpha-2\Delta = -\gamma$, giving $\Delta = \beta+\gamma$ and $\alpha+2\beta+\gamma = 2$, giving the above mentioned equality.

Near T_c, we note that free energy is not a function of the independent variables T and h, rather it is a function of the single scaled variable $\frac{h}{(\Delta T)^{\Delta}}$.

Hyperscaling Relation: We have defined the correlation length as $\xi \sim |T - T_c|^{-\nu}$. The idea is that within a volume ξ^d the spins are effectively all aligned and therefore can be taken as a single degree of freedom, rather than considering individual spins. We shall see, this is the idea which will finally lead to the renormalization scheme. Now the effective number of degrees of freedom $N_{\text{eff}} \sim \xi^d$. So, the free energy

$$f \sim \frac{F}{N_{\text{eff}}} \sim \xi^{-d} \sim (\Delta T)^{\nu d} .$$ (C13)

But, $f \sim (\Delta T)^{2-\alpha}$, implying

$$\alpha + \nu d = 2 .$$ (C14)

C.1.2
Formalism

Renormalization scheme is based on the fact that the effective number of degrees of freedom reduces as the critical point is approached. This is ensured by the presence of a finite correlation. So we can define an effective interaction which gives the same partition function, that is

$$Z(H_N) = Z(H'_{N'<N}) \, . \tag{C15}$$

If there is a hundred percent correlation, we could work with a single degree of freedom.

In a thermodynamic system, if the correlation length is large compared to the interatomic separation, then the experimental results for the system are obtained theoretically considering a single particle. Correlation length diverges as the critical point is approached. Hence, near the critical point we have to consider, instead of an Avogadro's number of particles, a single particle.

Let us define the operator $\mathcal{R}$ as

$$\mathcal{R} H_N = H'_{N'<N} \, . \tag{C16}$$

We keep operating this until correlation exists. Finally we reach a point where

$$\mathcal{R} H_{N*}^* = H_{N*}^* \, . \tag{C17}$$

After this, no further reduction in the degrees of freedom is possible. This is called the fixed point near which, the relation between the free energy densities can be written as

$$f(|\Delta T|, h) = \frac{N^*}{N} f\left(|\Delta T^*|, h^*\right), \tag{C18}$$

where $h^* = 0$ and hence $\Delta h = h$ itself.

Very near to the fixed point, $\mathcal{R}$ can always be written as a linear operator (linearization) and we can diagonalize it to write

$$\begin{pmatrix} |\Delta T'| \\ h' \end{pmatrix} = \mathcal{R} \begin{pmatrix} |\Delta T| \\ h \end{pmatrix}$$

where $|\Delta T'| = |T' - T^*|$, $|\Delta h'| = |h' - h^*| = h'$ and $|\Delta T| = |T - T^*|$, $|\Delta h| = |h - h^*| = h$. Both T and T', h and h' are very close to the critical point. In terms of the eigenvalues Λ_j of the linearized operator $\mathcal{R}$, one can write

$$h'_i = h_i^* + \sum_j \Lambda_j \left(h_j - h_j^*\right).$$

Therefore we can write

$$f(|\Delta T|, h) = \frac{N'}{N} f(\Lambda_T |\Delta T|, \Lambda_h h) \, .$$

Now, if the length scale is changed by a factor b the we have $\frac{N'}{N} = b^{-d}$, where d is the dimension. So,

$$f(|\Delta T|, h) = b^{-d} f(\Lambda_T |\Delta T|, \Lambda_h h) \, .$$

Also, note that $\mathcal{R}_{b_1}\mathcal{R}_{b_2} = \mathcal{R}_{b_1 b_2}$, that is it does not matter if we change the degrees of freedom in two steps or in a single step. So we have $\Lambda_{b_1}\Lambda_{b_2} = \Lambda_{b_1 b_2}$, which implies, $\Lambda \sim b^\lambda$ and λ are independent of the choice of b. Therefore we can write

$$f(|\Delta T|, h) = b^{-d} f\left(b^{\lambda_T}\Delta T, b^{\lambda_h}h\right).$$

Choosing now $b^{\lambda_T}|\Delta T| = 1$, as b can be arbitrary

$$f(|\Delta T|, h) = |\Delta T|^{d/\lambda_T} Y\left(\frac{h}{|\Delta T|^{\lambda_h/\lambda_T}}\right); \quad Y(x) \equiv f(1, x),$$

which is of the same form as that of the Widom scaling hypothesis (C12), with the identification

$$\frac{d}{\lambda_T} = 2 - \alpha \quad \frac{\lambda_h}{\lambda_T} = \Delta = \frac{1}{2}(2 - \alpha + \gamma), \tag{C19}$$

and scaling function $Y(x) = f(1, x)$.

C.1.3
RG for One-Dimension Ising Model

We now intend to explain the renormalization method for 1-d Ising model in presence of external field. The Hamiltonian can be written as

$$H = -J \sum_{i=1,2,3,\dots N} S_i S_{i+1} - h \sum_i S_i + \sum_i c. \tag{C20}$$

The last term (constant), although irrelevant at this point, is included because, a similar term will be generated by the RG transformations. We want to change the scale of the system by a factor $b = 2$. To do this, first we write the Hamiltonian in a slightly different form

$$H = -\sum_{i=2,4,6,\dots N}\left[J S_i (S_{i+1} + S_{i-1}) + h S_i + h\frac{S_{i+1} + S_{i-1}}{2} + 2c\right].$$

Hence, the partition function

$$Z = \text{Tr}_{\{S_i\}}$$

$$\times \prod_{i=2,4,6,\dots N} \exp\left[J S_i (S_{i+1} + S_{i-1}) + h S_i + \frac{h}{2}(S_{i+1} + S_{i-1}) + 2c\right].$$

We now perform a partial trace over the spins at even sites, getting

$$Z = \text{Tr}'_{\{s_i\}} \prod_{i=2,4,6,\dots} \left\{\exp\left[J (S_{i+1} + S_{i-1}) + h + \frac{h}{2}(S_{i+1} + S_{i-1}) + 2c\right]\right.$$

$$\left. + \exp\left[-J (S_{i+1} + S_{i-1}) - h + \frac{h}{2}(S_{i+1} + S_{i-1}) + 2c\right]\right\}.$$

We now rename the sites and write the partition function as

$$Z = \text{Tr}_{\{S_i\}} \prod_{i=1,2,\dots,N/2} \left\{ \exp\left[J\left(S_i + S_{i+1} \right) + h + \frac{h}{2}\left(S_i + S_{i+1} \right) + 2c \right] \right.$$

$$\left. + \exp\left[-J\left(S_i + S_{i+1} \right) - h + \frac{h}{2}\left(S_i + S_{i+1} \right) + 2c \right] \right\}.$$

Our demand however is that we can always write the partition function in the original form, of course in terms of the re-normalized coupling constants. Hence,

$$Z = \text{Tr}_{\{S_i\}} \prod_{i=1,2,\dots,N/2} \exp\left(J' S_i S_{i+1} + h' \frac{S_i + S_{i+1}}{2} + c' \right). \tag{C21}$$

But the above two equations should match for all combinations of S_i and S_{i+1}. Hence for $S_i = S_{i+1} = +1$

$$e^{J'+h'+c'} = e^{2J+2h+2c} + e^{-2J+2c}$$

for $S_i = S_{i+1} = -1$,

$$e^{-J'-h'+c'} = e^{-2J+2c} + e^{2J-2h+2c},$$

for $S_i = 1, S_{i+1} = -1$,

$$e^{-J'+c'} = e^{h+2c} + e^{-h+2c}. \tag{C22}$$

By use of simple algebra we get

$$e^{2h'} = e^{2h}\left[\frac{\cosh\left(2J + h \right)}{\cosh\left(2J - h \right)} \right]$$

$$e^{4J'} = \frac{\cosh\left(2J + h \right)\cosh\left(2J - h \right)}{\cosh^2 h}$$

$$e^{4c'} = 16e^{8c}\cosh\left(2J + h \right)\cosh\left(2J - h \right)\cosh^2 h.$$

Now we see that even if we had $c = 0$ we would get a c'. Defining the following

$$e^{-2h} = y \quad e^{-2h'} = y' \quad e^{-4J} = x \quad e^{-4J'} = x' \quad e^{-4c} = w \quad e^{-4c'} = w'.$$

We can write

$$x' = \frac{x\left(1 + y \right)^2}{\left(x + y \right)\left(1 + xy \right)}, \tag{C23}$$

$$y' = \frac{y\left(x + y \right)}{1 + xy}, \tag{C24}$$

$$w' = \frac{w^2 x y^2}{\left(1 + xy \right)\left(x + y \right)\left(1 + y^2 \right)}. \tag{C25}$$

Fixed points and flow diagram is shown in Figure C.2. If the system starts from any nonzero temperature, it flows to $x = 1$, that is the infinite temperature configuration. The fixed point $x = 0, y = 0$ corresponds to zero temperature and infinite field, when all the spins are aligned. Finally, the system shows critical behavior at $x = 0, y = 1$ fixed point, which corresponds to zero temperature and zero field.

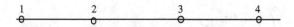

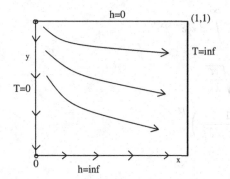

Figure C.1 RG steps: decimation and rescaling.

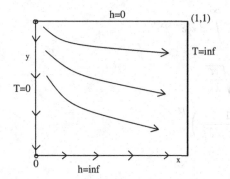

Figure C.2 Flow diagram for one-dimension Ising model.

C.1.4
Momentum Space RG for $4 - \epsilon$ Dimensional Ising Model

Let us consider again the Ising model without any external field. The Hamiltonian reads as

$$H = -J \sum_{\langle ij \rangle} S_i S_j \,. \tag{C26}$$

The partition function for the system can be written as

$$Z = \prod_i \text{Tr} \exp \left\{ \left(\frac{J}{k_B T} \right) \sum_{\langle ij \rangle} S_i S_j \right\}$$

$$= \prod_i \int \exp \left\{ \frac{J}{k_B T} \sum_{\langle ij \rangle} S_i S_j \right\} \delta(S_i^2 - 1) dS_i \,, \tag{C27}$$

where $K = J/k_B T$ and the delta function represents the density of states.

Now let us consider the following integral

$$I_1 \equiv \int_{-\infty}^{\infty} e^{-\bar{x}Ax} \, dx \,. \tag{C28}$$

Suppose A is diagonizable. Hence,

$$U^{-1}AU = \text{Diag}\{\lambda_1, \lambda_2, \dots\} \qquad \text{also} \qquad x' = Ux \,. \tag{C29}$$

Therefore,

$$\int_{-\infty}^{\infty} e^{-\bar{x}Ax} \, dx = \prod_i \int_{-\infty}^{\infty} e^{-\sum_i \lambda_i x_i'^2} \, dx_i'$$

$$= \int_{-\infty}^{\infty} dx_1' e^{-\lambda_1 x_1'^2} \int_{-\infty}^{\infty} dx_2' e^{-\lambda_2 x_2'^2} \dots$$

$$= \sqrt{\frac{\pi}{\lambda_1}} \sqrt{\frac{\pi}{\lambda_2}} \dots \sqrt{\frac{\pi}{\lambda_N}}$$

$$= \left(\sqrt{\pi}\right)^N \left(\frac{1}{\lambda_1 \lambda_2 \dots \lambda_N}\right)^{1/2}$$

$$= \frac{\pi^{N/2}}{\left(\prod_i \lambda_i\right)^{1/2}}$$

$$= \frac{\pi^{N/2}}{(\det A)^{1/2}} \,. \tag{C30}$$

Similarly,

$$I_2 \equiv \prod_i \int_{-\infty}^{\infty} e^{-\bar{x}Ax + Bx} \, dx_i$$

$$= \prod_i \int_{-\infty}^{\infty} e^{-(x - \frac{1}{2}A^{-1}B)A - (x - \frac{1}{2}A^{-1}B)} e^{\frac{1}{2}BA^{-1}B} \, dx_i$$

$$= \frac{\pi^{N/2}}{(\det A)^{1/2}} e^{\frac{1}{2}BA^{-1}B} \,. \tag{C31}$$

Also,

$$I_3 \equiv \prod_i \int_{-\infty}^{\infty} x_k e^{-\bar{x}Ax} dx_i$$

$$= \prod_i \int_{-\infty}^{\infty} x_k e^{-\bar{x}Ax+Bx} dx_i \Bigg|_{B\to 0}$$

$$= \frac{\partial}{\partial B_k}(I_2)\Bigg|_{B\to 0} = 0 , \tag{C32}$$

as $\exp\left[\frac{1}{2}BA^{-1}B\right]$ can be expanded as $1 + \frac{1}{2}BA^{-1}B + \dots$ and eventually $B \to 0$. Similarly,

$$I_4 \equiv \prod_{i=1}^{N} \int_{-\infty}^{\infty} x_k x_l dx_i e^{-\frac{1}{2}\sum_{\langle ij \rangle} x_i A_{ij} x_j}$$

$$= \frac{\partial}{\partial B_k}\frac{\partial}{\partial B_l}(I_2)\Bigg|_{B\to 0}$$

$$= \frac{\pi^{N/2}}{(\det A)^{1/2}}(A^{-1})_{kl} . \tag{C33}$$

We represent the double-delta function density of state coming from $\delta(S^2 - 1)$ in the integral representation of the trace in (C27) in the following way through the double-Gaussian

$$\rho(S) = e^{-\kappa(S^2-1)^2}\Bigg|_{\kappa\to 0} .$$

So the effective Hamiltonian ($Z = \mathrm{Tr}\, e^{-H_{\mathrm{eff}}/k_B T}$) will now have the form

$$H_{\mathrm{eff}} = -\frac{J}{k_B T}\sum_{\langle ij \rangle} S_i S_j - 2\kappa \sum_i S_i^2 + \kappa \sum_i S_i^4$$

$$= -\frac{J}{k_B T}\sum_{\langle ij \rangle}\left[S_i^2 + S_j^2 - (S_i - S_j)^2\right] - 2\kappa \sum_i S_i^2 + \kappa \sum_i S_i^4$$

$$= -\frac{J}{2k_B T}\left[4d\sum_i S_i^2 - \sum_{\langle ij \rangle}(S_i - S_j)^2\right] - 2\kappa \sum_i S_i^2 + \kappa \sum_i S_i^4$$

$$= -\left(\frac{2dJ}{k_B T} + 2\kappa\right)\sum_i S_i^2 + \kappa \sum_i S_i^4 + \frac{J}{2k_B T}(S_i - S_j)^2 ,$$

where in the second step we have expanded $S_i S_j$, and in the third step we have used the fact that $\sum_{\langle ij \rangle} S_i^2 = 2d \sum_i S_i^2 = \sum_{\langle ij \rangle} S_j^2$ with d being the dimensionality of the lattice. This Hamiltonian is called the Landau–Ginzburg–Wilson Hamiltonian

(H_{LGW}). Now we can make the Fourier expansion,

$$S_i = \sum_q e^{-iqr_i} S_q \,.$$

So,

$$\sum_i S_i^2 = \sum_i \sum_{qq'} e^{-i(q+q')r_i} S_q S_{q'}$$

$$= \sum_{qq'} \delta(q+q') S_q S_q'$$

$$= \sum_q S_q S_{-q} \,.$$

Also,

$$\sum_{\langle ij \rangle} S_i S_j = \sum_{\langle ij \rangle} \sum_{qq'} S_q S_{q'} e^{-iqr_i - iq'r_j}$$

$$= \sum_{\langle ij \rangle} \sum_{qq'} S_q S_q' e^{-i(q+q')r_i} e^{-iq'(r_i - r_j)}$$

$$= \sum_{\langle ij \rangle} \sum_{qq'} S_q S_{q'} e^{-i(q+q')r_i} \left(1 - \frac{q'^2 (r_i - r_j)^2}{2} \right)$$

$$= 2d \sum_{qq'} \sum_i e^{-i(q+q')r_i} \left(1 - \frac{q'^2 a^2}{2} \right) S_q S_{q'}$$

$$= 2d \sum_{qq'} \delta(q+q') \left(1 - \frac{q'^2 a^2}{2} \right) S_q S_{q'}$$

$$= 2d \sum_q \left(1 - \frac{q^2 a^2}{2} \right) S_q S_{-q} \,,$$

with a as the lattice constant here.
We had

$$H_{\text{LGW}} = C \sum_{\langle ij \rangle} (S_i - S_j)^2 + (T - T_c) \sum_i S_i^2 + C' \sum_i S_i^4 \,.$$

In momentum space

$$H_{\text{LGW}} = \sum_q (r^* + q^2) S_q S_{-q}$$

$$+ U \sum_{q_1 q_2 q_3 q_4} S_{q_1} S_{q_2} S_{q_3} S_{q_4} \delta(q_1 + q_2 + q_3 + q_4) \,, \tag{C34}$$

where $r^* \sim T - T_c^{\mathrm{MF}}$ and $U = C'$. So the partition function is

$$Z = \int e^{-H_{\mathrm{LGW}}(s_q)} \, dS_q \equiv \int_{S_q} e^{-H_{\mathrm{LGW}}(S_q)} . \qquad (C35)$$

Here $\int_{S_q}$ denotes the functional integral over S_q.

We can divide

$$S_q = S_q^0 + \tilde{S}_q , \qquad (C36)$$

where

$$\tilde{S}_q = S_q \Theta \left(|q| - \frac{1}{b} \right)$$

$$S_q^0 = S_q \Theta \left(\frac{1}{b} - |q| \right) . \qquad (C37)$$

We will integrate out $\tilde{S}_q$, which represents the small-scale fluctuations and will be left with $\tilde{H}_{\mathrm{eff}}(S_q^0)$. We demand that the form of the Hamiltonian does not change, only the parameters r and U get renormalized. Hence, $\tilde{H}_{\mathrm{eff}} \equiv \tilde{H}$ with

$$\tilde{H} = \sum_q (r' + q^2) S_q^0 S_{-q}^0$$

$$+ U' \sum_{q_1 q_2 q_3 q_4} S_{q_1} S_{q_2} S_{q_3} S_{q_4} \delta(q_1 + q_2 + q_3 + q_4) . \qquad (C38)$$

First let us evaluate Z when $H \equiv H(S_q)$. We consider the U-term as a perturbation, and hence

$$Z = \int_{S_q} e^{-\int_q (r + q^2) S_q S_{-q}}$$

$$\times \left(1 - U \int_{q_1} \int_{q_2} \int_{q_3} \int_{q_4} S_{q_1} S_{q_2} S_{q_3} S_{q_4} \delta(q_1 + q_2 + q_3 + q_4) \right) , \qquad (C39)$$

where $\int_{q_i}$ denotes the integral over q_i. One can write

$$e^{-\int_q (r + q^2) S_q S_{-q}}$$

$$= \exp \left[-\int_{q_1} \int_{q_2} (r + q^2)(S_{q_1}^0 + \tilde{S}_{q_1})(S_{q_2}^0 + \tilde{S}_{q_2}) \delta(q_1 + q_2) \right]$$

$$= \exp \left[-\int_q (r + q^2) S_q^0 S_{-q}^0 + \int_q (r + q^2) \tilde{S}_q \tilde{S}_{-q} \right] .$$

Cross terms will vanish due to the delta-function. So,

$$
Z = \int_{S_q} \exp\left\{ -\int_q (r + q^2) S_q^0 S_{-q}^0 \right.
$$

$$
- \int_q (r + q^2) \tilde{S}_q \tilde{S}_{-q} \left[1 + U \int_{q_1} \int_{q_2} \int_{q_3} \int_{q_4} (S_{q_1}^0 + \tilde{S}_{q_1}) \right.
$$

$$
\left. \left. \times (S_{q_2}^0 + \tilde{S}_{q_2})(S_{q_3}^0 + \tilde{S}_{q_3})(S_{q_4}^0 + \tilde{S}_{q_4}) \delta(q_1 + q_2 + q_3 + q_4) \right] \right\} . \quad \text{(C40)}
$$

Possible contributions from the different bracket-terms are

$$
X_1 = U \int_{q_1} \int_{q_2} \int_{q_3} \int_{q_4} S_{q_1}^0 S_{q_2}^0 S_{q_3}^0 S_{q_4}^0 \delta(q_1 + q_2 + q_3 + q_4)
$$

$$
X_2 = 6U \int_{q_1} \int_{q_2} \int_{q_3} \int_{q_4} S_{q_1}^0 S_{q_2}^0 \tilde{S}_{q_3} \tilde{S}_{q_4} \delta(q_1 + q_2) \delta(q_3 + q_4)
$$

$$
X_3 = U \int_{q_1} \int_{q_2} \int_{q_3} \int_{q_4} \tilde{S}_{q_4} \tilde{S}_{q_4} \tilde{S}_{q_4} \tilde{S}_{q_4} \delta(q_1 + q_2 + q_3 + q_4) . \quad \text{(C41)}
$$

The factor $(6 = {}^4C_2)$ comes from counting the number of ways to select two lower momenta out of four. Now, as we integrate the higher momenta, the X_3 term will be a constant and we shall not consider any effect of it from here on. The X_1 term will give the usual U term as is seen in the unrenormalized Hamiltonian. It is the X_2 term which will contribute to the correction on r. There will be no correction on U in this order. To get that we will consider a higher order correction later. Now,

$$
Z = \int_{S_q^0} \exp\left[-\int_q (r + q^2) S_q^0 S_{-q}^0 \right]
$$

$$
\times \int_{\tilde{S}_q} \exp\left[-\int_{q'} (r + q^2) \tilde{S}_q \tilde{S}_{-q} \right] [1 + X_1 + X_2 + X_3]. \quad \text{(C42)}
$$

Clearly, the higher momentum integration will give constant terms for all the terms except for the X_2 term. Integration of that term will be like

$$6U \int_{\tilde{S}_q} \exp\left[-\int_q (r+q^2)\tilde{S}_q\tilde{S}_{-q}\right]$$

$$\times \int_{q_1}\int_{q_2}\int_{q_3}\int_{q_4} S^0_{q_1} S^0_{q_2} \tilde{S}_{q_3} \tilde{S}_{q_4} \delta(q_1+q_2)\delta(q_3+q_4)$$

$$= 6U \int_{q_1}\int_{q_3}\int_{q_4} S^0_{q_1} S^0_{-q_1} \int_{\tilde{S}_q} \tilde{S}_{q_3}\tilde{S}_{q_4} \exp\left[\int_q (r+q^2)\tilde{S}_q\tilde{S}_{-q}\right] \delta(q_3+q_4) \,.$$

$$\tag{C43}$$

This term has the same form as I_4 (C33) only with a reduced number of variables. Identifying $A = \mathrm{diag}(r+q^2, r+q^2)$ we have the integral to be (apart from constants) $6U \int_q S^0_q S^0_{-q} \frac{1}{r+q^2}$. So, for the renormalized partition function we have (neglecting overall constants $\frac{\pi^{N/2}}{(\det A)^{1/2}}$ which will also come out from the X_3 term)

$$Z = \int_{S_q} \exp\left[-\int_q (r+q^2)S^0_q S^0_{-q}\right.$$

$$+ U\int_{q_1}\int_{q_2}\int_{q_3}\int_{q_4} S^0_{q_1} S^0_{q_4} S^0_{q_4} S^0_{q_4} \delta(q_1+q_2+q_3+q_4)$$

$$\left. + 6U \int_q S^0_q S^0_{-q} \frac{1}{r+q^2}\right]. \tag{C44}$$

Now comparing with (C38), we see that

$$r' = r + 6U\int_q \frac{1}{r+q^2} \quad \text{and} \quad U' = U\,. \tag{C45}$$

We thus see that we get a renormalized r but do not get any correction to U in the first-order. In the second-order, we shall have eight-spin terms and it can be shown that the renormalized U will be given by

$$U' = U - 36U^2 \int_{\tilde{q}_1} \frac{1}{(r+\tilde{q}^2)^2}\,, \tag{C46}$$

where $\tilde{q} \equiv q\Theta\left(q - \frac{1}{b}\right)$.

Now in H', q changes from 0 to $1/b$, but in H, q changes from 0 to 1. So, for comparison between H and H', we multiply by a scale factor b. Hence, after rescaling we will have

$$r' = C^2 b^d \left[r + 12 \int_{\tilde{q}} \frac{1}{r + \tilde{q}^2} \right]$$

$$U' = C^4 b^d \left[U - 36 U^2 \int_{\tilde{q}} \frac{1}{(r + \tilde{q}^2)^2} \right]. \tag{C47}$$

Now, on comparing H and H', $C^2 b^d b^{-2} q^2 = q^2$. So, $C^4 = b^{4-2d}$. Hence,

$$r' = b^2 \left[r + 12 \int_{\tilde{q}} \frac{1}{r + \tilde{q}^2} \right]$$

$$U' = b^{4-d} \left[U - 36 U^2 \int_{\tilde{q}} \frac{1}{(r + \tilde{q}^2)^2} \right]. \tag{C48}$$

The above equations imply that for $d > 4$, the U' term can be neglected and we would get a Gaussian model.

Let us write $r' = r_b$ and $U' = U_b$ and change the length scale by $n = (1 + \delta)$. Then,

$$r_{nb} = n^2 \left[r_b + 12 \int_{1/nb}^{1} \frac{1}{r_b + q^2} \right]$$

$$= (1 + \delta)^2 \left[r_b + 12 \int_{1/(1+\delta)b}^{1} \frac{1}{r_b + q^2} \right], \tag{C49}$$

and

$$U_{nb} = n^{4-d} \left[U_b - 36 U_b^2 \int_{1/b}^{1} \frac{1}{(r_b + q^2)^2} \right]$$

$$= (1 + \delta)^{4-d} \left[U_b - 36 U_b^2 \int_{1/(1+\delta)b}^{1} \frac{1}{(r_b + q^2)^2} \right]. \tag{C50}$$

We want to write the recurrence relations in the form of differential equations. Note that

$$\frac{dr_b}{db} = \lim_{\delta \to 0} \frac{r_{(1+\delta)b} - r_b}{\delta b}. \tag{C51}$$

Using this and a similar form for U_L we will have

$$b\frac{dr_b}{db} = 2r_b + \frac{12\Omega_d}{1 + r_b}$$

$$b\frac{dU_b}{db} = \epsilon U_b - 36\frac{U_b^2\Gamma_d^2}{(1 + r_b)^2}, \tag{C52}$$

with $\epsilon = 4 - d$ and Γ_d being the result of the integration of the angular variables. Now the fixed points are those, after which there will be no further change in Z, that is the derivatives in the above equations should be zero. Using it and dropping the subscript we can write

$$r^* + \frac{12\Gamma_d U^*}{1 + r^*} = 0$$

$$\epsilon U^* - 36\frac{\Gamma_d U^{*2}}{(1 + r^*)^2} = 0. \tag{C53}$$

Solving these, we will get two fixed points

$$r^* = 0 \qquad \text{and} \qquad U^* = 0 \tag{C54}$$

and

$$r^* = -\frac{\epsilon}{6 + \epsilon} \approx -\frac{\epsilon}{6}$$

$$U^* = \frac{\epsilon(6 + 2\epsilon)}{(6 + \epsilon)^2 6\Omega_d} \approx \frac{\epsilon}{36\Omega_d}. \tag{C55}$$

We take a point slightly away from the fixed point

$$r = r^* + \Delta r \qquad \text{and} \qquad U = U^* + \Delta U. \tag{C56}$$

Putting these in (C53) we get equations in terms of Δr and ΔU. Linearizing those equations, we get the following eigenvalue equations:

$$\begin{vmatrix} \lambda - 2 & -12\Omega_d \\ 0 & -(\lambda - \epsilon) \end{vmatrix} = 0.$$

The eigenvalues are: $\lambda_r = 2$ and $\lambda_U = \epsilon$. But we know $\nu = 1/\lambda_r = 0.5$. Again, for the other fixed point

$$\begin{vmatrix} \lambda - 2 + \epsilon/3 & -12\Omega_d(1 + \epsilon/6) \\ 0 & -(\lambda - \epsilon) \end{vmatrix} = 0$$

The eigenvalues here will be $\lambda_r = 2 - \epsilon/3$ and $\lambda_U = -\epsilon$. So, ν here for $d = 3$ will be ≈ 0.6, which is much improved.

Now from (C56) we see that $\Delta r \sim b^{\lambda_r}$ and $\Delta U \sim b^{\lambda_U}$. So, if there is any U term in the original Hamiltonian then we see that below dimension 4, at the first fixed point U blows up. But at the second fixed point ΔU is decaying. So, below dimension 4 the first fixed point is not observed, only the second fixed point is observed. There, the first fixed point is the repelling fixed point and the second fixed point is the steady fixed point. But above dimension 4, the U term blows up in the second fixed point and that decays in the first fixed point. Hence, above dimension 4, the first fixed point is the steady fixed point, which gives the mean field results (since, $r^* = 0$). The first fixed point is called the Gaussian fixed point and the second fixed point is called the Ising fixed point.

C.1.5
Real Space RG for Transverse Field Ising Chain

Here the basic idea of real space block renormalization is illustrated by applying it on an Ising chain in a transverse field. Taking the cooperative interaction along the x-axis and the transverse field along the z-axis, the Hamiltonian reads

$$H = -\Gamma \sum_{i=1}^{N} S_i^z - J \sum_{i=1}^{N-1} S_i^x S_{i+1}^x$$

$$= H_B + H_{IB} .$$ (C57)

With the following

$$H_B = \sum_{p=1}^{N/b} H_p ; \quad H_p = -\sum_{i=1}^{b} \Gamma S_{i,p}^z - \sum_{i=1}^{b-1} J S_{i,p}^x S_{i+1,p}^x$$

and

$$H_{IB} = \sum_{p=1}^{N/(b-1)} H_{p,p+1} ; \quad H_{p.p+1} = -J S_{b,p}^x S_{1,p+1} .$$

The above rearrangement of the Hamiltonian recasts the picture of N spins with nearest neighbor interaction into one in which there are $N/(b-1)$ blocks, each consisting of b number of spins. The part H_B represents the interaction between the

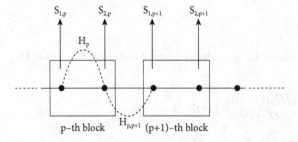

Figure C.3 The linear chain is broken up into blocks of size b (= 2 here) and the Hamiltonian (C57) can be written as the sum of block Hamiltonians H_p and the inter-block Hamiltonians $H_{p,p+1}$. The Hamiltonian H_p is diago-nalized exactly and the lowest lying two states are identified as the renormalized spin states in terms of which the inter-block Hamiltonian is rewritten to get the RG recursion relation.

spins within the blocks, while H_{IB} represents the interaction between the blocks through their terminal spins (Figure C.3).

Here we will consider $b = 2$, as shown in the figure. Now H_p has 4 eigenstates and one can express them in terms of the linear superposition of the eigenstates of $S_{1,p}^z \otimes S_{2,p}^z$; namely,

$$|\uparrow\uparrow\rangle \, , \quad |\downarrow\downarrow\rangle \, , \quad |\uparrow\downarrow\rangle \quad \text{and} \quad |\downarrow\uparrow\rangle \, .$$

Considering the orthonormality of the eigenstates, one may easily see that the eigenstates of H_p can be expressed as:

$$|0\rangle = \frac{1}{\sqrt{1 + \phi^2}} \left(|\uparrow\uparrow\rangle + \phi |\downarrow\downarrow\rangle \right)$$

$$|1\rangle = \frac{1}{\sqrt{2}} \left(|\uparrow\downarrow\rangle + |\downarrow\uparrow\rangle \right)$$

$$|2\rangle = \frac{1}{\sqrt{2}} \left(|\uparrow\downarrow\rangle - |\downarrow\uparrow\rangle \right)$$

$$|3\rangle = \frac{1}{\sqrt{1 + \phi^2}} \left(\phi |\uparrow\uparrow\rangle - |\downarrow\downarrow\rangle \right) .$$

Here ϕ is a coefficient required to be chosen properly, so that $|0\rangle$ and $|3\rangle$ are the eigenstates of H_p. One gets

$$H_p |0\rangle = H_p \left[\frac{1}{\sqrt{1 + \phi^2}} \left(|\uparrow\uparrow\rangle + \phi |\downarrow\downarrow\rangle \right) \right]$$

$$= \left[-\Gamma (S_1^z + S_2^z) - J(S_1^x S_2^x) \right] \frac{1}{\sqrt{1 + \phi^2}} \left(|\uparrow\uparrow\rangle + \phi |\downarrow\downarrow\rangle \right)$$

$$= \frac{1}{\sqrt{1 + \phi^2}} \left[-\Gamma(2 |\uparrow\uparrow\rangle - 2\phi |\downarrow\downarrow\rangle) - J |\downarrow\downarrow\rangle + J\phi |\uparrow\uparrow\rangle \right]$$

$$= -(2\Gamma + J\phi) \frac{1}{\sqrt{1 + \phi^2}} \left[|\uparrow\uparrow\rangle + \left(\frac{2\Gamma - J/\phi}{2\Gamma + J\phi} \right) \phi |\downarrow\downarrow\rangle \right] .$$

Thus, for $|0\rangle$ to be an eigenstate of H_p, one must have

$$-\frac{2\Gamma - J/\phi}{2\Gamma + J\phi} = 1$$

$$\Rightarrow J\phi^2 + 4\Gamma\phi - J = 0$$

$$\Rightarrow \phi = \frac{\pm\sqrt{4\Gamma^2 + J^2} - 2\Gamma}{J} .$$

To minimize the energy, we have to choose

$$\phi = \frac{\sqrt{4\Gamma^2 + J^2} - 2\Gamma}{J} . \tag{C58}$$

One can now see, applying H_p on its eigenstates

$$H_p \,|0\rangle = E_0\,|0\rangle\,, \quad E_0 = -\sqrt{4\Gamma^2 + J^2}$$
$$H_p \,|1\rangle = E_1\,|1\rangle\,, \quad E_1 = -J$$
$$H_p \,|2\rangle = E_2\,|2\rangle\,, \quad E_2 = +J$$
$$H_p \,|3\rangle = E_3\,|3\rangle\,, \quad E_3 = \sqrt{4\Gamma^2 + J^2}\,.$$

Now we define our new renormalized spin variables Ss, each replacing a block in the original Hamiltonian. We retain only the two lowest lying states $|0\rangle$ and $|1\rangle$ of a block and define corresponding $S_p'^z$ to have them as its two eigenstates, $|\uparrow\rangle = |0\rangle$ and $|\downarrow\rangle = |1\rangle$. We also define

$$S'^x = \frac{S_1^x \otimes \mathcal{I} + \mathcal{I} \otimes S_2^x}{2}\,,$$

where $\mathcal{I}$ is the 2×2 identity matrix. Since

$$\langle 0|\, S'^x\,|1\rangle = \frac{1+\phi}{\sqrt{2(1+\phi^2)}}\,,$$

we take our renormalized J to be

$$J' = J\frac{(1+\phi)^2}{2(1+\phi^2)}\,,$$

and since the energy gap between $|0\rangle$ and $|1\rangle$ must be equal to $2\Gamma'$ (this gap was 2Γ in the unrenormalized state), we set

$$\Gamma' = \frac{E_1 - E_0}{2} = \frac{\sqrt{4\Gamma^2 + J^2} + J}{2} = \frac{J}{2}\left[\sqrt{4\lambda^2 + 1} + 1\right],$$

where $a = \sqrt{4\lambda^2 + 1} - 2\lambda$, defining the relevant variable $\lambda = \Gamma/J$.
   The fixed points of the recurrence relation (rewritten in terms of λ) are

$$\lambda^* = 0$$
$$\lambda^* \to \infty$$
$$\text{and} \quad \lambda^* \approx 1.277\,.$$

Now if the correlation length goes as

$$\xi \sim (\lambda - \lambda_c)^\nu\,,$$

in the original system, then in the renormalized system we should have

$$\xi' \sim (\lambda' - \lambda_c)^\nu\,.$$

So,

$$\frac{\xi}{\xi'} = \left(\frac{\lambda' - \lambda_c}{\lambda - \lambda_c}\right)^{-\nu}$$

$$\Rightarrow \left(\frac{\xi'}{\xi}\right)^{-1/\nu} = \frac{d\lambda'}{d\lambda}\bigg|_{\lambda = \lambda_c \equiv \lambda*} .$$

Since the actual physical correlation length should stay the same as we renormalize, ξ' must be smaller by the factor b (that scales the length), then ξ (correlation length in the original scale), that is $\xi'/\xi = b$, or

$$b^{-1/\nu} = \left(\frac{d\lambda'}{d\lambda}\right)_{\lambda = \lambda_c \equiv \lambda*} \equiv \Omega \qquad \text{(say)},$$

hence,

$$\nu = \left(\frac{\ln \Omega}{\ln b}\right)_{\lambda = \lambda*} = \frac{\ln \Omega}{\ln 2} \approx 1.47 , \quad \text{(for } b = 2\text{)},$$

compared to the exact value $\nu = 1$ for the $(d + 1 =)2$ dimensional classical Ising system. Similarly, $E_g \sim \omega \sim (\text{time})^{-1} \sim \xi^{-z}$; $z = 1$. But for $b = 2$ we do not get $z = 1$. Instead, $\lambda'/\lambda \sim b^{-z}$ gives $z \approx 0.55$. An energy gap is

$$\Delta(\lambda) \sim |\lambda_c - \lambda|^s \sim \xi^{-z} \sim |\lambda_c - \lambda|^{\nu z} .$$

Hence, $s = \nu z = 0.55 \times 1.47 \approx 0.81$ (compared to the exact result $s = 1$). Results improve rapidly for large b values.

C.1.6
RG Method for Percolation

RG methods can be applied to the percolation theory to extract the critical point and critical exponents through simple calculations. For simplicity, we shall begin with the site percolation in one dimension and then go over to the treatments in two dimensions for both site and bond percolation.

C.1.6.1   Site Percolation in One Dimension
In one dimension, site percolation is a trivial problem. It is obvious that a spanning cluster would require all the sites to be occupied. So, $p_c = 1$ for this case. The connectedness or pair correlation function reads

$$C(r, p) = p^r = \exp\left(-r \ln\left(1/p\right)\right)$$

$$= \exp\left(-r/\xi\right), \tag{C59}$$

Figure C.4 Schematic representation of site percolation in one dimension: nonpercolating case.

with

$$\xi = \frac{1}{\ln (1/p)}$$

$$= -\frac{1}{\ln p}$$

$$= -\frac{1}{\ln (1 - (p_c - p))} \quad \text{(with} \quad p_c = 1\text{)}$$

$$= \frac{1}{(p_c - p)} .$$

Hence, we get,

$$\xi = (p_c - p)^{-1},$$

giving $\nu = 1$.

Now applying RG here, we will have the length scale scaled as $\xi' = \xi/b$. If the renormalized probability is p', we shall have

$$\left(p' - p_c\right)^{-\nu} = \frac{(p - p_c)^{-\nu}}{b}$$

$$\Rightarrow -\nu \ln \left(\frac{p' - p_c}{p - p_c}\right) = \ln \frac{1}{b} = -\ln b$$

$$\Rightarrow \nu = \frac{\ln b}{\ln \lambda} ,$$

with, $\lambda = \frac{p' - p_c}{p - p_c} \approx \left(\frac{dp'}{dp}\right)_{p_c}$.

Note that p denotes the probability of occupation of a given site. In the renormalized state, a single site is formed by b number of sites in the unrenormalized lattice. Hence, we have $p' = p^b$. So, for the fixed point

$$p^* = p^{*b} ,$$

giving the two fixed points, $p^* = 0$ and $p^* = 1$. Now performing the stability analysis close to the fixed point $p^* = 0$

$$p^* + \delta p' = \left(p^* + \delta p\right)^b$$

$$\Rightarrow \delta p' = (\delta p)^b .$$

p=0   p=1

Figure C.5 Flow diagram for one dimension site percolation.

In the next iteration it becomes, $\delta p'' = (\delta p)^{2b}$. Finally, the deviation will go to zero. Hence p is an irrelevant parameter for this fixed point. Therefore, we conclude that $p^* = 0$ is a stable fixed point.

Now, close to the other fixed point $p^* = 1$ we will have

$$1 - \delta p' = (1 - \delta p)^b$$
$$1 - \delta p' = 1 - b\delta p$$
$$\Rightarrow \delta p' = b\delta p$$

Clearly the deviation grows in every iteration and the system *flows away* from this fixed point. Hence, we conclude that $p^* = 1$ is an unstable fixed point. The flow diagram is shown in Figure C.5.

Now, we have $p' = p^b$, giving

$$\left.\frac{dp'}{dp}\right|_{p_c} = \left(bp^{b-1}\right)_{p_c} = b. \tag{C60}$$

So,

$$\nu = \frac{\ln b}{\ln b} = 1. \tag{C61}$$

C.1.6.2 Site Percolation in Two Dimension Triangular Lattice

Consider a triangular lattice. Take alternate triangles and replace the three sites by a super-site (Figure C.6). The super-site is occupied if two or more sites in those three sites were occupied. So, we can write the renormalized probability as

$$p' = p^3 + 3p^2(1 - p). \tag{C62}$$

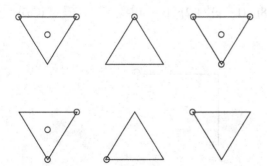

Figure C.6 Site percolation in two dimensions.

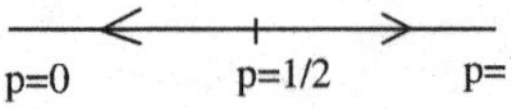

Figure C.7 Flow diagram for two dimension site percolation.

The first term corresponds to all sites being present, the factor 3 in the second term refers to the fact that there are three ways to keep one site empty. For the fixed point we write

$$p^* = p^{*3} + 3p^{*2}(1 - p^*),$$

which on solving gives, $p^* = 1, \frac{1}{2}, 0$. Clearly, $p^* = 1/2$ is a nontrivial fixed point. Let us perform stability analysis around it

$$\frac{1}{2} + \delta p' = \left(\frac{1}{2} + \delta p\right)^3 + 3\left(\frac{1}{2} + \delta p\right)^2 \left(1 - \left(\frac{1}{2} + \delta p\right)\right), \qquad (C63)$$

implying,

$$\delta p' = \frac{3}{2}\delta p. \qquad (C64)$$

So, $\lambda = \frac{dp'}{dp} = 3/2$, giving, $\nu = \frac{\ln b}{\ln \lambda} = \frac{\frac{1}{2}\ln 3}{\ln 3/2} \sim 1.355$, which is close to the assumed exact value $4/3$. We have used here $b = \sqrt{3}$, because $N' = N/3$ and $N' = N/b^d$. Here the dimensionality $d = 2$, giving $b = \sqrt{3}$, where N and N' is the number of sites in the normal and super lattice, respectively.

C.1.6.3   Bond Percolation in Two Dimension Square Lattice

Consider the eight-bond structure (Figure C.8), which can be taken as a unit cell for a square lattice. Let us replace this eight-bond structure by the two-bond structure, which can also be taken as a unit cell. Suppose we are concerned about percolation in the vertical direction. Then we can ignore the bond BG and DH for finding the contributions to connectivity between AE or equivalently, $A'E'$. Of course on the large scale, those bonds could also contribute to the vertical connectivity, but here we make this approximation. So, the contributions to the connection probability of AE can be written as

$$p' = p^5 + 5p^4(1 - p) + 8p^3(1 - p)^2 + 2p^2(1 - p)^3, \qquad (C65)$$

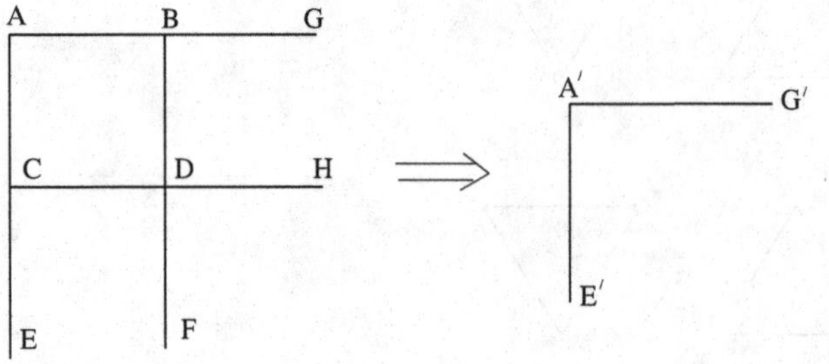

Figure C.8 Eight-bond structure: rescaling to two-bond structure.

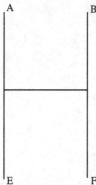

A   B

E   F   **Figure C.9** Five-bond structure: open bonds neglected.

where the first term corresponds to the occupation of all the five bonds, the factor 5 corresponds to the fact that there are five ways of removing one bond and still have vertical connectivity. The other factors come from similar arguments. For the fixed point we have

$$p^* = p^{*5} + 5p^{*4}\left(1 - p^*\right) + 8p^{*3}\left(1 - p^*\right)^2 + 2p^{*2}\left(1 - p^*\right)^3 . \quad (C66)$$

It can be seen that the physically relevant fixed points are $p = 0, \frac{1}{2}, 1$ of which, $p = \frac{1}{2}$ is nontrivial. Now

$$\lambda = \left.\frac{dp'}{dp}\right|_{p=1/2} = \frac{13}{8} . \quad (C67)$$

So,

$$\nu = \frac{\ln 2}{\ln \frac{13}{8}} \sim 1.428 , \quad (C68)$$

where we have used $N' = N/4 = N/b^d$ and $d = 2$, $b = 2$. Note that the assumed exact value is $\frac{4}{3}$ and the estimation approximately matches.

Appendix D
Spin Glasses and Optimization Problems: Annealing

D.1
Spin Glasses [14]

Spin glasses are magnetic systems with randomly competing (frustrated) interactions. Frustration is a situation where all the spins present in the system cannot energetically satisfy every bond associated to them. Here the frustration arises due to competing (ferromagnetic and anti-ferromagnetic) quenched random interactions between the spins. As a result huge barriers arise ($\mathcal{O}(N)$, N = system size) in the free energy landscape of the system. In thermodynamic limit, the height of such barriers occasionally go to infinity. These barriers strongly separate different configurations of the system so that once the system gets stuck in a deep valley in between two barriers, it practically gets trapped around that configuration for a macroscopically large time. Because of frustration, the ground state is largely degenerate; degeneracy being on the order of exp(N). As discussed above, these different ground state configurations are often separated by $\mathcal{O}(N)$ barriers, so that once the system settles into one of them, it cannot visit the others equally often over the course of time, as predicted by the Boltzmann probability factor. The system thus becomes 'nonergodic' and may be described by a nontrivial order parameter distribution in the thermodynamic limit (unlike the nonfrustrated cooperative systems, where the distribution becomes trivially delta function-like). The spins in such a system thus gets frozen in random orientation below a certain transition temperature. Although there is no long range magnetic order, that is the space average of the spin moments vanishes, the time average of any spin is nonzero below the transition (spin-glass) temperature. This time average is treated as a measure of spin freezing or the spin glass order parameter.

D.1.1
Models

Several spin glass models have been studied extensively using both analytical and computer simulation techniques. The Hamiltonian for such models can be written

Econophysics. Sitabhra Sinha, Arnab Chatterjee, Anirban Chakraborti, and Bikas K. Chakrabarti
Copyright © 2011 WILEY-VCH Verlag GmbH & Co. KGaA, Weinheim
ISBN: 978-3-527-40815-3

as

$$H = -\sum_{i<j} J_{ij} S_i^z S_j^z, \tag{D1}$$

where $S_i^z = \pm 1$, denotes the Ising spins, interacting with random quenched inter-actions J_{ij}, which differs in various models. Some examples are:

a. In the Sherrington–Kirkpatrick (S–K) model, J_{ij} are long-ranged and are dis-tributed with a Gaussian probability (centered around zero), as given by

$$P(J_{ij}) = \left(\frac{N}{2\pi J^2}\right)^{1/2} \exp\left(-\frac{N J_{ij}^2}{2 J^2}\right). \tag{D2}$$

b. In the Edward–Anderson (EA) model, the J_{ij} are short-ranged (between near-est neighbors only), but once again distributed with Gaussian probability (D2).
c. In another kind of model, the J_{ij} are again short-ranged, but having a binary $(\pm J)$ distribution with probability p

$$P(J_{ij}) = p\,\delta(J_{ij} - J) + (1 - p)\delta(J_{ij} + J). \tag{D3}$$

D.1.2
Critical Behavior

Since the disorder in the spin system is quenched, one has to perform configura-tional averaging (denoted by overhead bar) over $\ln Z$, where $Z = \mathrm{Tr}\exp(-H/k_B T)$ is the partition function of the system. To evaluate $\langle Z \rangle$, one usually employs a replica trick based on the representation $\ln Z = \lim_{n\to 0}[(Z^n - 1)/n]$. Now for the classical Hamiltonian (with all commuting spin components), $Z^n = \prod_{a=1}^{n} Z_a = Z\left(\sum_{a=1}^{n} H_a\right)$, where H_a is the ath replica of the Hamiltonian H in (D1) and Z_a is the corresponding partition function. The spin freezing can then be measured in terms of replica overlaps and the Edwards–Anderson order parameter takes the form

$$q = \frac{1}{N}\sum_{i=1}^{N} \overline{\langle S_i^z(t) S_i^z(0) \rangle}\Big|_{t\to\infty} \approx \frac{1}{N}\sum_{i=1}^{N} \overline{\langle S_{i\alpha}^z S_{i\beta}^z \rangle},$$

where α and β corresponds to different replica.

Extensive Monte Carlo studies, together with the analytical solutions for the mean field of S–K and E–A models have revealed the nature of the spin glass transition. It appears that the lower critical dimension d_l^c for E–A model, below which transition ceases to occur (with transition temperature T_c becoming zero), is between 2 and 3: $2 < d_l^c < 3$. The upper critical dimension d_u^c, at and above which mean field results (e.g., those of S–K model) apply, appears to be 6. Within these dimensions ($d_l^c < d < d_u^c$), the spin glass transitions occur (for Hamiltonian with short-ranged interactions) and the transition behavior can be characterized by various exponents. Although the linear susceptibility shows a cusp at the transition point, the nonlinear susceptibility $\chi_{SG} = (1/N) \sum_r g(r)$, where $g(r) = (1/N) \sum_i \overline{\langle S_i^z S_{i+r}^z \rangle^2}$, diverges at the spin glass point

$$\chi_{SG} \sim (T - T_c)^{-\gamma_c}, \quad g(r) \sim r^{-(d-2+\eta_c)} f\left(\frac{r}{\xi}\right); \quad \xi \sim |T - T_c|^{-\nu_c}. \quad (D4)$$

Here ξ denotes the correlation length, which determines the length scaling in the spin correlation function $g(r)$ (f in $g(r)$ denotes the scaling function). Numerical simulation gives $\nu_c = 1.3 \pm 0.1, 0.08 \pm 0.15, 1/2$ and $\gamma_c = 2.9 \pm 0.5, 1.8 \pm 0.4, 1$ for $d = 2, 3$ and 6, respectively. One can define the characteristic relaxation time τ through the time dependence of spin autocorrelation

$$q(t) = \overline{\langle S_i^z(t) S_i^z(0) \rangle} \sim t^{-x} \tilde{q}\left(\frac{t}{\tau}\right); \quad \tau \sim \xi^z \sim |T - T_c|^{-\nu_c/z_c}, \quad (D5)$$

where $x = (d - 2 + \eta_c)/2z_c$, and z_c denotes the classical dynamical exponent. Numerical simulations give $z_c = 6.1 \pm 0.3$ and 4.8 ± 0.4 in $d = 3$ and 4, respectively. Of course, such large values of z_c (particularly in lower dimensions) also indicates the possibility of the failure of power-law variation (D5) of τ with $(T - T_c)$ and rather suggests a Vogel–Fulcher-like variation: $\tau \sim \exp[A/(T - T_c)]$. In the $\pm J$ spin glasses (type (c) above), some exact results are known along the 'Nishimori Line', and the nature of the phase transition there is precisely known.

D.1.3
Replica Symmetric Solution of the S–K Model

The Hamiltonian of the Sherrington–Kirkpatrick (S–K) model in the presence of an external field is given by

$$H = -\sum_{ij} J_{ij} S_i S_j - h \sum_i S_i,$$

where J_{ij} follows the distribution

$$P(J_{ij}) = \left(\frac{N}{2\pi\Delta^2}\right)^{1/2} \exp\left(\frac{-N J_{ij}^2}{a\Delta^2}\right), \quad (D6)$$

which was first studied by Ishi and Yamamoto.

We shall briefly review the replica symmetric solution of the classical S–K model in a longitudinal field, the Hamiltonian of which is

$$H = -\sum_{\langle ij \rangle} J_{ij} S_i^z S_j^z - h \sum_i S_i^z, \quad (D7)$$

where J_{ij} follows the Gaussian distribution given by (D6).

The replica trick relies on the exact relation

$$\ln \bar{Z}\{x\} = \lim_{n \to 0} \frac{1}{n}(\bar{Z}^n\{x\} - 1) = \lim_{n \to 0} \frac{\partial}{\partial n}[\bar{Z}^n\{x\}]. \tag{D8}$$

For positive integral values of n one can express Z^n as n identical replicas of the system

$$Z^n\{x\} = \prod_{a=1}^{n} Z_a\{x\} = \prod_{a=1}^{n} \exp[-H^a/k_B T]$$

$$= \exp\left[-\sum_{a=1}^{n} H^a/k_B T\right], \tag{D9}$$

where Z_a is the partition function of the ath replica and $H^a = -\sum_{a=1}^{n} \sum_{\langle ij \rangle} J_{ij} S_{ia}^z \cdot S_{ja}^z - h \sum_{a,i} S_{ia}^z$. Using the replica trick, one obtains for the configuration averaged n-replicated partition function $\bar{Z}^n$ as

$$\bar{Z}^n = \sum_{S_{ia}=\pm 1} \int_{-\infty}^{\infty} P(J_{ij}) d J_{ij} \exp\left[\frac{1}{k_B T} \sum J_{ij} \sum S_{ia}^z S_{ja}^z + \frac{h}{k_B T} \sum S_{ia}^z\right].$$

Performing the Gaussian integral, using the Hubbard–Stratonovich transformation and finally using the method of steepest descent to evaluate integrals for thermodynamically large systems, one obtains the free energy per site f as

$$-\frac{f}{k_B T} = \lim_{n \to 0}\left[\frac{\Delta^2}{4 k_B T}\left(1 - \frac{1}{n}\sum_{a,\beta} q_{a,\beta}^2 + \frac{1}{n} \ln \mathrm{Tr}(\exp(L))\right)\right], \tag{D10}$$

where $L = (J/k_B T)^2 \sum_{a,\beta} q_{a,\beta} S_a^z S_\beta^z + (1/k_B T) \sum_{a=1}^{n} S_a^z$ and $q_{a,\beta}$ is self-consistently given by the saddle point condition $(\partial f/\partial q_{a,\beta}) = 0$. Considering the replica symmetric case $(q_{a,\beta} = q)$, one finds

$$-\frac{f}{k_B T} = \frac{\Delta^2}{2 k_B^2 T^2}(1 - q) + \frac{1}{\sqrt{2\pi}} \int_{-\infty}^{\infty} dr e^{r^2/2} \ln\left[2 \cosh\left\{\frac{h^{\mathrm{eff}}(r)}{k_B T}\right\}\right],$$

where r is the excess static noise arising from the random interaction J_{ij}, and the spin glass order parameter q is self-consistently given by

$$q = \frac{1}{\sqrt{2\pi}} \int_{-\infty}^{\infty} dr e^{-r^2/2} \tanh^2\left(\frac{h^{\mathrm{eff}}(r)}{k_B T}\right), \tag{D11}$$

and $h^{\mathrm{eff}}(r) = \Delta \sqrt{qr} + h$ can be interpreted as a local molecular field acting on a site. Different sites have different fields because of disorder and the effective distribution of $h^{\mathrm{eff}}(r)$ is Gaussian with mean 0 and variance $\Delta^2 q$.

D.2
Optimization and Simulated Annealing [15–18]

Optimization deals with the problem of finding the minimum of a given cost function (the relationship between the total cost of production and the quantity of a product produced). In combinatorial optimization problems the cost function depends on a large number of variables and hard problems are those for which the computational time is not bound by any polynomial in the problem size. Solving hard optimization problems is a challenging task. Several techniques have been developed to obtain the solution(s) of such problems. Here we will discuss some of the problems and also the techniques that have already been implemented.

Apart from many day-to-day engineering problems, many problems in physics have turned out to be, or can be transformed to, optimization problems, for example conformations of polymers in random media, optimization of wirings on printed circuit boards, and ground state of spin glasses.

An optimization problem can be described mathematically in the following way: let $\sigma = (S_1, S_2, \ldots, S_N)$ be a vector with N elements, which can take values from a domain $X^N : S_i \in X$. The domain X can be either discrete, for instance $X = 0, 1$ or $X = Z$ the set of all integers (in which case it is an integer optimization problem) or X can be continuous, for instance $X = R$, the real numbers. The ultimate goal is to minimize a given cost or energy function $\mathcal{H}(S_1, S_2, \ldots, S_N)$ with respect to N variables $S_1, S_2, \ldots, S_N$ subject to some constraints. The task is to find a set of values for these variables (a configuration) for which the function $\mathcal{H}(S_i)$ has the minimum value (see Figure D.1). In some optimization problems the set of feasible configurations from which an optimum is to be chosen is a finite set (for finite N) and the variables are discrete. These problems are combinatorial in nature. Certain problems can also be reduced to combinatorial problems.

Now we will discuss the *time complexity* of an algorithm which describes its speed. An optimization problem is said to belong to class P (P for polynomial), if it can be solved in polynomial time (i.e., the evaluation time varies as some polynomial in N) using polynomially bound resources (computer space, processors, etc.). The existence of such a polynomial bound on the evaluation time is interpreted as the "easiness" of the problem. Many important optimization problems seem to fall outside this class, such as the traveling salesman problem.

There is another important class of problems which can be solved in polynomial time by nondeterministic machines. This class is the nondeterministic polynomial (NP) class. P is included completely in the NP class, since a deterministic Turing machine is a special case of nondeterministic Turing machine. Unlike a deterministic machine, which takes a specific step deterministically at each instant (and hence follows a single computational path), a nondeterministic machine has a host of different "allowed" steps at its disposal at every instant. At each instant it explores all allowed steps and if any of them leads to the goal, the job is considered complete. Thus it explores in parallel many paths (whose number varies roughly exponentially with time) and checks if any of them reaches the goal.

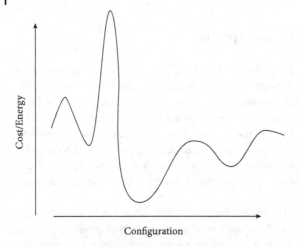

Figure D.1 A typical cost function variation of a computationally hard problem. It has a large number of local minima.

Among the NP problems, there are certain problems, known as NP-complete problems, such that any NP problem can be "reduced" to this problem type using a polynomial algorithm. If one has a routine to solve an NP-complete problem of size N, then using that routine one can solve any NP problem at the cost of an extra overhead in time that varies only polynomially with N. Problems in this class are considered to be hard, since so far a general nondeterministic machine cannot be simulated by a deterministic Turing machine (or any sequential computer with polynomially bound resources) without an exponential growth of execution time. In fact, not withstanding some enthusiasm with quantum annealing (see later), it is widely believed (though not proved yet) that it is impossible to do so (i.e., $P \neq NP$) in principle. However, assuming this to be true, one can show that there are indeed problems in the NP class that are neither NP-complete nor P.

D.2.1
Some Combinatorial Optimization Problems

There are a large number of combinatorial optimization problems, such as the traveling salesman problem, vehicle routing problem, minimum spanning tree problem, spin glass problem, etc. In this section we will discuss only the first and the last ones.

D.2.1.1 The Traveling Salesman Problem (TSP)
The traveling salesman problem (TSP) is a simple example of a combinatorial optimization problem and perhaps the most famous one. Given a set of cities and the intercity distance metric, a traveling salesman must find the shortest tour in which he visits all the cities and comes back to his starting point. It is a nondeterministic polynomial complete (NP-complete) problem.

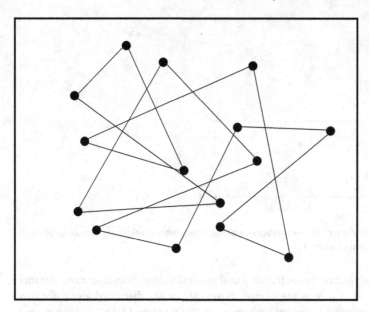

Figure D.2 A typical move of a TSP problem on a $2d$ continuum. The black dots represent cities and the lines joining them indicate the moves from one city to another.

The problem was first formulated as a mathematical problem during the 1930s. In the TSP, the most naive algorithm for finding the optimal tour is to consider all the $(N - 1)!/2$ possible tours for N number of cities and check for the shortest of them (Figure D.2). Working this way, the fastest computer available today would require more time than the current age of the universe to solve a case with about 30 cities. The typical case behavior is difficult to characterize for the TSP though it is believed to require exponential time to solve in the worst case. For this reason the TSP serves as a prototype problem for the study of the combinatorial optimization problems in general.

In the normal TSP, we have N number of cities distributed in some continuum space and we determine the average optimal travel distance per city $\bar{l}_E$ in the Euclidean metric (with $\Delta r_E = \sqrt{\Delta x^2 + \Delta y^2}$), or $\bar{l}_C$ in the Cartesian metric (with $\Delta r_C = |\Delta x| + |\Delta y|$). Since the average distance per city (for fixed area) scales with the number of cities N as $1/\sqrt{N}$, we find that the normalized travel distance per city $\Omega_E = \bar{l}_E \sqrt{N}$ or $\Omega_C = \bar{l}_C \sqrt{N}$ becomes the optimized constants and their values depend on the method used to optimize the travel distance. Numerically, $\Omega_E \approx 0.71$ and $\Omega_C \approx 0.90$.

Bounds for Optimized Travel Distance   Approximate analytic methods have also been employed to estimate the upper and lower bounds of Ω_E and Ω_C. The travel distance is expressed as a function of a single variable and the distance is optimized with respect to that variable. In the one-dimensional case any directed tour will solve such a problem. In two dimensions, we consider a unit square lattice

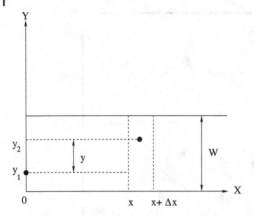

Figure D.3 Calculating the average distance between two nearest neighbors (shown by black dots) along a strip of width W.

with the random city concentration p and generalize the optimized travel distance $\ln \sqrt{p}$ (instead of $\sqrt{N}$ as mentioned above). Ω_E or Ω_C discussed above (for continuum) correspond to the case where $p \to 0$ ($\Omega_E(p) = \Omega_C(p) = 1$ for $p \to 1$ trivially). In a mean field-like approach for the solution of the TSP, one can reduce it (approximately) to a one-dimensional problem, where the square (country) is divided into strips of width W and within each strip, the salesman visits the cities in a directed way. The total travel distance is then optimized with respect to W.

Let the strip width be W and the probability density of cities be p (= N for unit area). We have a city at $(0, y_1)$ (Figure D.3). The probability that the next city is between distances x and $x + \Delta x$, is $p W \Delta x$. So the probability that there is no city at distance Δx is $(1 - p W \Delta x)$ and no city in such n consecutive strips of length Δx, where $x = n\Delta x$, is $(1 - p W \Delta x)^n \sim e^{(-p W x)}$. The probability that there is a city between y and $y + \Delta y$, is $\Delta y / W$. Hence the probability that there is no other city within distance y is $(1 - y/W)$. The average distance between any two consecutive cities is therefore

$$\bar{l}_E = 2 \int_{x=0}^{\infty} \int_{y=0}^{W} \sqrt{x^2 + y^2}\, p\, W\, dx\, e^{-(p W x)} \frac{dy}{W} \left(1 - \frac{y}{W}\right). \tag{D12}$$

The factor 2 arises to take care of the fact that y can be both positive and negative. Substituting the variables with $u = p W x$ and $v = y/W$ gives

$$\bar{l}_E = 2 \int_{u=0}^{\infty} \int_{v=0}^{1} \frac{1}{p W} \sqrt{u^2 + p^2 W^4 v^2}\, e^{-u} (1 - v)\, du\, dv. \tag{D13}$$

Introducing two dimensionless quantities $\Omega_E = \sqrt{p}\,\bar{l}_E$ and $\tilde{W} = \sqrt{p}\, W$ yields

$$\Omega_E = \frac{2}{\tilde{W}} \int_{u=0}^{\infty} \int_{v=0}^{1} \sqrt{u^2 + \tilde{W}^4 v^2}\, e^{-u} (1 - v)\, du\, dv.$$

Using Monte Carlo integration to evaluate the above integral, we get the minimum $\Omega_E \sim 0.92$ at normalized strip width of $\tilde{W} \sim 1.73$.

In the Cartesian metric the average distance between any two consecutive cities is

$$\bar{l}_C = 2 \int\limits_{x=0}^{\infty} \int\limits_{y=0}^{W} (x+y)\, p\, W\, dx\, e^{-(p\,W\,x)} \frac{dy}{W} \left(1 - \frac{y}{W}\right).$$

Introducing $u = p\,W\,x$ and $v = y/W$ yields

$$\bar{l}_C = 2 \int\limits_{u=0}^{\infty} \int\limits_{v=0}^{1} \frac{1}{p\,W} \sqrt{u + p\,W^2 v}\, e^{-u}(1-v)\, du\, dv.$$

Introducing the dimensionless quantities $\Omega_C = \sqrt{p}\,\bar{l}_C$ and $\tilde{W} = \sqrt{p}\,W$ gives

$$\Omega_C = \frac{2}{\tilde{W}} \int\limits_{u=0}^{\infty} \int\limits_{v=0}^{1} \sqrt{u + \tilde{W}^2 v}\, e^{-u}(1-v)\, du\, dv.$$

Using Monte Carlo integration to evaluate the above integral, we get the minimum $\Omega_C \sim 1.15$ at the normalized strip width of $\tilde{W} \sim 1.73$.

Let us now estimate a lower bound of the minimum travel distance per city. Let the distance between any two cities be denoted by l. Then the probability that there is a city between l and $l + dl \sim (p-1)2\pi l dl \sim 2p\pi l dl$. Now, the probability that there is no other city in the distance $l \sim (1 - \pi l^2)^{p-2} \sim e^{-(p-2)\pi l^2} \sim e^{-p\pi l^2}$. Therefore, $P(l)dl = (2p\pi l)e^{-p\pi l^2} dl$ with $\int P(l)dl = 1$. Hence the average nearest neighbor distance is

$$\bar{l}_E = \int\limits_{0}^{\infty} l^2 e^{-p\pi l^2}\, dl = \frac{1}{2}\frac{1}{\sqrt{p}}.$$

Therefore, a simple lower bound for Ω_E is $1/2$.

Assuming now a random orientation of the Euclidean nearest neighbor distance l_E or Ω_E with respect to the Cartesian axes, one can easily estimate the corresponding lower bound for Ω_C as $\Omega_C = 2\Omega_E \langle |\cos\theta| \rangle = (4/\pi)\Omega_E = 2/\pi$.

D.2.2
Details of a few Optimization Techniques

There are some excellent deterministic algorithms for solving certain optimization problems exactly. These algorithms are, however, small in number and strictly problem specific. For NP or harder problems, only approximate results can be found using these algorithms in polynomial time. Even if one can solve a NP-complete problem up to a certain approximation using some polynomial algorithm, that does not ensure that one can solve all other NP problems using the same algorithm up to the said approximation in polynomial time. Thus, one has to look for some heuristic algorithms, which are based on some stochastic iterative movements. In this chapter we will discuss three such techniques namely the local search, simulated annealing and quantum annealing.

Local Search In this algorithm one begins with a random configuration C_0 and makes some local changes in the configuration following some rules (stochastic or deterministic) to generate a new configuration C_1 and calculates the change in cost. If the cost is lowered then the change is accepted, otherwise the old one is retained. In the next step another such local change is attempted and thus an iterative procedure is carried on which steadily reduces the cost locally. When no further lowering is possible by any other local moves, the algorithm is stopped. In many optimization problem, we will end up with a local minima in such a search and will consequently be far from the global minima. An exhaustive search for an N city TSP would require $N!$ order of searches. In improved algorithms such as the *nearest neighbor algorithm, greedy algorithm* or *pair exchange heuristic*, the search time is somewhat lower.

Simulated (or Thermal or Classical) Annealing In many optimization problems there occur many local minima in the cost-configuration landscape which are far away from the global one. Local search algorithms may lead to one of those minima and thus we end up with a wrong ground state. Starting with another new configuration may lead to another such local minima and after a large number of trials we have to sort out the best one. One way of escaping these local minima is to introduce some noise or fluctuations in the searching process so that the searching is not uni-directional (towards lower energy configuration) but also a finite probability towards higher energy states. This fluctuation may be classical or quantum.

When the fluctuation is imposed as a temperature it is called *simulated annealing*. *Annealing* is a common technique used in metallurgy, where a material is heated and then cooled down slowly to increase the size of its crystals and reduce their defects. The heating helps in exploring the higher energy states (escaping from local minima) and the slow cooling gradually leads the system towards lower energy states than the previous ones. In the simulated annealing technique an "artificial" temperature T is introduced in the problem such that the transition probability from a configuration C_i to a configuration C_f is given by $\min\{1, \exp-(\Delta_{if}/T)\}$, where $\Delta_{if} = E_f - E_i$, with E_k denoting the cost or energy of the configuration C_k. The temperature T is reduced slowly from a high initial value following some annealing schedule. Here both 'uphill' as well as 'downhill' moves are allowed, but the allowed moves are increasingly downhill as T approaches zero. This method was independently described by S. Kirkpatrick, C. D. Gelatt and M. P. Vecchi in 1983. The TSP was one of the first problems to which simulated annealing was applied. If the temperature is reduced logarithmically,

$$T(t) \geq N/\ln t , \tag{D14}$$

where t denotes the cooling time and N the system size, the global minimum is attained with certainty in the limit $t \to \infty$. But such a temperature schedule will take a longer time than finding it by exhaustive search and is thus irrelevant in practical applications. Researchers have also used faster annealing schedules, for example $T(t) \sim C^t$, where $C < 1$. This can be realized by performing a fixed number of trials at each temperature, after which the temperature is reduced by a typical factor of 0.98. In such *exponential cooling* schemes, the temperature will reach a value sufficiently close to zero after a polynomially bounded time and no further uphill movements are possible (freezing sets in).

Quantum Annealing [19]   The simulated annealing technique can suffer severe set backs in the case of nonergodic systems. In such systems the degeneracy increases exponentially with the system size. Moreover the local minima are separated by $O(N)$-sized energy barriers, and at any finite temperature thermal fluctuations are not enough to relax the system to the global minimum by crossing those large barriers. This problem can be overridden (in many cases) by implementing quantum fluctuations, called *quantum annealing*.

In quantum annealing, one adds a quantum kinetic term (say $\mathcal{H}'(t)$) with the classical one (say $\mathcal{H}_0$, that has to be optimized) and reduces the quantum term from a very high initial value to zero. One can then solve the time-dependent Schrödinger equation

$$i\hbar \frac{\partial \psi}{\partial t} = \left[\mathcal{H}_0 + \mathcal{H}'(t)\right] \psi = \mathcal{H}_{\text{tot}} \psi , \tag{D15}$$

for the wave function $\psi(t)$ of the entire system $\mathcal{H}_{\text{tot}} = \mathcal{H}_0 + \mathcal{H}'(t)$. The kinetic term $\mathcal{H}'(t)$ helps the system in tunneling through the barriers and thus come out of the local traps (Figure D.4). Initially the tunneling term is much higher than the classical part, so the ground state is realizable trivially. The reduction of the

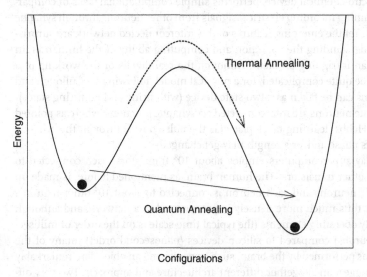

Figure D.4 Schematic diagram of the advantage of quantum annealing over classical annealing.

kinetic term is done adiabatically; it assures that the system eventually settles (at $t \to \infty$, $\mathcal{H}'(t) \to 0$) in one of the eigenstates of $\mathcal{H}_0$ which is hopefully the ground state. The tunneling term makes the apparently large barriers transparent to the system and thus the system can visit any configuration with a finite probability.

Now the crucial point in this method is the choice of $\mathcal{H}'(t)$, which can be written as $\Gamma(t)\mathcal{H}_{\text{kin}}$, where Γ is the parameter that controls the quantum fluctuations. Initially Γ is kept high so that the $\mathcal{H}_{\text{kin}}$ term dominates. Following a certain annealing schedule this Γ term is brought to zero and as mentioned in the previous paragraph if the decrement of Γ is slow enough, the adiabatic theorem of quantum mechanics assures that the system will always remain at the instantaneous ground state of the evolving Hamiltonian $\mathcal{H}_{\text{tot}}$. As Γ is finally brought to zero, $\mathcal{H}_{\text{tot}}$ will coincide with $\mathcal{H}_0$ and the desired ground state is reached.

Quantum annealing can be done in several ways:

1. Quantum Monte Carlo annealing
2. Quantum annealing using real-time adiabatic evolution
3. Annealing of a kinetically constrained system
4. Experimental realizations.

D.3
Modeling Neural Networks

The human brain is formed out of an interconnected network of roughly 10^{10} to 10^{12} relatively simple computing elements called neurons. Each neuron, although a complex electro-chemical device, performs simple computational tasks of comparing and summing incoming electrical signals from other neurons through synaptic connections. Yet the emerging features of this interconnected network are surprising, and understanding the cognition and computing ability of the human brain is certainly an intriguing problem. Although the real details of the working of a neuron can be quite complicated, for a physical model, following McCullough and Pitts, neurons can be taken as a two-state device (with firing and nonfiring states). Such two-state neurons are interconnected via synaptic junctions, where as pointed out first by Hebbs, learning takes place (as the pulse travels through the synaptic junctions, its phase and/or strength may get changed).

Present day supercomputers employ about 10^8 transistors, each connected to about 2 to 4 other transistors. The human brain, as mentioned before, is made up of about 10^{10} neurons and each neuron is connected to about 10^4 other neurons. The brain is thus much more densely interconnected as a network, and although it uses slowly operating elements (the typical time-scale is on the order of milliseconds for neurons) compared to silicon devices (nanosecond order), many of the computations performed by the brain, such as perceiving an object, are remarkably faster and suggest an altogether different architecture and approach. Two very different features of neural computations by the brain very largely outperform present day computers (though they perform arithmetic and logical operations with ever increasing speed): (a) associative memory or retrieval from partial information, and (b) pattern or rhythm recognition and inductive inference capability.

By associative memory, we mean that the brain can recall a pattern or sequence from partial information (partially erased pattern). In terms of the language of the dynamics of any network, this means that by learning various patterns, the neural network forms corresponding (distinct) attractors in the "configuration" space. If the partial information is within the basin of attraction of the pattern then the dynamics of the network takes it to the (local) fixed point or attractor corresponding to the pattern, and the network is then said to recognize the pattern. A look at the neural network in this way suggests that the "learning" of a large (macroscopic) number of patterns in such a network means creation of a large number of attractors or fixed points of dynamics corresponding to the various patterns (uncorrelated by any symmetry operation) to be learned. The pattern or rhythm recognition capability of the brain is responsible for inductive knowledge. By looking at a part of a sequence, say a periodic function of time, the brain can extrapolate or predict (by adaptively changing the synaptic connections) the next few steps or the rest of the pattern. By improved training, expert systems can comprehend more complicated (quasi-periodic or quasi-chaotic) patterns or time sequences (e.g., in medical diagnosis).

D.3.1
Hopfield Model of Associative Memory [20]

In the Hopfield model, a neuron i is represented by a two-state Ising spin at that site i. The synaptic connections are represented by a spin–spin interaction and they are taken to be symmetric. This symmetry of synaptic connections allows one to define an energy function. Synaptic connections are constructed following Hebb's rule of learning, which says that for p patterns the synaptic strength for the pair (i, j) is

$$J_{ij} = \frac{1}{N} \sum_{\mu}^{p} \zeta_i^{\mu} \zeta_j^{\mu} , \tag{D16}$$

where $\zeta_i^{\mu}, i = 1, 2, \ldots, N$, denotes the μth pattern learned ($\mu = 1, 2, \ldots, p$). Each can take values ± 1. The parameter N is the total number of neurons each connected in a one-to-all manner and p is the number of patterns to be learned. For a system of N neurons, each with two states, the energy function is

$$H = -\sum_{i>j}^{N} J_{ij} \sigma_i \sigma_j . \tag{D17}$$

The expectation is that, with J_{ij}s constructed as in (D16), the Hamiltonian or the energy function (D17) will ensure that any arbitrary pattern will have higher energy than those for the patterns to be learned; they will correspond to the (local) minima in the (free) energy landscape. Any pattern then evolves following the dynamics

$$\sigma_i (t + 1) = \text{sgn} (h_i (t)) , \tag{D18}$$

where $h_i(t)$ is the internal field on the neuron i, given by

$$h_i = \sum_j J_{ij}\sigma_j(t).$$ (D19)

Here, a fixed point of dynamics or attractor is guaranteed, that is after a certain (finite) number of iterations t^*, the network stabilizes and $\sigma_i(t^* + 1) = \sigma_i(t^*)$. Detailed analytical as well as numerical studies (see [20]) shows that the local minima for H in (D17) indeed correspond to the patterns fed to the system to be learned in the limit when memory loading factor $\alpha\ (= p/N)$ tends to zero. They are less than $\sim 3\%$ off from the patterns fed to the system to be learned when $\alpha < \alpha_c \sim 0.142$. Above this loading, the network goes to a confused state where the local minima in the energy landscape do not have any significant overlap with the patterns that are fed to be learned by the network.

Appendix E
Nonequilibrium Phenomena

E.1
Nonequilibrium Phenomena

E.1.1
Fluctuation Dissipation Theorem

When a thermodynamic system in equilibrium is perturbed by the application of an external perturbation, for example, in a magnetic field h on a magnet or in temperature T, the linear coefficients like the susceptibility χ ($= dm/dh$, m denoting the equilibrium magnetization) or specific heat C ($= dU/dT$, U denoting the average internal energy of the system) characterize the thermodynamic response of such systems. One can easily show that some appropriate fluctuations in equilibrium, in fact, determine such linear responses.

Let us consider a magnetic Hamiltonian like (B3), where the average magnetization m is given by

$$m = \frac{1}{N} \sum_i \langle S_i \rangle = \frac{\mathrm{Tr}\, S_i e^{-H/k_B T}}{\mathrm{Tr}\, e^{-H/k_B T}} , \tag{E1}$$

and susceptibility χ as

$$\chi = \frac{dm}{dh}\bigg|_{h \to 0} = \frac{1}{N} \sum_i \frac{d}{dh} \langle S_i \rangle$$

$$= \frac{1}{N} \sum_i \frac{d}{dh} \frac{\mathrm{Tr}\, S_i e^{-H/k_B T}}{\mathrm{Tr}\, e^{-H/k_B T}}\bigg|_{h \to 0} ,$$

where $H = -J \sum_{\langle ij \rangle} S_i S_j - h \sum_i S_i$. One can rewrite χ as

$$\chi = \frac{1}{N k_B T} \sum_{ij} \left[\frac{\mathrm{Tr}\, S_i S_j e^{-H/k_B T}}{\mathrm{Tr}\, e^{-H/k_B T}} - \frac{\mathrm{Tr}\, S_i e^{-H/k_B T} \mathrm{Tr}\, S_j e^{-H/k_B T}}{\left[\mathrm{Tr}\, e^{-H/k_B T} \right]^2} \right] , \tag{E2}$$

Econophysics. Sitabhra Sinha, Arnab Chatterjee, Anirban Chakraborti, and Bikas K. Chakrabarti
Copyright © 2011 WILEY-VCH Verlag GmbH & Co. KGaA, Weinheim
ISBN: 978-3-527-40815-3

or,

$$\chi = \frac{1}{Nk_B T} \sum_{ij} [\langle S_i S_j \rangle - \langle S_i \rangle \langle S_j \rangle]; \qquad \langle S_i \rangle = m = \langle S_j \rangle$$

$$\equiv \frac{1}{Nk_B T} \Delta m ,$$

where $\langle \dots \rangle$ corresponds to equilibrium average when $h = 0$ and Δm corresponds to fluctuation in m.

Similarly, if (internal) energy here is expressed as

$$U = \frac{J}{Nk_B T} \sum_{\langle ij \rangle} \langle S_i S_j \rangle ,$$

when the external field (h) is zero, one can express the specific heat C as $\frac{dU}{dT}$. As $g(r_{ij}) = \langle S_i S_j \rangle = \mathrm{Tr} \, S_i S_j e^{-H/k_B T}$, one can similarly express C by ΔU, where

$$\Delta U = \langle (U - \langle U \rangle)^2 \rangle . \tag{E3}$$

These are examples of the fluctuation-dissipation theorem, by which one can express the linear response coefficients of a thermodynamic system under (small) perturbation by the corresponding equilibrium fluctuations.

E.1.2
Fokker–Planck Equation and Condition of Detailed Balance

Under dynamic situations, the population density ρ_α (say, in any particular N spin configuration α) will have the simple dynamical equation

$$\frac{d\rho_\alpha}{dt} = -\sum_\beta (\omega_{\beta\alpha} \rho_\alpha - \omega_{\alpha\beta} \rho_\beta) , \tag{E4}$$

where $\omega_{\alpha\beta}$ refers to rate of change of the probability of the state β to α. The above equation can also formally be obtained from the Fokker–Planck diffusion equation relating diffusion to velocity fluctuations. In equilibrium $\rho_\alpha \sim \exp(-E_\alpha/k_B T)$, where E_α refers to the energy of the state α, and hence

$$\frac{\omega_{\alpha\beta}}{\omega_{\beta\alpha}} = \exp \left[-(E_\alpha - E_\beta)/k_B T \right] , \tag{E5}$$

or the detailed balance condition is satisfied.

E.1.3
Self-Organized Criticality (SOC) [21–23]

Self-organized criticality (SOC) is a property of a class of dynamical systems, which evolves collectively to a self-organized state. The SOC is different from the static critical behavior of different thermodynamical systems, like magnets or percolating systems, in the sense that unlike in those cases, the system evolves naturally towards the self-organized critical point and stays there self-tuned. There is no need to tune any external parameter, like the temperature or the magnetic field in the case of magnets, or dilution concentration in the case of percolation. More mathematically, SOC is shown by a class of dynamical systems, and it has an *attractor* as its critical point.

A simple example may be a pile of sand which grows on a horizontal plate due to random deposition of sand grains. The dynamics are such that if the local slope at the point of the surface of the pile grows beyond a critical value, a local avalanche occurs and the sand grains are redistributed. This reduces the local slope anywhere to values less than or equal to the critical value. Eventually, the conical pile ceases to grow, as the additional sand grains ultimately fall off the plate through various sizes of avalanches. This self-organized state of the pile, with statistical fluctuations over the average angle of repose, is an attractor of its dynamics. These self-organized states are often critical, in the sense that self-similarities are observed in spatial and temporal structures over all possible scales, leading to fractal-like power law behaviors. It is suggested, and partly observed, that the avalanche size in a sand pile, at its self-organized angle of repose, follows a power law distribution. It appears that many random dynamical systems, undergoing global failure, approach such critical states in a self-organized way, through the stress distributions due to local failures. We intend to discuss one such model in detail.

E.1.3.1 The BTW Model and Manna Model

Bak and others proposed a sand pile model on a square lattice which captures correctly the properties of a natural sand pile. At each lattice site (i, j), there is an integer variable $h_{i,j}$, which represents the height of the sand column at that site. A unit of height (one sand grain) is added at a randomly chosen site at each time step and the system evolves in discrete time. The dynamics starts as soon as any site (i, j) has a height equal to the threshold value ($h_{th} = 4$): that site topples, that is $h_{i,j}$ becomes zero there, and the heights of the four neighboring sites increase by one unit

$$h_{i,j} \rightarrow h_{i,j} - 4, h_{i\pm1,j} \rightarrow h_{i\pm1,j} + 1, h_{i,j\pm1} \rightarrow h_{i,j\pm1} + 1 . \qquad \text{(E6)}$$

If, due to this toppling at site (i, j), any neighboring site become unstable (its height reaches the threshold value), the same dynamic follows. Thus the process continues till all sites become stable ($h_{i,j} < h_{th}$ for all (i, j)). When toppling occurs at the boundary of the lattice, extra heights are off the lattice and are removed from the system. With continuous addition of unit height (sand grain) at random sites of the lattice, the avalanches (toppling) become correlated over longer and longer ranges and the average height (h_{av}) of the system grows with time. Gradually the correlation length (ξ) becomes on the order of the system size L as the system attains the critical average height $h_c(L)$. On average, the additional height units start leaving the system and the average height remains stable there (see Figure E.1). The distributions of the avalanche sizes and the corresponding life times follow robust power laws, hence the system becomes critical here.

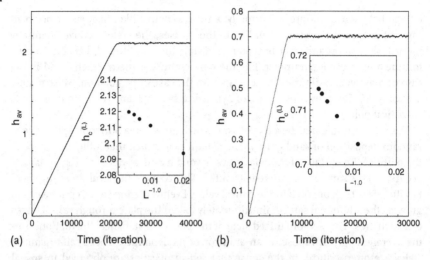

Figure E.1 Growth of BTW (a) and Manna (b) sand piles. Inset shows the system size dependence of the critical height.

We can perform a finite size scaling fit $h_c(L) = h_c(\infty) + CL^{-1/\nu}$ (by setting $\xi \sim | \; h_c(L) - h_c(\infty) \; |^{-\nu} = L$), where C is a constant, with $\nu \simeq 1.0$, which gives $h_c \equiv h_c(\infty) \simeq 2.124$ (see inset of Figure E.1). A similar finite size scaling fit with $\nu = 1.0$ gives $h_c(\infty) \simeq 2.124$ in earlier large-scale simulations.

Manna proposed the stochastic sand pile model by introducing randomness in the dynamics of sand pile growth in two dimensions. Here, the critical height is 2. Therefore at each toppling, two rejected grains choose their host among the four available neighbors randomly with equal probability. After constant adding of sand grains, the system ultimately settles at a critical state having height h_c. A similar finite size scaling fit $h_c(L) = h_c(\infty) + CL^{-1/\nu}$ gives $\nu \simeq 1.0$ and $h_c \equiv h_c(\infty) \simeq$ 0.716 (see inset of Figure E.1). This is close to an earlier estimate $h_c \simeq 0.71695$, made in a somewhat different version of the model. The avalanche size distribution has power laws similar to the BTW model, however, the exponent seems to be different, compared to that of the BTW model.

E.1.3.2 Subcritical Response: Precursors

We are going to investigate the behavior of BTW and Manna sand piles when they are away from the critical state ($h < h_c$). At an average height h_{av}, when all sites of the system have become stable (dynamics have stopped), a fixed number of height units h_p (pulse of sand grains) is added at any central point of the system. Just after this addition, the local dynamics start and it takes a finite time or iterations to return back to the stable state ($h_{i,j} < h_{\text{th}}$ for all (i, j)) after several toppling events. We measure the response parameters: $\Delta \rightarrow$ number of toppling, $\tau \rightarrow$ number of iteration and $\xi \rightarrow$ correlation length which is the distance of the furthest toppled site from the site where h_p has been dropped.

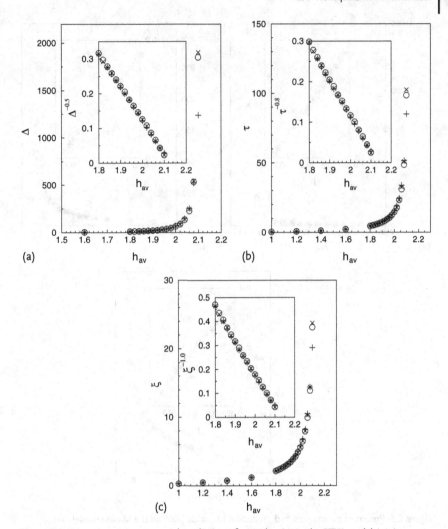

Figure E.2 Precursor parameters and prediction of critical point in the BTW model (a–c).

(A) In BTW Model We choose $h_p = 4$ for the BTW model to ensure toppling at the target site. We found that all the response parameters follow a power law as h_c is approached: $\Delta \propto (h_c - h_{av})^{-\lambda}$, $\tau \propto (h_c - h_{av})^{-\mu}$, $\xi \propto (h_c - h_{av})^{-\nu}$, $\lambda \cong 2.0$, $\mu \cong 1.2$ and $\nu \cong 1.0$. Now if we plot $\Delta^{-1/\lambda}$, $\tau^{-1/\mu}$ and $\xi^{-1/\nu}$ with h_{av} all the curves follow a straight line and they should touch the x-axis at $h_{av} = h_c$. Therefore by a proper extrapolation we can estimate the value of h_c; we find $h_c = 2.13 \pm .01$ (Figure E.2), which agrees well with direct estimates of the same.

(B) In Manna Model Obviously we have to choose $h_p = 2$ for the Manna model to ensure toppling at the target site. Then we measured all the response parameters and they seem to follow a power law as h_c is approached: $\Delta \propto (h_c - h_{av})^{-\lambda}$, $\tau \propto$

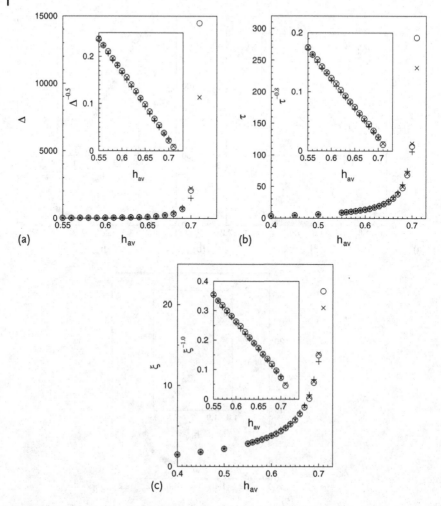

Figure E.3 Precursor parameters and prediction of critical point in the Manna model (a–c).

$(h_c - h_{av})^{-\mu}$, $\xi \propto (h_c - h_{av})^{-\nu}$, $\lambda \cong 2.0$, $\mu \cong 1.2$ and $\nu \cong 1.0$. As in the BTW model, all the curves follow a straight line if we plot $\varDelta^{-1/\lambda}$, $\tau^{-1/\mu}$ and $\xi^{-1/\nu}$ with h_{av}, and proper extrapolations estimate the value of $h_c = 0.72 \pm .01$, which is again a good estimate (Figure E.3).

Simulation results suggest that although BTW and Manna models belong to different universality classes with respect to their properties at the critical state, both models show similar subcritical response or precursors. A proper extrapolation method can estimate the respective critical heights of the models quite accurately.

E.1.4
Dynamical Hysteresis [24]

The detailed nature of the dynamic response of extended systems (having many interacting degrees of freedom), under time-dependent fields, is being investigated intensively these days. Considerable efforts have been made, in particular using computer simulations, to investigate the nature of the above mentioned dynamic phase transition and hysteresis in Ising models. As discussed before, simple Ising systems contain ferromagnetically interacting spin degrees of freedom, each with binary (up/down or ± 1) spin states. Let us consider a simple ferromagnetic system represented by an Ising model with nearest neighbor ferromagnetic coupling, which is put under an oscillating external field. Such a system can be represented by the Hamiltonian

$$H = -\sum_{\langle ij \rangle} J_{ij} S_i S_j - h(t) \sum_i S_i ,\tag{E7}$$

with

$$h(t) = h_0 \sin \omega t ,\tag{E8}$$

where the symbols have the usual meanings as before. Note that for truly long-range interactions, the sum in (E7) extends over all the pairs and the value of J_{ij} decreases inversely with the number of sites in the system. The system is in contact with an isothermal heat bath at temperature T. For simplicity all J_{ij} (> 0) are taken to be a constant J ($= 1$), and the temperature T is measured in the units of J, setting the Boltzmann constant to unity. In order to keep H/T dimensionless, h_0 is also measured in units of J.

Specifically, one considers quantities like the loop area $A = (\oint m\,dh)$ and the dynamic order parameter $Q = \frac{\omega}{2\pi} \oint m\,dt$. The mean field equation of motion for the average magnetization m can be written as

$$\frac{dm}{dt} = -m + \tanh\left(\frac{m + h(t)}{T}\right) .\tag{E9}$$

This simple nonlinear equation is indeed capable of capturing a number of important features of dynamic hysteresis and of the dynamic transition. However, the lack of fluctuation of the thermodynamic average value m in the above equation is responsible for the loss of some very significant features. For example a nonzero hysteresis loop area can be found in the static limit.

The observed variation of the loop area A with frequency ω follows the generic form discussed earlier: A decreases for both low and high values of ω. However, it may be noted that A does not quite vanish in the zero frequency limit. The exact variation can, in fact, be fitted to a form

$$A = A_0 + h_0^\alpha \omega^\beta g\left(\frac{\omega}{h_0^\gamma}\right) ,\tag{E10}$$

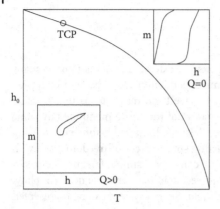

Figure E.4 Schematic diagram of the dynamical phase boundary in the field amplitude (h_0) and temperature (T) plane at a fixed nonzero frequency. The line represents the phase boundary (seperating the phases with $Q > 0$ and $Q = 0$). The small circle represents the tricritical point (TCP). Insets demonstrate the breaking of the symmetry of the dynamical hysteresis m–h loop due to dynamical transition.

with the scaling exponents α, β and γ and with scaling function g having a suitable nonmonotonic form such that $g(x) \to 0$ as $x \to 0$ or ∞. Here A_0 is the loop area in the zero frequency limit.

Linearizing the mean field (E9), for small m and h_0 ($T > T_c = 1$), one gets

$$\frac{dm}{dt} = \epsilon m + \frac{h(t)}{T}, \quad \epsilon = \frac{T-1}{T}, \tag{E11}$$

the steady state solution of which can be written as $m(t) = m_0 \cos(\omega t - \phi)$ for $h(t) = h_0 \cos \omega t$. A direct substitution then gives $m_0 = h_0 / T \sqrt{\epsilon^2 + \omega^2}$ and $\phi = \sin^{-1}(\omega / \sqrt{\epsilon^2 + \omega^2})$. For the loop area A in this linearized limit, one gets

$$A = \oint m\, dh \sim h_0^2 g(\omega) / T, \qquad g(\omega) = \frac{\omega}{\epsilon^2 + \omega^2}.$$

The general solution of (E9) would give $m(t) = m_b + m_0 \cos(\omega t - \phi)$ (the above linear solution gives the symmetry breaking term $m_b = 0$) and that gives the dynamic transition discussed above (see, e.g., Figure E.4).

E.1.5
Dynamical Transition in Fiber Bundle Models [25]

The fiber bundle (see Figure E.5) consists of N fibers or Hook springs, each having identical spring constant κ. The bundle supports a load $W = N\sigma$ and the breaking threshold $(\sigma_{\text{th}})_i$ of the fibers are assumed to be different for different fibers (i). For the equal load sharing model we consider here, the lower platform is absolutely rigid, and therefore no local deformation and hence no stress concentration occurs anywhere around the failed fibers. This ensures equal load sharing, that is the intact fibers share the applied load W equally and the load per fiber increases

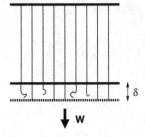

Figure E.5 The fiber bundle consists initially of N fibers attached in parallel to a fixed and rigid plate at the top and a downwardly movable platform, from which a load W is suspended at the bottom. In the equal load sharing model considered here, the platform is absolutely rigid and the load W is consequently shared equally by all the intact fibers.

as more and more fibers fail. The strength of each of the fibers $(\sigma_{th})_i$ in the bundle is given by the stress value it can bear, and beyond which it fails. The strength of the fibers are taken from a randomly distributed normalized density $\rho(\sigma_{th})$ within the interval 0 and 1 such that

$$\int_0^1 \rho(\sigma_{th})\,d\sigma_{th} = 1\,.$$

The equal load sharing assumption neglects 'local' fluctuations in stress (and its redistribution) and renders the model as a mean field one.

The breaking dynamics start when an initial stress σ (load per fiber) is applied on the bundle. The fibers having strength less than σ fail instantly. Due to this rupture, the total number of intact fibers decreases and rest of the (intact) fibers have to bear the applied load on the bundle. Hence, effective stress on the fibers increases and this compels some more fibers to break. These two sequential operations, namely the stress redistribution and further breaking of fibers continue till an equilibrium is reached, where either the surviving fibers are strong enough to bear the applied load on the bundle or all fibers fail.

This breaking dynamic can be represented by recursion relations in discrete time steps. For this, let us consider a very simple model of fiber bundles where the fibers (having the same spring constant κ) have a white or uniform strength distribution $\rho(\sigma_{th})$ up to a cut-off strength normalized to unity, as shown in Figure E.6: $\rho(\sigma_{th}) = 1$ for $0 \leq \sigma_{th} \leq 1$ and $= 0$ elsewhere. Let us also define $U_t(\sigma)$ to be the fraction of fibers in the bundle that survive after (discrete) time step t, counted from the time $t = 0$ when the load is applied (time step indicates the number of stress redistributions). As such, $U_t(\sigma = 0) = 1$ for all t and $U_t(\sigma) = 1$ for $t = 0$ for any σ; $U_t(\sigma) = U^*(\sigma) \neq 0$ for $t \to \infty$ and $\sigma < \sigma_c$, the critical or failure strength of the bundle, and $U_t(\sigma) = 0$ for $t \to \infty$ if $\sigma > \sigma_c$.

Therefore $U_t(\sigma)$ follows a simple recursion relation (see Figure E.6)

$$U_{t+1} = 1 - \sigma_t\,; \quad \sigma_t = \frac{W}{U_t N}$$

$$\text{or,} \quad U_{t+1} = 1 - \frac{\sigma}{U_t}\,. \tag{E12}$$

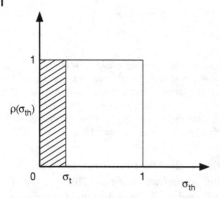

Figure E.6 The simple model considered here assumes uniform density $\rho(\sigma_{th})$ of the fiber strength distribution up to a cut-off strength (normalized to unity). At any load per fiber level σ_t at time t, the fraction σ_t fails and $1 - \sigma_t$ survives.

At the equilibrium state $(U_{t+1} = U_t = U^*)$, the above relation takes a quadratic form of U^*

$$U^{*^2} - U^* + \sigma = 0 .$$

The solution is

$$U^*(\sigma) = \frac{1}{2} \pm (\sigma_f - \sigma)^{1/2}; \sigma_f = \frac{1}{4} .$$

Here σ_f is the critical value of initial applied stress beyond which the bundle fails completely. The solution with (+) sign is the stable one, whereas the one with (−) sign gives the unstable solution. The quantity $U^*(\sigma)$ must be real valued as it has a physical meaning: it is the fraction of the original bundle that remains intact under a fixed applied stress σ, when the applied stress lies in the range $0 \le \sigma \le \sigma_f$. Clearly, $U^*(0) = 1$. Therefore the stable solution can be written as

$$U^*(\sigma) = U^*(\sigma_f) + (\sigma_f - \sigma)^{1/2} ; \quad U^*(\sigma_f) = \frac{1}{2} \quad \text{and} \quad \sigma_f = \frac{1}{4} . \quad \text{(E13)}$$

For $\sigma > \sigma_f$ we cannot get a real-valued fixed point as the dynamics never stop until $U_t = 0$, when the bundle breaks completely.

(a) At $\sigma < \sigma_f$
It may be noted that the quantity $U^*(\sigma) - U^*(\sigma_f)$ behaves like an order parameter that determines a transition from a state of partial failure ($\sigma \le \sigma_f$) to a state of total failure ($\sigma > \sigma_f$)

$$O \equiv U^*(\sigma) - U^*(\sigma_f) = (\sigma_f - \sigma)^\beta ; \quad \beta = \frac{1}{2} . \quad \text{(E14)}$$

To study the dynamics away from criticality ($\sigma \to \sigma_f$ from below), we replace the recursion relation (E12) by a differential equation

$$-\frac{dU}{dt} = \frac{U^2 - U + \sigma}{U} .$$

Close to the fixed point we write $U_t(\sigma) = U^*(\sigma) + \epsilon$ (where $\epsilon \to 0$). Thus using $U^{*2} - U^* + \sigma = 0$ gives

$$-\frac{d\epsilon}{dt} \approx \epsilon \frac{2U^* + 1}{U^*} \sim (\sigma_f - \sigma)^{1/2} \tag{E15}$$

or

$$\epsilon = U_t(\sigma) - U^*(\sigma) \approx \exp(-t/\tau) ,$$

where $\tau \sim (\sigma_f - \sigma)$. Near the critical point we can write

$$\tau \propto (\sigma_f - \sigma)^{-\alpha} ; \quad \alpha = \frac{1}{2} . \tag{E16}$$

Therefore the relaxation time diverges following a power law as $\sigma \to \sigma_f$ from below.

One can also consider the breakdown susceptibility χ, defined as the change of $U^*(\sigma)$ due to an infinitesimal increment of the applied stress σ

$$\chi = \left| \frac{dU^*(\sigma)}{d\sigma} \right| = \frac{1}{2}(\sigma_f - \sigma)^{-\gamma} ; \quad \gamma = \frac{1}{2}$$

from (E14). Hence the susceptibility diverges as the applied stress σ approaches the critical value $\sigma_f = \frac{1}{4}$. Such a divergence in χ had already been observed in numerical studies.

(b) At $\sigma = \sigma_f$
At the critical point ($\sigma = \sigma_f$), we observe a different dynamic critical behavior in the relaxation of the failure process. From the recursion relation (E12), it can be shown that decay of the fraction $U_t(\sigma_f)$ of unbroken fibers that remain intact at time t follows a simple power-law decay

$$U_t = \frac{1}{2}\left(1 + \frac{1}{t+1}\right), \tag{E17}$$

starting from $U_0 = 1$. For large t ($t \to \infty$), this reduces to $U_t - 1/2 \propto t^{-\delta}$; $\delta = 1$; a strict power law which is a robust characterization of the critical state.

The universality class of the model has been checked taking two other types of fiber strength distributions: (i) linearly increasing density distribution and (ii) linearly decreasing density distribution within the (σ_{th}) limit 0 and 1. One can show that while σ_f changes with different strength distributions ($\sigma_f = \sqrt{4/27}$ for case (i) and $\sigma_f = 4/27$ for case (ii)), the critical behavior remains unchanged: $\alpha = 1/2 = \beta = \gamma$, $\delta = 1$ for all these equal load sharing models.

References

1 Huang, K. (1987) *Statistical Mechanics*, John Wiley and Sons, New York.

2 Chowdhury, D. and Stauffer, D. (2000) *Principles of Equilibrium Statistical Mechanics*, Wiley-VCH, Berlin.

3 Kittel, C. (1963) *Quantum theory of Solids*, Wiley, New York.

4 Chakrabarti, B.K., Sen, P., and Dutta, A. (1996) *Quantum Ising Phases and Transitions*, Springer, Heidelberg.

5 Peierls, R. (1979) *Surprises in Theoretical Physics*, Princeton University Press, Princeton.

6 Temperley, H.N.V. (1956) *Phys. Rev.*, **103**, 1.

7 Fisher, M.E. and Sykes, M.F. (1959) *Phys. Rev.*, **114**, 45.

8 Fisher, M.E. (1968) *Phys. Rev.*, **176**, 257.

9 Harris, A.B. (1974) *J. Phys. C*, **7**, 1671.

10 de Gennes, P.G. (1979) *Scaling concepts in polymer physics*, Cornell University Press, Ithaca, New York.

11 Stauffer, D. and Aharony, A. (1994) *Introduction to percolation theory*, Taylor and Francis.

12 Mandelbrot, B. (1982) *Fractal Geometry of Nature*, Freeman, San Francisco.

13 Wilson, K. and Kogut, J. (1974) *Phys. Rep.*, **12C**, 2.

14 Binder, K. and Young, A.P. (1986) *Rev. Mod. Phys*, **58**, 801.

15 Garey, M.R. and Johnson, D.S. (1979) *Computers and Intractability: A Guide to the Theory of NP-Completeness*.

16 Breadwood, J., Halton, J.H., and Hammersley, J.M. (1959) *Proc. Camb. Phil. Soc.*, **55**, 299.

17 Armour, R.S. and Wheeler, J.A. (1983) *Am. J. Phys.*, **51**(5), 405.

18 Kirkpatrick, S., Gelatt Jr., C.D., and Vecchi, M.P. (1983) *Science*, **220**, 671.

19 Das, A. and Chakrabarti, B.K. (2008) *Rev. Mod. Phys.*, **80**, 1061.

20 Hertz, J., Krogh, A., and Palmer, R.G. (1991) *Introduction to the Theory of Neural Computation*, Addison-Wesley, Redwood City.

21 Bak, P., Tang, C., and Wiesenfeld, K. (1987) *Phys. Rev. Lett.*, **59**, 381.

22 Bak, P., Tang, C., and Wiesenfeld, K. (1998) *Phys. Rev. A*, **38**, 364.

23 Dhar, D. (1999) *Physica A*, **270**, 69.

24 Chakrabarti, B.K. and Acharyya, M. (1999) *Rev. Mod. Phys.*, **71**, 847.

25 Pradhan, S., Hansen, A. and Chakrabarti, B.K. (2010) *Rev. Mod. Phys.*, **82**, 499.

Some Extensively Used Notations in Appendices

p	concentration of impurity
P	pressure
v	velocity
V	volume
N	number of particles
h	magnetic field
H	Hamiltonian
$\hbar$	Planck's constant
m	magnetization
M	mass
q	momentum
Q	heat
u	displacement vector
U	internal energy
W	work
S	entropy
T	absolute temperature
k_B	Boltzmann constant
Ω	number of microstates/optimized travel distance
g	degeneracy factor
Z	partition function
z	lattice coordination number
F	free energy
f	free energy per degree of freedom
J	interaction strength
S_i	spin at site i
χ	susceptibility
ξ	correlation length
Γ	transverse field
$C, C^\dagger$	annihilation, creation operators
a	lattice constant
d	space dimension (Euclidean)

Econophysics. Sitabhra Sinha, Arnab Chatterjee, Anirban Chakraborti, and Bikas K. Chakrabarti
Copyright © 2011 WILEY-VCH Verlag GmbH & Co. KGaA, Weinheim
ISBN: 978-3-527-40815-3

D	fractal dimension
C	specific heat
ϵ	microstate energy
α	specific heat exponent
β	order parameter exponent
γ	susceptibility exponent
ν	correlation length exponent
η	anomalous dimension

Index

Econophysics. Sitabhra Sinha, Arnab Chatterjee, Anirban Chakraborti, and Bikas K. Chakrabarti
Copyright © 2011 WILEY-VCH Verlag GmbH & Co. KGaA, Weinheim
ISBN: 978-3-527-40815-3